Renewable Energy Systems from Biomass

Renewable Energy Systems from Biomass

Efficiency, Innovation, and Sustainability

Edited by
Vladimir Strezov
Hossain M. Anawar

CRC Press
Taylor & Francis Group
Boca Raton London New York

CRC Press is an imprint of the
Taylor & Francis Group, an **informa** business

CRC Press
Taylor & Francis Group
6000 Broken Sound Parkway NW, Suite 300
Boca Raton, FL 33487-2742

First issued in paperback 2022

© 2019 by Taylor & Francis Group, LLC
CRC Press is an imprint of Taylor & Francis Group, an Informa business

No claim to original U.S. Government works

ISBN-13: 978-1-498-76790-3 (hbk)
ISBN-13: 978-1-03-233872-9 (pbk)
DOI: 10.1201/9781315153971

Library of Congress Cataloging-in-Publication Data

Names: Strezov, Vladimir, editor. | Anawar, Hossain M., editor.
Title: Renewable energy systems from biomass : efficiency, innovation and
sustainability / editors: Vladimir Strezov, Hossain M. Anawar.
Description: Boca Raton : Taylor & Francis, 2019. | Includes bibliographical
references and index.
Identifiers: LCCN 2018043069 | ISBN 9781498767903 (hardback : alk. paper)
Subjects: LCSH: Biomass energy. | Sustainability.
Classification: LCC TP339 .R493 2019 | DDC 662/.88--dc23
LC record available at https://lccn.loc.gov/2018043069

Visit the Taylor & Francis Web site at
http://www.taylorandfrancis.com

and the CRC Press Web site at
http://www.crcpress.com

Contents

Preface

The world faces a range of sustainability challenges due to its fossil-fuel dependence for energy production. One of the major environmental problems at present is climate change due to greenhouse gas emissions from fossil fuels. Furthermore, fossil fuels have limited reserves and deplete with use, which illustrates the importance of introducing and accelerating technological developments for wider adoption and use of renewable energy sources. Biomass offers an important opportunity to substitute for fossil fuels, as it is the only carbon-based, renewable energy source; its carbon is net greenhouse gas neutral, because the CO_2 released from combustion or processing of biomass is the same as the CO_2 fixed from the atmosphere by the plant, through photosynthesis. Biomass offers other opportunities through production of liquid, gaseous, or high-energy-density solid fuels with already available processing technologies, such as pyrolysis, gasification followed by Fischer–Tropsch synthesis, hydrothermal processing, fermentation of high-sugar-containing crops, or transesterification of high lipid containing biomass. These technologies were reviewed and discussed in detail previously (Strezov and Evans, 2015).

There still are, however, challenges in realizing the full potential of biomass fuels for energy conversion. First, most of the biomass processing technologies for production of liquid fuels for transportation are based on the first-generation energy crops, such as sugar cane, corn, wheat, soybean, and rapeseed, which require high nutrients, water, and high-quality agricultural land for cultivation, thereby creating the food versus energy debate. The second (lignocellulosic biomass, agricultural, forestry, and other organic wastes) or third generation of biomass fuels (micro- and macro-algae) can provide solutions to this debate and substantially improve the sustainability of energy production and energy conversion of these biomass resources.

As the technology for sustainable use of biomass resources develops, there is an opportunity to design the fourth generation of biomass technologies, which will not only solve the current problems from the excessive use of fossil fuels to sustainably generate electricity or produce high-energy-density fuels and petrochemicals, but also solve other environmental problems, such as improving quality of marginal soils, providing carbon sequestration, accelerating adoption of other renewable energy forms (e.g., solar and wind) in agricultural and rural areas, remediating contaminated land and wastewater, contributing beneficially to development of the hydrogen economy, and integrating biomass into green-star-energy-rated building environments. This can be achieved through an integrated approach to design renewable energy production and utilization systems from biomass. This book aims to provide a discussion on the biomass utilization systems that have been developed or are in development to more efficiently use biomass resources and further contribute to improved environmental and sustainability benefits.

REFERENCE

Strezov, V., and T.J. Evans. *Biomass Processing Technologies*. Boca Raton, FL: CRC Press, 2015.

Editors

Professor Vladimir Strezov is a Professor in the Department of Environmental Sciences, Faculty of Science and Engineering, Macquarie University, Australia. He holds a PhD in chemical engineering and a bachelor of engineering (with honors) in mechanical engineering. Prior to commencing academic work at Macquarie University, he was a researcher at the Department of Chemical Engineering, the University of Newcastle, and at BHP Research in Newcastle, Australia. Professor Strezov leads a research group at Macquarie University that is working on renewable and sustainable energy, industrial ecology, and control of environmental pollution, and is designing sustainability metrics of industrial operations. Professor Strezov is an advisory panel member for the Australian Renewable Energy Agency (ARENA) and Fellow of the Institution of Engineers Australia. He is associate editor of the *Journal of Cleaner Production* and editorial member for the journals *Sustainability, Environmental Progress & Sustainable Energy*, and *International Journal of Chemical Engineering and Applications*. Professor Strezov is author of more than 200 articles and two books: *Biomass Processing Technologies*, with T. J. Evans (2014), and *Antibiotics and Antibiotics Resistance Genes in Soils*, with M. Z. Hashmi and A. Varma (2017).

Dr. Hossain M. Anawar is currently working in the Department of Environmental Sciences, Faculty of Science and Engineering, Macquarie University NSW, Australia. Dr. Anawar holds a bachelor of science (honors) and master of science in chemistry, a second master of science in environmental chemistry and geoscience, and a PhD in environmental biogeochemistry. He has been conducting research in different internationally recognized universities and research institutes for more than 14 years in Japan, Spain, Czech Republic, Portugal, South Africa, Botswana, and Australia. Dr. Anawar held the position of researcher at different levels in the above academic institutes. He was lecturer in the Department of Environmental Sciences and Management at Independent University, Bangladesh (IUB). Dr. Anawar has an internationally-reputed research record in innovative chemical, biogeochemical, environmental technological, microbial, and nano-technological approaches to understanding the effects of contaminants within the aquatic, plant, and soil environments. His current research work focuses on renewable energy sources, recovery of sustainable and economic renewable energy systems, resources, and materials. His other area of research priority is sustainable environmental management of waste materials and life cycle assessment. His research emphasizes the development of technological solutions for mining, contaminated land rehabilitation, resource recovery, waste to resources, nanomaterial contaminants, environmental assessment, and remediation. Dr. Anawar has been awarded several internationally reputed government and industry-funded scholarships and research grants. He has published more than 60 articles in peer-reviewed international journals.

Contributors

Hossain M. Anawar
Department of Environmental Sciences
Faculty of Science and Engineering
Macquarie University
Sydney, New South Wales, Australia

Hannah Hyunah Cho
Department of Environmental Sciences
Faculty of Science and Engineering
Macquarie University
Sydney, New South Wales, Australia

Tim Evans
Department of Environmental Sciences
Faculty of Science and Engineering
Macquarie University
Sydney, New South Wales, Australia

Tao Kan
Department of Environmental Sciences
Faculty of Science and Engineering
Macquarie University
Sydney, New South Wales, Australia

Yani Kendra
Department of Environmental Sciences
Faculty of Science and Engineering
Macquarie University
Sydney, New South Wales, Australia

Xiaofeng Li
Department of Environmental Sciences
Faculty of Science and Engineering
Macquarie University
Sydney, New South Wales, Australia

Peter Nelson
Department of Environmental Sciences
Faculty of Science and Engineering
Macquarie University
Sydney, New South Wales, Australia

Suraj Adebayo Opatokun
Department of Environmental Sciences
Faculty of Science and Engineering
Macquarie University
Sydney, New South Wales, Australia

Margarita Rosa Albis Salas
Department of Environmental Sciences
Faculty of Science and Engineering
Macquarie University
Sydney, New South Wales, Australia

Vladimir Strezov
Department of Environmental Sciences
Faculty of Science and Engineering
Macquarie University
Sydney, New South Wales, Australia

Graham Town
School of Engineering
Faculty of Science and Engineering
Macquarie University
Sydney, New South Wales, Australia

Haftom Weldekidan
Department of Environmental Sciences
Faculty of Science and Engineering
Macquarie University
Sydney, New South Wales, Australia

1 Current Status of Renewable Energy Systems from Biomass
Global Uses, Acceptance, and Sustainability

Hossain M. Anawar and Vladimir Strezov

CONTENTS

1.1 APPLICATION OF BIOENERGY SYSTEMS: CURRENT AND FUTURE PROSPECTS

High prices and limited reserve of fossil fuels, environmental pollution, and climate change have pushed the increase of investment in renewable energy solutions by 17% to $270.2 billion in 2014. Collection and utilization of distributed biomass resources after employing suitable biomass to energy conversion technology can meet the future energy demand in the developing countries (Naqvi et al., 2018). Hybrid or integrated renewable energy systems are effective technologies to explore in respect to resilience, environmental and economic benefits, and sustainability (Moomaw et al., 2011; Bartolucci et al., 2018) to generate electricity, heat, or biogas. The local renewable energy sources include farm-based and indigenous agricultural waste, bioenergy plants, crop residues, and animal wastes. The increased use of renewable energy can be available after technological development, long-term planning, implementation of integration strategies, and appropriate investments. The appropriate thermal or biochemical conversion technologies should be developed to convert biomass resources for cooking and liquid fuel systems (e.g., biogas, ethanol, methane, dimethyl ether [DME], methanol, bioethanol, and hydrogen). The biofuels can be blended with petroleum-based fuels to meet vehicle engine fuel specifications. Biofuels reduce emissions of CO_2, hydrocarbons, and SO_x emissions, resulting in improved environmental performance (Shuba and Kifle, 2018).

The biomass combined heat and power are more convenient and beneficial for increased bioenergy use in an integrated district heating (DH) and cooling system than bioheat boilers. Furthermore, biomass can be used in biorefineries and heat pumps in individual, central, and district heating (Hagos et al., 2017). The dimethyl ether biorefinery is more economical than Fischer–Tropsch (FT) biodiesel biorefinery system for energy production. By using clean energy technologies, the bioenergy and geosequestration can provide a low-carbon energy system (heat, power, electricity), energy efficiency, material recycling, and CO_2 capture and storage in the food (rice processing, sugar) and fiber processing industries (pulp and paper) (Sims et al., 2011; Spataru et al., 2014).

1.2   CONTRIBUTION OF ENERGY CROPS TO PRODUCTION OF BIOFUELS

Large-scale cultivation of energy crops and commercial production of biofuels are feasible when the sufficient amount of land areas is available by avoiding the food versus energy conflict and environmental side effects (Femeena et al., 2018; Li and Chen, 2018). The energy crop *Jatropha curcas* L. (Jatropha) has high potential for biofuel production and environmental benefits, but this crop has also issues of habitat suitability for cultivation and potential environmental problems (Hu, 2017). The marginal lands are the preferred option for large-scale biomass production that can reduce competition for land between food and energy crops (Jiang et al., 2018). Approximately 13.6 million ha of land that is not suitable for agricultural food production are recommended for switchgrass cultivation in seven US Great Plains states (Li and Chen, 2018). The study, based on the random utility theory and three model crops (switchgrass, miscanthus, and willow), indicates that the owners of marginal lands have potential interest to plant energy crops; however, owners with no marginal lands do not want to compromise the price of the crops (Jiang et al., 2018). However, the farmers are sometimes not interested in planting energy crops, due to their lower popularity and economic factors. Successful growth in cellulose-based energy crops largely depends on acceptance by local communities (Baumber, 2018). The perennial grasses or C_4-type crops, such as switchgrass (*Panicum virgatum* L.), miscanthus (*Miscanthus × giganteus Greef et Deu.*), willow, and reed canary, are widely recommended as biofuel feedstock. These plants require less fertilizers compared to other crop plants, can grow in marginal/infertile lands, and produce higher biomass than C_3 crops (Koçar and Civaş, 2013; Femeena et al., 2018). Femeena et al. (2018) developed a framework for cropping pattern with the use of corn stover (crop residue) and cultivation of switchgrass and *Miscanthus* for biofuel production. The energy crop insurance can reduce the total costs of meeting the one billion gallon mandate by 1.3%, whereas establishment cost subsidy can reduce these costs by 34% (Miao and Khanna, 2017). Corn, sugar beet, palm oil, soybean, rapeseed, and wheat are used for biofuels and energy production in most of the developed countries, including European countries (Koh and Wilcove, 2008) and the United States (Runge and Senauer, 2007; Tilman et al., 2009), especially bioethanol from sugarcane and corn and biodiesel from oilseed plants (Perdue et al., 2017). Table 1.1 demonstrates the biofuel targets of several countries (TEPGE, 2012; Koçar and Civaş, 2013).

The utilization of second-generation or cellulosic biofeedstocks and crop residues, such as corn stover, can stop the grain-based ethanol production and food versus fuel conflict (Femeena et al., 2018). However, the extensive use of crop residues for fuel production might require more fertilizer application for nutrient recovery in soil. Therefore, Hu (2017) suggested developing cultivation policies for *Jatropha curcas* L. The cultivation of micro-algae can not only produce the third-generation biofuels but also provide multiple environmental benefits, such as wastewater treatment and CO_2 mitigation (Shuba and Kifle, 2018).

TABLE 1.1
Biofuel Targets of Several Countries

Countries	Years	Target	Feedstock
United States	2012	28 billion L ethanol	Corn, soybean oil, sorghum, cellulosic sources in the future
	2013	1 billion L of cellulosic ethanol	
	2020	25% ethanol	
	2005	2% biodiesel	
Brazil	2012	25% ethanol and B2	Soybean, sugarcane, palm oil
	2013	B5 (2.4 billion biodiesel)	
	2020	B20	
EU	2005	2%	Rapeseed, sunflower, wheat, sugar beet, barley
	2010	5.75%	
	2020	10%	
China	2010	1.5–2.0 million L biodiesel	Corn, cassava, sweet potato, rice, jatropha
	2020	10% ethanol (=8.5 million tones) 10.6–12.0 million biodiesel	
Thailand	2012	10% biodiesel	Cassava, molasses, sugarcane, soybean, coconut, jatropha, peanut
Canada	2010	5% ethanol	Wheat
	2012	2% biodiesel	
India	2012	5% biofuel	Molasses, sugarcane in the future, jatropha
	2017	10% biofuel	
Australia	2010	350 million L of biofuel	Wheat, sugarcane, molasses, palm oil, cotton oil
	2012	10% ethanol and 10% biodiesel	
	2017	20% ethanol and 20% biodiesel	
Japan	2010	360 million L biofuel	Imported ethanol, rice bran
	2020	6 billion L biofuel	
	2030	10% biofuel	

Source: Reprinted from *Renew. Sustain. Energy Rev.*, 28, Koçar, G., and Civaş, N., An overview of biofuels from energy crops: Current status and future prospects, 900–916, Copyright 2013, with permission from Elsevier.

1.3 TECHNOLOGICAL DEVELOPMENT

The technological improvement can help the cost-effective conversion of lignocellulosic biomass to ethanol (Li and Chen, 2018). The different energy crops and the different parts of plant have various compositions of lignocellulosic feedstock. Single-treatment technology, such as anaerobic digestion or thermochemical pretreatment, cannot effectively convert biomass components into biofuels and bio-based products. Therefore, combined application of the above technologies could efficiently convert biomass components into biofuels and other useful things (Surendra et al., 2018). The conversion of biomass to liquid fuels—especially transport fuel, using the FT process—can represent a significant technological development for producing clean and carbon-neutral bioenergy (Ail and Dasappa, 2016).

1.4 RENEWABLE ENERGY AND LOCAL SUSTAINABILITY

Given the sustainability criteria, a contrasting study between a smart-energy system and a traditional, non-integrated renewable energy system suggests that the first has high potential in terms of electricity production, sustainable development, socioeconomic and environmental benefits, and technological availability, while the second consumes 50% more biomass than the recommended level (sustainable level). However, the costs of both systems are approximately similar (del Rio and Burguillo, 2008; Bačeković and Østergaard, 2018). Sustainability criteria of renewable energy systems for sustainable development should consider not only the environmental problems, but also socioeconomic benefits, such as diverse energy supply and stronger regional and rural developments, developing domestic industry and higher employment (del Rio and Burguillo, 2008; Richardson, 2013; Cavicchi, 2018). Regional development policy and effective frameworks should be developed, including the sustainable renewable energy systems, different sources of socioeconomic benefits, and environmental safety. Cavicchi (2018) studied the triple bottom line of sustainability (social, economic, and environmental benefits) of bioenergy development in Norway. Although the forest-based bioenergy development increased rapidly, its continuous sustainable development was subsequently hindered by conflicting local interests, power relations, and market dynamics.

The production of nitrogen fertilizers and their extensive use in cultivation of bioenergy crops significantly contribute to fossil energy consumption and greenhouse gas (GHG) emissions. Sastre et al. (2016) used life-cycle assessment to evaluate the sustainability of different bioenergy pathways, where the soil nitrogen balance can help to maintain soil fertility, remove negative effects, and maintain sustainable development.

1.5 CURRENT STATUS OF GLOBAL BIOENERGY POTENTIAL

The global use of renewable energy has increased rapidly, reaching 19.2% of the global final energy consumption in 2014 (REN21, 2016; Scarlat et al., 2018). In the European Union (EU), bioenergy contributes significantly to the renewable energy source in the energy mix, and it is expected that, by 2020, the share will reach more than 60% of renewable energy and approximately 12% of the final energy consumption (Scarlat et al., 2015).

The bioenergy systems incorporate technological, societal, cultural, economic, and environmental considerations (Zabaniotou, 2018) and offer a circular waste-based bioeconomy. For effective success, the bioeconomy should include local knowledge, public health, and community resilience. However, the waste-based, global bioenergy sector can be restricted by recovery of new biomaterials from the same sources. Therefore, sustainable bioenergy systems can be integrated with the cascade biorefinery models. Otherwise, it can provide waste management solutions by stand-alone, decentralized systems. The biomass feedstock supply chains can affect future policy targets for worldwide development of bioenergy through competitiveness, reliability, and sustainability (Gabrielle et al., 2014). The biomass feedstocks are related to agricultural crop type, agronomic practices, dry matter yields, agricultural input requirements and environmental impacts, soil, and climate conditions. The use of grass–legume mixtures or residues from biomass conversion processes can improve the recovery of biomass feedstock. Further improvement in bioenergy feedstock supply can be obtained through research findings on multi-crop and multi-site experiments, optimization of management practices and innovative cropping systems, use of alternative land under future climate changes, direct and indirect effects of bioenergy development on land-use change, investigation of the effect of perennial stands on biodiversity, and improvement of methodologies to assess social impacts of the bioenergy projects.

1.5.1 Current Status of Bioenergy Resources in Asian Countries

Renewable energy policies and individual and government initiatives encourage harnessing of the renewable energy resources and implementation of renewable energy systems in Asian countries. Governments have taken initiatives to deploy renewable energy in households and industrial sectors, and to partially replace fossil fuels in South Asian countries (Shukla et al., 2017). A few examples of Asian countries and their renewable energy statuses are described below.

FIGURE 1.1   Jana Landfill biogas generation project at Puchong, Malaysia. (Reprinted from *Renew. Sustain. Energy Rev.*, 40, Mekhilef, S. et al., Malaysia's renewable energy policies and programs with green aspects, 497–504, Copyright 2010, with permission from Elsevier.)

Malaysia: As a country of tropical and humid climate, Malaysia has opportunity for utilizing multiple sources of renewable energy. Given the increasing prices of fossil fuel and impacts on climate, the government has taken different initiatives to encourage industries and individuals and exploit renewable energy systems in power applications (Mekhilef et al., 2014). As an example, its Small Renewable Energy Power Program has installed the Jana Landfill biogas generation project at Puchong (Figure 1.1).

Jordan: Jaber et al. (2015) used strength, weakness, opportunities, and threat (SWOT) analysis to determine the status of renewable energy sources and systems in Jordan. There are a couple of obstacles that might reduce the employment of renewable energy systems in Jordan. Among them, the most notable are lack of available allocated financing programs, the future price of electricity, and investment for public awareness and training projects. Government and private sectors should take new initiatives to remove these obstacles and invest to develop potential and viable technology for different renewable energy options.

Nepal: The burning of biomass, woods, crop residues, and animal dung is used for energy generation in domestic usage, meeting about 86% of the national energy consumption, while less than 50% of the population has access to electricity, and the increasing rate of importing petroleum is a major burden for the country's economy. The geographical, technical, political, and economical reasons have hindered employment in the renewable energy sector in Nepal. However, there is a significant potential for developing renewable energy technologies in this country (Surendra et al., 2011).

India and China: The current renewable energy policies have introduced different types of technologies to meet the increasing demand of energy and reduce environmental pollution in India and China. A few studies have depicted the status, opportunities, barriers, and potential of renewable energy in India and China (Gera et al., 2013; Jia et al., 2015; Saravanan et al., 2018). Substantial efforts are underway to harness the different sources of renewable energy in these countries (Gera et al., 2013; Saravanan et al., 2018). The renewable energy policies emphasize biofuel production from micro-algae and marketing in sustainable energy supply. The governments have announced programs to provide financial help for bio-based fuel production and the blending of ethanol with gasoline and diesel with biodiesel. A number of sources, such as sweet sorghum, neem seed, mahua seed, sugarcane molasses, and jatropha, have been assessed as potential materials for producing biofuels in India (Sharma and Kumar, 2018).

1.5.2 RENEWABLE ENERGY IN THE UNITED STATES: CURRENT STATUS AND FUTURE PROSPECTS

The forest log, wood-pellet and residues, and short rotation woody crops (SRWC), such as *Pinus taeda* L. (loblolly pine), are the most important feedstocks for renewable energy production in the southeastern United States (Dale et al., 2017; Perdue et al., 2017). The forest industry in the Southeast is producing a large amount of wood pellets that are mainly exported to the EU countries to replace coal in power plants. The question of sustainability, ecosystem services, and environmental issues revolves around this mass-scale production of forest-based wood pellets (Dale et al., 2017). The science and technology-based management of SRWC, forests, and different forestry-based products, including bioenergy and bio-based fuel, can protect and improve the forest ecosystem services and socioeconomic benefits, and generate income for the landowners, protect soil and water quality, and save wildlife and biodiversity (Butler et al., 2017; Cornwall, 2017). The sustainable harnessing of biomass energy from forest should consider the moderate level of logging, new plantation, conservative and managed forest, systematic monitoring, and improved management of forest. The US Federal Biomass Research and Development Technical Advisory Committee has planned to replace current US petroleum consumption with biofuels by 30% by 2030. Although large areas of land are considered to be available for bioenergy production, the geospatial analyses of land suggest large reductions in the estimates of potential land areas available for bioenergy production (Merry et al., 2017). Sanford et al. (2016) studied the six-year average production potential and biomass yield of seven model bioenergy cropping systems in both southcentral Wisconsin and southwest Michigan. Out of these, corn had the highest production, followed by giant miscanthus and then switchgrass. The less-productive cropping systems were native grasses ≈ restored prairie ≈ early successional ≈ and hybrid poplar. However, the bioenergy crop yields of the native grass, prairie, and hybrid poplar can be improved to a great extent, provided that simple changes are adopted in agronomic management (e.g., harvest timing and harvest equipment modification).

1.5.3 BIOENERGY IN CANADA

Bioenergy production from forest residue, wood fiber, pulp mill residual fiber, crop residue, etc. has high potential to generate thermal and electrical energy and mitigate climate change in Canada, as shown in Figure 1.2 (Dymond and Kamp, 2014; Smyth et al., 2016, 2017; Liu et al., 2018).

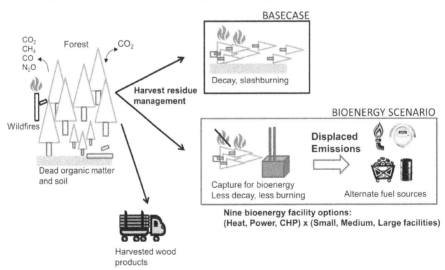

FIGURE 1.2 Schematic of C flows in the Base Case and Bioenergy Scenario. The Bioenergy Scenario differs from the Base Case forest management assumptions by reducing slashburning (where applicable) and utilizing harvest residues for bioenergy. (Reprinted from Smyth, C. et al., *GCB Bioenergy*, 9, 817–832, 2017.)

The bioenergy production from biomass did not increase the extensive and intensive harvesting of forests (Dymond and Kamp, 2014). The use of these residues for bioenergy production could contribute significantly when they displace the highest-emitting fuels in the fuel mix for heat and electricity. However, negative mitigation potential is shown when biomass displaces low-emission hydroelectricity in some areas. Therefore, bioenergy needs an integrated assessment, using a systems approach, at regional and national levels. Canada has vast areas of marginal land to produce energy crops and bioenergy, while both fertile and marginal lands show potential to produce food crops (Liu et al., 2018). Excessive use of crop residue can affect the soil quality and soil carbon reserve. Therefore, their use for bioenergy production should be done in a sustainable way.

1.5.4 BIOGAS PRODUCTION AND FUTURE PROSPECTS IN EUROPE

The EU climate and energy framework for 2030 has set EU-wide targets and policy objectives for reduction of GHG emissions by 40% compared to 1990 levels, at least a 27% share of renewable energy consumption, and energy savings of at least 27% (COM[2014] 15 final, 2014). The 2030 policy will work towards the long-term 2050 GHG reduction targets (COM[2016] 767 final/2, 2017). The 27% target of renewable energy might be increased to 35% by 2030 according to the European Parliament. The use of food or feed crops in biofuel production will be limited to 3.8% in 2030. The current EU renewable energy framework will contribute significantly to biogas/biofuel/bioenergy production for electricity, heat and transport use, economic and environmental benefits, waste management, GHG reductions, and climate control. In the EU, biogas production has increased to 18 billion m^3 of methane (654 PJ) in 2015, which is one-half of the global biogas production (Scarlat et al., 2018). Table 1.2 shows the energy production from biogas in Europe in 2015 (IEA, 2016).

TABLE 1.2
Energy Production from Biogas in Europe in 2015

	Electricity Capacity (MW)	Average Capacity (kW)	Electricity Production (GWh)	Heat Production (TJ)	Derived Heat (TJ)
EU					
Belgium	183	897	955	4,272	388
Bulgaria	20	1,818	120	182	24
Czech Republic	368	664	2,611	6,491	623
Denmark	102	671	485	3,265	2,099
Germany	4,803	443	33,073	69,047	9,285
Estonia	11	611	50	286	112
Ireland	53	1,828	202	370	0
Greece	49	1,750	230	661	0
Spain	224	1,612	982	2,474	0
France	320	446	1,783	6,859	1,432
Croatia	28	1,217	177	219	219
Italy	1,336	859	8,212	10,469	8,604
Cyprus	10	769	51	214	51
Latvia	60	1,017	391	1,256	892
Lithuania	21	583	86	403	91
Luxembourg	12	400	62	390	80
Hungary	69	972	293	667	131
Malta	3	1,500	7	30	6
Netherlands	239	892	1,036	0	48
Austria	194	437	624	2,036	145

(Continued)

TABLE 1.2 (*Continued*)
Energy Production from Biogas in Europe in 2015

	Electricity Capacity (MW)	Average Capacity (kW)	Electricity Production (GWh)	Heat Production (TJ)	Derived Heat (TJ)
Poland	216	780	906	3,703	436
Portugal	66	1,031	294	336	0
Romania	14	1,273	61	303	156
Slovenia	32	1,231	132	383	304
Slovakia	91	650	541	2,122	473
Finland	0	0	358	1,600	763
Sweden	95	337	62	2,150	274
UK	1,488	2,845	7,189	6,641	0
Switzerland	74	116	303	1,342	1,199
Iceland	0	0	0	37	0
Norway	17	138	7	834	118
Macedonia	4	1,333	20	37	0
Serbia	5	714	23	45	0
Moldova	3	750	15	159	11
Ukraine	18	1,125	10	282	360
European Union (EU)	10,107	609	60,973	126,829	26,636
Europe	10,228	588	61,351	129,565	28,324

Source: Reprinted from Scarlat, N. et al., *Renew. Energy*, 129, 457–472, 2018.

The technical efficiency level of the bioenergy industry, assessed to enhance bioenergy production through proper use of available resources, is higher in developing countries than in developed countries in the EU28 region. The pure technical efficiency levels are more influenced by technical efficiency (Alsaleh et al., 2017). The technical efficiency and cost efficiency of the bioenergy industry are significantly affected by capital input, labor input, gross domestic product, inflation, and interest rate in the developing and developed countries of the EU28 region. Alsaleh and Abdul-Rahim (2018) determined the impact of country-specific and macroeconomic determinants of cost efficiency rates in the bioenergy industry in the EU28 zone. The cost efficiency rates of the bioenergy industry are equal among the developing and developed countries in the EU28 zone.

1.5.5 BIOENERGY IN AUSTRALIA

National and regional level biomass production can provide significant amounts of bioenergy feedstocks for production of electricity, heat, and biofuels (solid, liquid, or gas) in Australia for current and future generations (Farine et al., 2012; RIRDC, 2014; Kosinkova et al., 2015; Crawford et al., 2016). The biomass potential bioenergy feedstocks available in Australia are crop stubble, native grasses, forest harvesting or wood-processing pulpwood and residues, private native forests, bagasse, and new short-rotation tree crops. An estimated 80 Mt per year of biomass feedstock can be obtained from three major sources: crop stubble (27.7 Mt per year), grasses (19.7 Mt per year), and forest plantations (10.9 Mt per year). However, this figure can augment to 100–115 Mt per year over the next 20–40 years. The new plantings of short-rotation trees are the major sources of the increase (14.7 Mt per year by 2030 and 29.3 Mt per year by 2050). This estimate of potential biomass does not include oilseeds, algae, and regrowth of vegetation naturally on cleared land. The energy facility or biorefinery industry should be set up based on the distribution, location, spatial density, and seasonal supply of biomass. Therefore,

Kosinkova et al. (2015) estimated the region-specific availability of second- and third-generation feedstocks and identified the most appropriate bioenergy solutions and supply chains for each region. Three states in Australia, New South Wales, Queensland, and Victoria (NSW, QLD, and VIC), have high potential for the second-generation biofuels, while Western Australia and Northern Territory (WA, NT) are the most suitable regions for micro-algae cultivation, according to land use opportunity cost and climate.

The bioenergy market development can provide the owners of private native forests with economic benefits to adopt the silvicultural management necessary to promote growth of the retained forests and a sustainable supply of ecological and economic benefits in the future (Hayward et al., 2015; Ngugi et al., 2018). Extensive use of crop residues for the bioenergy industry could negatively affect the soil quality of organic carbon (SOC), nitrogen pools, soil erosion, soil moisture, and soil fertility (Powlson et al., 2011). However, a significant amount of crop residue can be sustainably harvested when adequate agronomic management practices are applied, and crop residue is harvested from croplands with high primary productivity and low SOC decomposition rate (Wilhelm et al., 2007). Zhao et al. (2015) quantified the limits of sustainable harvest for wheat residue in the Australian agricultural sites and reported that, with fertilization of up to 75 kg N ha^{-1}, the amount of crop residue that can be sustainably harvested in southwestern and southeastern Australia can reach 75% and 50%, respectively. Higher fertilization rates do not contribute to further increase in harvest rates of sustainable residue.

1.5.6 BIOENERGY IN AFRICA

Renewable energy can contribute to sustainable development of the African countries that have significant potential of using forest biomass, agriculture, sugar mill molasses, cane fibers, and residues for harnessing bioenergy/biofuel, such as heat, electricity, and ethanol, depending on availability of raw resources in different regions (Leal et al., 2016). However, the adoption of modern bioenergy technologies should be available in Africa to improve energy production and utilization efficiency, reducing negative health effects and GHG emissions (Lynd et al., 2015). The integrated production of food crops, bioenergy crops, and livestock, and sustainable use of forest wood, crop residue, and manure, along with adaptation and development of modern technologies (e.g., anaerobic digestion, composting, and pyrolysis), business models, and government subsidies, can contribute significantly to the bioenergy/biofuels sector of Africa (Mohammed et al., 2015; Smith et al., 2015; Akbi et al., 2017; Maqhuzu et al., 2017).

1.5.7 BIOENERGY POTENTIAL IN LATIN AMERICA

Brazil is a country of high potential for producing renewable energy and is currently driving the large amount of ethanol (flex cars) production (Meisen and Hubert, 2010). The sources, extent, and potential of renewable energy resources vary significantly in the Latin American countries. Brazil can play an important role in uniting these energy sources and driving the continent-wide grid to meet the energy demand, reducing the dependence on the fossil fuels, increasing the renewable energy business model, and improving the socioeconomic conditions of the continent.

Although the bioenergy projects demonstrate potential benefits for socioeconomic development, they sometimes are not evaluated properly, due to the inherent challenges and complexity of the project (Nogueira et al., 2017). Often, the bioenergy project is evaluated using the SWOT matrix. The integrated approach is necessary to obtain maximum output of the bioenergy production. The neotropical palm Acrocomia aculeata can provide biomass feedstock for bioenergy production and potentially can be cultivated in Latin America, especially Central America, including the Caribbean, northern Colombia, and Venezuela, as well as southern Brazil and eastern Paraguay, under current abiotic environmental conditions (Plath et al., 2016). However, given future ecological scenarios, this plant should be cultivated cautiously and in a sustainable manner.

1.6 CONCLUSIONS

The global use of renewable energy has increased rapidly, reaching 19.2% of global final energy consumption in 2014. Hybrid or integrated renewable energy systems are more effective technologies to explore local, renewable energy sources in terms of their resilience, environmental and economic benefits, and sustainability criteria to generate electricity, heat, or biogas. Corn, sugar beet, palm oil, soybean, rapeseed, and wheat are used for biofuels and energy production in most of the developed countries. The potential bioenergy feedstocks are farm-based and indigenous agricultural waste, bioenergy plants, crop residues, and animal wastes. Widely known energy crops that have high potential for biofuel production and environmental benefits are *Jatropha curcas* L. (Jatropha), switchgrass, miscanthus, and willow. Micro-algae can produce biofuels and provide environmental benefits, such as wastewater treatment and CO_2 mitigation. The marginal lands are the preferred option for large-scale biomass production that can reduce the competition for land between food and energy crops. Energy policy and framework development are necessary for cropping pattern of food crops, sustainable and moderate use of crop residue for bioenergy production, cultivation of energy crops, multi-crop and multi-site experiments, optimization of management practices, and innovative cropping systems.

The appropriate thermal or biochemical conversion techniques should be developed to convert biomass resources for biogas, ethanol, methane, DME, methanol, bioethanol, and hydrogen energy systems. The use of several technologies together, such as anaerobic digestion, thermochemical pre-treatment, and FT processes could efficiently convert biomass components in biofuels and other useful components. The biomass combined heat and power are more convenient and beneficial for increased bioenergy use in integrated DH and cooling systems than bioheat boilers. The smart energy system has more potential than a traditional, non-integrated renewable energy system with respect to electricity production, sustainable development, socioeconomic and environmental benefits, and technological availability. Renewable energy policies and individual and government initiatives encourage harnessing the renewable energy resources and implementation of renewable energy systems in Asia, Australia, the United States, and Canada, and in European, African, and Latin American countries.

REFERENCES

Ail, S. S., and S. Dasappa. 2016. Biomass to liquid transportation fuel via Fischer Tropsch synthesis—Technology review and current scenario. *Renewable and Sustainable Energy Reviews* 58: 267–286.

Akbi, A., M. Saber, M. Aziza, and N. Yassaa. 2017. An overview of sustainable bioenergy potential in Algeria. *Renewable and Sustainable Energy Reviews* 72: 240–245.

Alsaleh, M., and A. S. Abdul-Rahim. 2018. Determinants of cost efficiency of bioenergy industry: Evidence from EU28 countries. *Renewable Energy* 127: 746–762.

Alsaleh, M., A. S. Abdul-Rahim, and H. O. Mohd-Shahwahid. 2017. Determinants of technical efficiency in the bioenergy industry in the EU28 region. *Renewable and Sustainable Energy Reviews* 78: 1331–1349.

Bačeković, I., and P. A. Østergaard. 2018. A smart energy system approach vs a non-integrated renewable energy system approach to designing a future energy system in Zagreb. *Energy* 155: 824–837.

Bartolucci, L., S. Cordiner, V. Mulone, V. Rocco, and J. L. Rossi. 2018. Hybrid renewable energy systems for renewable integration in microgrids: Influence of sizing on performance. *Energy* 152: 744–758.

Baumber, A. 2018. Energy cropping and social licence: What's trust got to do with it? *Biomass and Bioenergy* 108: 25–34.

Butler, S. M., B. J. Butler, and M. Markowski-Lindsay. 2017. Family forest owner characteristics shaped by life cycle, cohort, and period effects. *Small-Scale Forestry* 16: 1–18.

Cavicchi, B. 2018. The burden of sustainability: Limits to sustainable bioenergy development in Norway. *Energy Policy* 119: 585–599.

COM. 2014. 15 final, A policy framework for climate and energy in the period from 2020 to 2030. Communication from the Commission to the European Parliament, the Council, the European economic and social Committee and the Committee of the Regions 2014, Brussels, Belgium.

COM. 2016. 767 final/2, Proposal for a Directive of the European Parliament and of the Council on the promotion of the use of energy from renewable sources (recast), 2017, Brussels, Belgium.

Cornwall, W. 2017. Is wood a green source of energy? Scientists are divided. *Science* 355: 18–21.

Crawford, D. F., M. H. O'Connor, T. Jovanovic, A. Herr, R. J. Raison, D. A. O'Connell, and T. Baynes. 2016. A spatial assessment of potential biomass for bioenergy in Australia in 2010, and possible expansion by 2030 and 2050. *GCB Bioenergy* 8: 707–722. doi:10.1111/gcbb.12295.

Dale, V. H., K. L. Kline, E. S. Parish, A. L. Cowie, R. Emory, R. W. Malmsheimer, R. Slade et al. 2017. Status and prospects for renewable energy using wood pellets from the southeastern United States. *GCB Bioenergy* 9: 1296–1305. doi:10.1111/gcbb.12445.

del Rio, P., and M. Burguillo. 2008. Assessing the impact of renewable energy deployment on local sustainability: Towards a theoretical framework. *Renewable and Sustainable Energy Reviews* 12: 1325–1344.

Dymond, C. C., and A. Kamp. 2014. Fibre use, net calorific value, and consumption of forest-derived bioenergy in British Columbia, Canada. *Biomass and Bioenergy* 70: 217–224.

Farine, D. R., D. A. O'Connell, R. J. Raison, B. M. May, M. H. O'Connor, D. F. Crawford, A. Herr et al. 2012. An assessment of biomass for bioelectricity and biofuel, and for greenhouse gas emission reduction in Australia. *Global Change Biology Bioenergy* 4: 148–175.

Femeena, P. V., K. P. Sudheer, R. Cibin, and I. Chaube. 2018. Spatial optimization of cropping pattern for sustainable food and biofuel production with minimal downstream pollution. *Journal of Environmental Management* 212: 198–209.

Gabrielle, B., L. Bamière, N. Caldes, S. De Cara, G. Decocq, F. Ferchaud, E. Pelzer, C. Loyce, Y. Perez, J. Wohlfahrt, and G. Richard. 2014. Paving the way for sustainable bioenergy in Europe: Technological options and research avenues for large-scale biomass feedstock supply. *Renewable and Sustainable Energy Reviews* 33: 11–25.

Gera, R. K., H. M. Rai, Y. Parvej, and H. Soni. 2013. Renewable energy scenario in India: Opportunities and challenges. *Indian Journal of Electrical and Biomedical Engineering* 1(1): 10–16.

Hagos, D. A., A. Gebremedhin, and T. F. Bolkesjø. 2017. The prospects of bioenergy in the future energy system of Inland Norway. *Energy* 121: 78–91.

Hayward, J. A., D. A. O'Connell, R. J. Raison, A. C. Warden, M. H. O'Connor, H. T. Murphy, T. H. Booth et al. 2015. The economics of producing sustainable aviation fuel: A regional case study in Queensland, Australia. *Global Change Biology Bioenergy* 7: 497–511.

Hu, J. 2017. Decreasing desired opportunity for energy supply of a globally acclaimed biofuel crop in a changing climate. *Renewable and Sustainable Energy Reviews* 76: 857–864.

International Energy Agency (IEA). 2016. Medium-term renewable energy market report 2016. Market Analysis and Forecasts to 2021, Paris, France.

Jaber, J. O., F. Elkarmi, E. Alasis, and A. Kostas. 2015. Employment of renewable energy in Jordan: Current status, SWOT and problem analysis. *Renewable and Sustainable Energy Reviews* 49: 490–499.

Jia, Y., Y. Gao, Z. Xu, K. P. Wong, L. L. Lai, L. L. Y. Xue, Z. Y. Dong, and D. J. Hill. 2015. Powering China's sustainable development with renewable energies: Current status and future trend. *Electric Power Components and Systems* 43(8–10): 1193–1204. doi:10.1080/15325008.2015.1009585.

Jiang, W., K. Y. Zipp, and M. Jacobson. 2018. Economic assessment of landowners' willingness to supply energy crops on marginal lands in the northeastern of the United States. *Biomass and Bioenergy* 113: 22–30.

Koçar, G., and N. Civaş. 2013. An overview of biofuels from energy crops: Current status and future prospects. *Renewable and Sustainable Energy Reviews* 28: 900–916.

Koh, L. P., and D. S. Wilcove. 2008. Is oil palm agriculture really destroying tropical biodiversity? *Conservation Letters* 1: 60–64.

Kosinkova, J., A. Doshi, J. Maire, Z. Ristovski, R. Brown, and T. J. Rainey. 2015. Measuring the regional availability of biomass for biofuels and the potential for microalgae. *Renewable and Sustainable Energy Reviews* 49: 271–1285.

Leal, M. R. L. V., J. G. D. B. Leite, M. F. Chagas, R. da Maia, and L. A. B. Cortez. 2016. Feasibility assessment of converting sugar mills to bioenergy production in Africa. *Agriculture* 6: 45.

Li, R., and J. Chen. 2018. Planning the next-generation biofuel crops based on soil-water constraints. *Biomass and Bioenergy* 115: 19–26.

Liu, J., T. Huffman, and M. Green. 2018. Potential impacts of agricultural land use on soil cover in response to bioenergy production in Canada. *Land Use Policy* 75: 33–42.

Lynd, L. R., M. Sow, A. F. Chimphango, L. A. Cortez, C. H. Brito Cruz, M. Elmissiry, M. Laser et al. 2015. Bioenergy and African transformation. *Biotechnology for Biofuels* 8: 18.

Maqhuzu, A. B., K. Yoshikawa, and F. Takahashi. 2017. Biofuels from agricultural biomass in Zimbabwe: Feedstock availability and energy potential. *Energy Procedia* 142: 111–116.

Mekhilef, S., M. Barimani, A. Safari, and Z. Salam. 2014. Malaysia's renewable energy policies and programs with green aspects. *Renewable and Sustainable Energy Reviews* 40: 497–504.

Meisen, P., and J. Hubert. 2010. Renewable energy potential of Brazil. Global Energy Network Institute (GENI), September 2010, Brazil.

Merry, K., P. Bettinger, D. Grebner, J. Siry, and Z. Ucar. 2017. Assessment of potential agricultural and short-rotation forest bioenergy crop establishment sites in Jackson County, Florida, USA. *Biomass and Bioenergy* 105: 453–463.

Miao, R., and M. Khanna. 2017. Costs of meeting a cellulosic biofuel mandate with perennial energy crops: Implications for policy. *Energy Economics* 64: 321–334.

Mohammed, Y. S., N. Bashir, and M. W. Mustafa. 2015. Overuse of wood-based bioenergy in selected sub-Saharan Africa countries: Review of unconstructive challenges and suggestions. *Journal of Cleaner Production* 96: 501–519.

Moomaw, W., F. Yamba, M. Kamimoto, L. Maurice, J. Nyboer, K. Urama, and T. Weir. 2011. Introduction. In *IPCC Special Report on Renewable Energy Sources and Climate Change Mitigation* [O. Edenhofer, R. Pichs-Madruga, Y. Sokona, K. Seyboth, P. Matschoss, S. Kadner, T. Zwickel et al. (Eds.)], Cambridge University Press, Cambridge, UK.

Naqvi, S. R., S. Jamshaid, M. Naqvi, W. Farooq, and W. Afzal. 2018. Potential of biomass for bioenergy in Pakistan based on present case and future perspectives. *Renewable and Sustainable Energy Reviews* 81: 1247–1258.

Ngugi, M. R., V. J. Neldner, S. Ryan, T. Li, J. Lewis, P. Norman, and M. Mogilski. 2018. Estimating potential harvestable biomass for bioenergy from sustainably managed private native forests in Southeast Queensland, Australia. *Forest Ecosystems* 5: 6.

Nogueira, L. A. H., L. G. A. de Souza, L. A. B. Cortez, and M. R. L. V. Leal. 2017. Sustainable and integrated bioenergy assessment for Latin America, Caribbean and Africa (SIByl-LACAf): The path from feasibility to acceptability. *Renewable and Sustainable Energy Reviews* 76: 292–308.

Perdue, J. H., J. A. Stanturf, T. M. Young, X. Huang, and Z. Guo. 2017. Profitability potential for *Pinus taeda* L. (loblolly pine) short-rotation bioenergy plantings in the southern USA. *Forest Policy and Economics* 83: 146–155.

Plath, M., C. Moser, R. Bailis, P. Brandt, H. Hirsch, A. M. Klein, D. Walmsley, and H. von Wehrden. 2016. A novel bioenergy feedstock in Latin America? Cultivation potential of Acrocomia aculeata under current and future climate conditions. *Biomass and Bioenergy* 91: 186–195.

Powlson, D. S., M. J. Glendining, K. Coleman, and A. P. Whitmore. 2011. Implications for soil properties of removing cereal straw: Results from long-term studies. *Agronomy Journal* 103: 279–287.

REN21. 2016. Renewables 2016 Global Status Report, Renewable Energy Policy Network for the 21st Century. http://www.ren21.net/resources/publications/.

Republic of Turkey. 2012. Ministry of Food, Agriculture and Livestock Institute of Agricultural Economics and Policy Development (TEPGE). Article no. 204: 34.

Richardson, T. 2013. Overcoming barriers to sustainability in bioenergy research. *Environmental Science & Policy* 33: 1–8.

RIRDC. 2014. Opportunities for Primary Industries in the Bioenergy Sector: National Research, Development and Extension Strategy. Priority Area RD&E Implementation Plan. Publication No 14/056, RIRDC, Canberra, Australia.

Runge, C. F., and B. Senauer. 2007. How biofuels could starve the poor. *Foreign Affairs* 86: 41–53.

Sanford, G. R., L. G. Oates, P. Jasrotia, K. D. Thelen, and R. D. Jackson. 2016. Comparative productivity of alternative cellulosic bioenergy cropping systems in the North Central USA. *Agriculture, Ecosystems & Environment* 216: 344–355.

Saravanan, A. P., T. Mathimani, G. Deviram, K. Rajendran, and A. Pugazhendhi. 2018. Biofuel policy in India: A review of policy barriers in sustainable marketing of biofuel. *Journal of Cleaner Production* 193: 734–747.

Sastre, C. M., J. Carrasco, R. Barro, Y. González-Arechavala, and P. Ciria. 2016. Improving bioenergy sustainability evaluations by using soil nitrogen balance coupled with life cycle assessment: A case study for electricity generated from rye biomass. *Applied Energy* 179: 847–863.

Scarlat, N., J. F. Dallemand, and F. Fahl. 2018. Biogas: Developments and perspectives in Europe. *Renewable Energy* 129: 457–472. doi:10.1016/j.renene.2018.03.006.

Scarlat, N., J. F. Dallemand, F. Monforti-Ferrario, M. Banja, and V. Motola. 2015. Renewable energy policy framework and bioenergy contribution in the European Union—An overview from National Renewable Energy Action Plans and Progress Reports. *Renewable & Sustainable Energy Reviews* 51: 969–985. doi:10.1016/j.rser.2015.06.062.

Sharma, M., and A. Kumar. 2018. Promising biomass materials for biofuels in India's context. *Materials Letters* 220: 175–177.

Shuba, E. S., and D. Kifle. 2018. Microalgae to biofuels: 'Promising' alternative and renewable energy, review. *Renewable and Sustainable Energy Reviews* 81: 743–755.

Shukla, A. K., K. Sudhakar, and P. Baredar. 2017. Renewable energy resources in South Asian countries: Challenges, policy and recommendations. *Resource-Efficient Technologies* 3: 342–346.

Sims et al. 2011. Integration of renewable energy into present and future energy systems. In *IPCC Special Report on Renewable Energy Sources and Climate Change Mitigation* [O. Edenhofer, R. Pichs-Madruga, Y. Sokona, K. Seyboth, P. Matschoss, S. Kadner, T. Zwickel et al. (Eds.)], Cambridge University Press, Cambridge, United Kingdom and New York, NY, USA.

Smith, J. U., A. Fischer, P. D. Hallett, H. Y. Homans, P. Smith, Y. Abdul-Salam, H. H. Emmerling, and E. Phimister. 2015. Sustainable use of organic resources for bioenergy, food and water provision in rural Sub-Saharan Africa. *Renewable and Sustainable Energy Reviews* 50: 903–917.

Smyth, C., G. Rampley, T. C. Lemprière, O. Schwab, and W. A. Kurz. 2016. Estimating product and energy substitution benefits in national-scale mitigation analyses for Canada. *Global Change Biology Bioenergy* (Published online). doi:10.1111/gcbb.12389.

Smyth, C., W. A. Kurz, G. Rampley, T. C. Lempriere, and O. Schwab. 2017. Climate change mitigation potential of local use of harvest residues for bioenergy in Canada. *GCB Bioenergy* 9: 817–832. doi:10.1111/gcbb.12387.

Spataru, C., E. Zafeiratou, and M. Barrett. 2014. An analysis of the impact of bioenergy and geosequestration in the UK Future Energy System. *Energy Procedia* 62: 733–742.

Surendra, K. C., S. K. Khanal, P. Shrestha, and B. Lamsal. 2011. Current status of renewable energy in Nepal: Opportunities and challenges. *Renewable and Sustainable Energy Reviews* 15: 4107–4117.

Surendra, K. C., R. Ogoshi, H. M. Zaleski, A. G. Hashimoto, and S. K. Khanal. 2018. High yielding tropical energy crops for bioenergy production: Effects of plant components, harvest years and locations on biomass composition. *Bioresource Technology* 251: 218–229.

Tilman, D., R. Socolow, J. A. Foley, J. Hill, E. Larson, L. Lynd, S. Pacala, J. Reilly, T. Searchinger, C. Somerville, and R. Williams. 2009. Beneficial biofuels—The food, energy, and environment trilemma. *Science* 325(5938): 270–271.

Wilhelm, W. W., J. M. F. Johnson, D. L. Karlen, and D. T. Lightle. 2007. Corn stover to sustain soil organic carbon further constrains biomass supply. *Agronomy Journal* 99: 1665–1667.

Zabaniotou, A. 2018. Redesigning a bioenergy sector in EU in the transition to circular waste-based Bioeconomy-A multidisciplinary review. *Journal of Cleaner Production* 177: 197–206.

Zhao, G., B. A. Bryan, D. King, Z. Luo, E. Wang, and Q. Yu. 2015. Sustainable limits to crop residue harvest for bioenergy: Maintaining soil carbon in Australia's agricultural lands. *GCB Bioenergy* 7: 479–487. doi:10.1111/gcbb.12145.

2 Modeling of Sustainable Energy System from Renewable Biomass Resources in Response to Technical Development, Lifecycle Assessment, Cost, and Availability

Hossain M. Anawar and Vladimir Strezov

CONTENTS

2.1 INTRODUCTION

Biomass has been one of the main energy providers in rural areas for long time. The fourth-largest source of global energy in the past decade was from biomass, accounting for 10%–14% of the final energy consumption after coal (12%–14%), natural gas (14%–15%), and electricity (14%–15%) (Balat and Ayar, 2005). Biomass as a worldwide source of energy creates significant socio-economic and environmental benefits, and contributes to improved sustainability. The worldwide use of bioenergy as a renewable energy source is attracting more attention due to the decreasing reserves of fossil fuels, increasing population growth, and global warming. There are different types of bioenergy feedstocks, such as bioenergy crops (jatropha, switchgrass, miscanthus, willow, and others), rapeseed, sugarcane, poplar, Eucalyptus camaldulensis, agricultural waste, and crop residue. Depending on the properties of biomass, they have various yield potential and energy conversion efficiencies.

Although both field experiments and crop growth models can be used to quantify the biomass yield of energy crops, the results of field experiments from one area cannot be extrapolated to other areas to determine the biomass production, because the biomass growth-controlling factors, such as climate, soil conditions, and essential inputs, including management, are not the same in different areas (Nair et al., 2012; Jiang et al., 2017). The crop growth models are presumed to be successful for the theoretical estimation of biomass yield of different energy crops at the local and regional levels. The accuracy of modeling results depends on how accurately the models use the biomass growth-controlling parameters, such as climate; soil quality; availability and limitations of water, nutrients and other agronomic inputs; effect of pests; diseases; and weeds (van den Broek et al., 2001; Nair et al., 2012; Jiang et al., 2017).

Different methods are applied to convert biomass to bioenergy and biomaterials through the integrated management of biomass supply. The most important methods are fermentation, anaerobic digestion, gasification, direct combustion, and pyrolysis (Saidur et al., 2011). The selection of methods depends on the property of biomass and availability of technology. The supply chain of biomass energy is controlled and affected by availability of agricultural and other variable sources of biomass (Iakovou et al., 2010).

The conversion of biomass into gaseous or liquid fuels and biomaterials occurs through gasification techniques that consist of multiple inherent transformation processes (Basu, 2010). Some studies have been reported that modeled the (1) biomass production potential of energy crops and (2) success and failure of a biomass gasifier (Arnavat et al., 2010; Baruah and Baruah, 2014). These models can be used to indicate the recovery percentage of fuels from a particular feedstock using a biomass gasifier. Therefore, the modeling of biomass potential of energy crops and a biomass gasifier is presumed to create significant advancement in bioenergy fields. However, these models do not always produce the accurate output/results. The growers of energy crops and the owners of the forest play a significant role in deciding the type of crops and harvesting of crops and forest that provide biomass supply for energy production. Therefore, inclusion of the stakeholders in biomass supply models is necessary to assess the accurate biomass production potential.

2.2 BIOMASS PRODUCTIVITY OF ENERGY CROPS AND THEIR MODELING

The yields of biomass from energy crops are typically assessed by modeling because of the scale of contribution of bioenergy to the global energy supply (Jiang et al., 2017). The energy crop models, in combination with the process-based models, can predict the sustainable production of energy crops and economic and environmental issues. Crop models are being used to predict the yields of different energy crops and crop rotations based on the site-specific data, input, and environmental parameters for long time (Bauböck, 2014). The bioenergy system is a complex process that needs substantial data and knowledge to design and analyze the system before it is applied in the energy production. Field-level experiments are expensive and need significant time and area to obtain the crop-specific and site-specific yield data. However, the mathematical models can be applied in all fields of bioenergy systems to design and optimize the system. The mathematical models can simulate the biomass productivity and biomass yield potential, conversion of biomass into energy and biomaterials in a biorefinery system, economics of bioenergy supply chain logistics, and environmental effects of bioenergy system and resource recovery/lifecycle assessment models (Wang et al., 2015). The development and validation of models encounters some challenges, such as model hypothesis, assumptions, ideas, and data input. Therefore, these models should include the more accurate field-level data and estimating uncertainty to simulate the biomass yield of energy crops.

2.2.1 Types of Energy Crop Models

Up to the present time, different studies presented more than 20 models to simulate the yields of different energy crops (Table 2.1). Van den Broek et al. (2001) used the SILVicultural Actual (SILVA) model to simulate the biomass yield of *Eucalyptus camaldulensis*. Other studies modeled the biomass production potential of switchgrass, Miscanthus, maize, poplar, willow, and sugarcane (Jiang et al., 2017). Based on different principles or approaches, three main types of mechanistic plant-growth models, such as the radiation model, water-controlled crop model, and integrated model, were used to estimate the biomass yields (Bauböck, 2014; Jiang et al., 2017). The above 20 mechanistic models are founded on a few biological approaches, such as light interception, conversion of intercepted light into biomass, and partition of biomass to the different plant parts. The radiation models (EPIC, ALMANAC, APSIM, ISAM, MISCANMOD, MISCANFOR, SILVA, DAYCENT, APEX and SWAT) are based on a radiation use efficiency approach (RUE). The radiation use efficiency

TABLE 2.1
Characteristics of Selected Energy Crop Models

Categories	Model	Scale	Energy Crops Covered	Crop Model	Reference
Radiation model	EPIC	Field	Switchgrass, *Miscanthus*	Generic, dynamic	Williams et al. (1984)
	ALMANAC	Field	Switchgrass, *Miscanthus*	Generic, dynamic	Kiniry et al. (1992)
	APSIM	Field	Sugarcane	Generic, dynamic	Keating et al. (1999)
	ISLAM	0.1°, country[1]	Switchgrass	Crop specific	Jain et al. (2010)
	MISCANMOD	Field	*Miscanthus*	Crop specific	Khanna et al. (2008)
	MISCANMOD	Field	*Miscanthus*	Crop genotype specific	Hastings et al. (2009)
	SILVA	Commercial	*Eucalyptus camaldulensis*	Crop specific	Van den Broek et al. (2001)
	DAYCENT	Field, Regional	*Miscanthus*, switchgrass	Generic, dynamic	Davis et al. (2012)
	APEX	Channel system, watershed	Switchgrass	Generic, dynamic	Gassman et al. (2009)
	SWAT	Watershed, ecosystem	*Miscanthus*	Generic, dynamic	Ng et al. (2010)
Water-controlled crop model	AquaCrop model	Field	Switchgrass	Generic, dynamic	Ahmadi et al. (2015)
			Miscanthus, maize		Stričević et al. (2015)

(Continued)

TABLE 2.1 (*Continued*)
Characteristics of Selected Energy Crop Models

Categories	Model	Scale	Energy Crops Covered	Crop Model	Reference
Integrated model—	CANEGRO	Field	Sugarcane	Specific, dynamic	Inman-Bamber (1991)
photosynthesis and respiration	CANEGRO	Field	Sugarcane	Specific, dynamic	Inman-Bamber (1991)
approach	3PG	Stand level	Hybrid poplar, willow	Generic growth model dynamic	Landsberg and Waring (1997)
	CropSyst	Field	Maize	Crop specific, dynamic	Stöckle et al. (2003)
	DSSAT	Field	Maize		Fosu et al. (2012)
Integrated model—	SECRETS	Stand-ecosystem	*Miscanthus*, poplar	Generic growth model	Sampson and Ceulemans (2000)
biochemical	LPJmL	Ecosystem	Sugarcane	Generic	Bondeau et al. (2007)
approach	Agro-BGC	Ecosystem	Switchgrass	Generic, dynamic	Di Vittorio et al. (2010)
	Agro-IBIS	Ecosystem	Sugarcane, *Miscanthus*	Generic, dynamic	Kucharik (2003)
	WIMOVAC	Field	Switchgrass, *Miscanthus*	Generic, dynamic	Miguez et al. (2009)
	DNDC	Field, ecosystem	*Miscanthus*	Generic, dynamic	Li et al. (1992)
	DRAINMOD-GRASS	Ecosystem	Switchgrass, *Miscanthus*	Generic, dynamic	Tian et al. (2016)
	AgTEM		Switchgrass, *Miscanthus*, maize	Generic, dynamic	Qin et al. (2012)

Source: Reprinted from *J. Integr. Agric.*, 16, Jiang, R. et al., Modeling the biomass of energy crops: Descriptions, strengths and prospective, 1197–1210, Copyright 2017, with permission from Elsevier.

approach is more commonly and widely used in crop models. The crop model based on water use benefit is the AquaCrop model that emphasizes crop water use (Bauböck, 2014; Jiang et al., 2017). The CANEGRO, 3PG, CropSyst and DSSAT integrated models are based on the photosynthesis and respiration approaches, while the SECRETS, LPJmL, Agro-BGC, Agro-IBIS, and WIMOVAC/BioCro, DNDC, DRAINMOD-GRASS, and AgTEM models use biochemical approaches.

The models WOFOST and CROPGRO use the carbon-based growth engine tools (Bauböck, 2014). The yield data of crops grown in the standard conditions cannot indicate the future changes in crop yield under the climate change. Therefore, the crop model of BioSTAR can estimate and predict the biomass productivity of energy crops and food crops at the local and regional scale, using the climate and soil-related site data (Bauböck, 2014).

2.2.2 PERENNIAL BIOENERGY CROPS: MODELING OF YIELD AND CROP MARKET

The perennial energy crops (e.g., switchgrass) that produce large quantity of biomass are some of the highest-potential bioenergy feedstocks for sustainable and renewable energy production, because these grasses and trees can grow in poor soil substrates with low concentrations of water and nutrients and are beneficial to the environment (Heaton et al., 2008). The mechanistic models are developed to simulate the yields of perennial energy crops depending on the genotypes, plant species, environment, climate, locations, and agronomic management. Some models are developed for a specific crop type that cannot be used for other plants, while generic models are used for all plant types. The mechanistic plant-growth models are very useful to simulate the plant biomass yields and have wider applications, including diverse crop types, environments, geographical locations, climate, and managements. However, the empirical models developed based on field data (Peters, 1980) of diverse crop types, environments, geographical locations, climate, and managements can help to develop mechanistic models (Jager et al., 2010). Although the empirical models developed for switchgrass demonstrated uncertainties, they provide some useful information for further research and development.

The deployment of energy crops, such as Miscanthus and short-rotation coppice in the UK fields, is lower than the expected level to meet the renewable energy target of 15% of total energy consumption by 2020 (DECC, 2011). The supply of biomass feedstocks from crop residues and energy crops is not sufficient to meet the renewable energy target (Department for Transport, 2012), although government financial incentives are trying to accelerate the renewable energy sources. Alexander et al. (2013) studied the UK perennial energy crop market using the agent-based modeling (ABM) that explained the contingent interaction of supply and demand, and the spatial and temporal dynamics of energy crop adoption. The following features are important for the energy crop market (Alexander et al., 2014) and therefore should be incorporated in the ABM model: (1) farmer choice for food crops and energy crops; (2) crop selection by local soil, climate and other factors; (3) individual farmer's choice and behavior change; (4) transportation cost (Borjesson and Gustavsson, 1996; Dunnett et al., 2008); and (5) investment for setting power plant that needs demands of energy and supply of biomass feedstocks at reasonable price for long time of the project (Hellmann and Verburg, 2011; MacDonald, 2011). This model can incorporate the nonlinear behaviors of market dynamics (Anon, 2010) and the complex system of the developing energy crop market. The short-rotation coppice (SRC) willow will contribute little to the proportion of the anticipated perennial energy crop target, while Miscanthus will have a significant contribution to the biomass energy under different climate scenarios (Alexander et al., 2014).

2.2.3 Prospective and Principles of Energy Crop Models

There are different types of energy crop models that can predict the biomass growth potential of energy crops. The biomass growth potential of energy crops, including herbaceous and woody energy crops, has been successfully modeled by 14 different models (Nair et al., 2012). Using the particular field data and Miscanthus as a model crop plant, up to six process-based models were developed that successfully predicted the biomass growth potential of energy crops (Robertson et al., 2015).

There are different models developed for energy crop modeling using different biological processes and methods. In one of the radiation models, solar radiation is considered the most important factor for crop production, along with temperature and water (Monteith, 1977). The water-crop model is water-driven, based on the biomass water productivity that separates evapotranspiration into crop transpiration and soil evaporation (Zhang et al., 2013; Stričević et al., 2015). This model principally simulates crop biomass and its yield under specific water supplies of rainfed, supplemental, deficit, and full irrigation conditions (Steduto et al., 2009; Mabhaudhi et al., 2014). The integrated model can be typically described by two principle approaches, such as the (1) photosynthesis and respiration approach, and (2) biochemical approach. Two typical models (CANEGRO model for the photosynthesis and respiration approach; WIMOVAC for the biochemical approach) are used to describe each principle for the integrated model.

2.3 SUPPLY CHAIN OF BIOMASS TO ENERGY AND ITS MODELING

The source and supply of biomass feedstock is not always the same for bioenergy production. Biomass supply chain can supply the biomass resources efficiently for biorefinery industries (Mafakheri and Nasiri, 2014). The supply chain model and farm-scale economic models can explain the annual profit and uptake of energy crops compared to traditional food crops, using the theory of supply chain economics (Bauen et al., 2010; Sherrington and Moran, 2010; Alexander et al. 2014). The availability of biomass resources, demand of bioenergy, effective utilization of bioenergy, and setting up of a bioenergy industry chain can significantly improve the bioenergy sector (Lu and Zhao, 2013).

2.4 BIOMASS SUPPLY CHAIN MODELING

The bioenergy supply chain network model can present a cost-effective biofuel supply chain by reducing the production cost and improving the biomass' inherent quality (Castillo-Villar et al., 2016). For example, the low-quality biomass feedstocks, higher ash and moisture contents of biomass, and harvest residues reduce the biomass quality and increase the supply chain cost. The sugar contents and particle size of feedstocks also play a significant role in the quality of biomass feedstocks. Therefore, if quality of biomass feedstocks is not high, supply cannot meet bioenergy targets (Kenney et al., 2013). There are five categories of biomass supply chain modeling, as shown in Figure 2.1.

2.4.1 HARVESTING OF BIOMASS

The land distribution and planning of harvesting and biomass collection are decided based on bioenergy demand, land supply, climate scenarios, and biomass soil/moisture contents. Given spatial restrictions due to supply of land and productivity, the following two models have been developed for scheduling of biomass harvest (Murray, 1999): Unit Restriction Model (URM) and Area Restriction Model (ARM). Two adjacent blocks of land are not harvested at the same time in URM, while in ARM, each block of land can be harvested no more than once in each planning period. An integer programming model can identify and explain the decisions regarding the land selection for harvesting in response to bioenergy demands (Gunnarsson et al., 2004). In addition, a mixed integer programming (MIP) model considers several factors and can reduce the total cost of a biomass supply chain to the minimum level (Eksioglu et al., 2009).

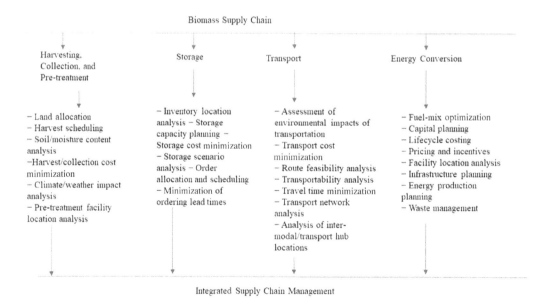

FIGURE 2.1 A taxonomy of the models developed for biomass supply chain operations management. (Reprinted from Mafakheri, F., and Nasiri, F. 2014. Modeling of biomass-to-energy supply chain operations: Applications, challenges and research directions. *Energy Policy* 67: 116–126. Copyright 2014, with permission from Elsevier.)

2.4.2 Biomass Pre-treatment

Some biomass pre-treatment processes facilitate the biomass-to-energy conversion by decreasing crystallinity of cellulose, increasing biomass surface area, removing hemicellulose, and breaking the lignin barrier (Chaula et al., 2014). The most important and common processes are physical separation, mechanical, and chemical processes. Furthermore, the pre-treatment processes commonly used in bioenergy production are drying, torrefaction, carbonization, pelletization, chopping, shredding, and grinding. These pre-treatment processes augment the energy conversion efficiency and decrease the cost of biomass supply chain (Mafakheri and Nasiri, 2014). Furthermore, a geographic information system (GIS)-based model can predict the cost efficiency, environmental impacts, and carbon footprint of different pre-treatment processes and facility areas that are used for energy production (Chiueh et al., 2012).

2.4.3 Biomass Storage

In biomass supply chains, decisions of biomass storage analyze the site/area, capacity and planning of storage. The properties of biomass resources and constraints of transportation options influence the appropriate location for biomass storage facilities. Allen et al. (1998) and Huisman et al. (1997) suggested the storage of biomass on the production field to decrease the cost of transportation. Kanzian et al. (2009) and Tatsiopoulos and Tolis (2003) developed a dynamic, discrete event simulation (Nilsson and Hansson, 2001) and linear programming models that suggested the locations of biomass storage between biomass production sites and the energy plant. A mixed-integer optimization model also explained the effect of including inter-modal storage facilities (Eksioglu et al., 2010). The dynamic programming approach and a linear programming model considered the location of biomass storage near the biomass energy plant and cost of biomass field to storage, thereby minimizing the total storage cost, depending on availability (Cundiff et al., 1997; Papadopoulos and Katsigiannis, 2002).

2.4.4 Biomass Transport

The transportation cost of biomass with heavy weight and low energy density discourages people from producing bioenergy using biomass resources (Castillo-Villar, 2014). The linear programming and MIP are used to develop most of the bioenergy supply chain models that consider the type, amount, and properties of biomass materials; availability of biomass; logistics facilities; and energy demand (Busato and Berruto, 2008). The effective optimization model can reduce the transportation cost and solve the bioenergy supply chain problems using the metaheuristic algorithmic approaches. Some studies used the GIS-based model to simulate the most suitable biomass delivery plan, minimum cost of transportation, environmental issues, and carbon footprint (Graham et al., 2000; Gronalt and Rauch, 2007; Frombo et al., 2009; Perpina et al., 2009).

Forsberg (2000) proposed a lifecycle analysis approach to identify the carbon footprint and greenhouse gas emissions in biomass transport. The various models are developed in the transport phase of the biomass supply chain, with the objectives of exploring feasible alternative routes, the best transport chain, and minimum transport cost, delivery time, and environmental effects in the biomass supply chain.

2.4.5 Biomass Energy Conversion

The location of a biomass conversion plant, selection of conversion technology, and operations and investment are important factors for the bioenergy investors. A hybrid GIS-linear programming approach (Velazquez-Marti and Fernandez-Gonzalez, 2010) and the mixed integer linear programming models (Zhu et al., 2011) were developed that could analyze feasible locations and find optimal locations of biomass conversion facilities. The MIP model was the best one to decide the optimal locations based on their advantages and disadvantages (Johnson et al., 2012).

The selection of biomass conversion technology is an important part of the biomass supply chains because they determine the energy conversion efficiency, type of biomass resources, processing of biomass materials, environmental issues, and total cost of biomass supply chains (McKendry, 2002). A bi-objective, multi-period mixed-integer linear programming optimization model can identify the conversion pathways and technologies, lifecycle costs, and carbon footprint, and provide the suboptimal least-cost supply chain (You and Wang, 2011). Figure 2.1 shows the various types of biomass supply chain models (Mafakheri and Nasiri, 2014).

2.5 CHALLENGES AND ISSUES

Integrated biofuel production is presumed to be more sustainable, using multiple sources of biomass resources, such as biomass waste, energy crops, animal waste, forest residue, and municipal waste. The optimization of biomass supply chain and decision support in bioenergy production can reduce the carbon and societal footprint, and environmental adverse effects. However, there is complexity and uncertainty in decisions of the producers and other stakeholders regarding the adoption of available resources, conversion technologies, and biomass supply chain. Some decisions might be currently sustainable. However, they might not be sustainable in the long term. Therefore, Seay and Badurdeen (2014) attempted to solve these problems using the discrete event simulation model. The following six types of challenges and issues exist with the biomass supply chain modeling influencing the operations: technical, financial, social, environmental, policy/regulatory, and institutional/organizational.

2.5.1 TECHNOLOGICAL MATTERS

The technological challenges of biomass supply chains are efficiencies of resource and supply chain, and production efficiency rates. Biomass harvesting for energy production without an effective, simultaneous planting program can lead to future scarcities of biomass resource supply for energy plants (Adams et al., 2011). The current biomass supply, management, and storage system are not sufficient to supply the biomass resources for large bioenergy production factories. Therefore, the supply chain efficiency, careful inventory planning, and optimal storage system, protecting the quality and quantity of biomass materials, are necessary for development of sustainable bioenergy systems (Hoogwijk et al., 2003; Gold and Seuring, 2011; Kurian et al., 2013).

2.5.2 FINANCIAL ISSUES

The total cost associated with the different components of biomass supply chain is the main financial issue of bioenergy production (Diamantopoulou et al., 2011). Other financial problems associated with the bioenergy supply chains are some uncertainties, such as inefficient conversion technologies, lack of profit and investments, a volatile energy market, and food crisis (Adams et al., 2011). The stochastic optimization model can optimize the total cost and financial risk of a biomass supply chain (Gebreslassie et al., 2012).

The farmers' choice of contract farming and close co-operation among farmers can promote the bioenergy supply chain. A new supply chain design is needed to increase the income of farmers, diversification in farming, and conversion of infertile, fallow land into fertile lands (Cembalo et al., 2014). Cembalo et al. suggested the *Arundo donax* as a potential bioenergy crop, because it can produce high biomass, mitigate soil erosion, and produce revenue compared to wheat. The adoption of a "minimum price guarantee by government" initiative can significantly reduce the "cost of the contract" and the negative effect of a long-term contract duration for the bioenergy company. It also encourages farmers in contract bioenergy farming.

2.5.3 Social Issues

The benefits of renewable energy sources are the enhancement of new jobs, socio-economic development in the regional areas, new income generation, and the meeting of energy demand in rural areas (Thornley et al., 2008). Sometimes, problems arise due to lack of (1) mutual participation of biomass producers and investors in the bioenergy projects, (2) realization of social benefits, and (3) mitigation of negative effects and carbon footprint. Furthermore, social benefits may not have been perceived locally and the negative impacts of biomass supply chains and power plants on local environments and land uses may not have been appropriately understood and mitigated (Upreti, 2004). The minimization of possible conflicts with food supply is another social challenge in biomass energy supply chain planning and management (Tilman et al., 2009).

2.5.4 Environmental Issues

Renewable energy use has some important environmental benefits, such as carbon footprint reduction, waste recycling, resource recovery, and waste management, provided that bioenergy production processes consider sustainability issues (Banos et al., 2011). Biomass supply chains have some negative ecological and other challenges due to transportation activities and space requirements (Awudu and Zhang, 2012) and unsustainable sources of feedstock. The discrete-event model (Mobini et al., 2011) and Arena-based simulation model (Zhang et al., 2012) were developed to predict the carbon footprint in forest biomass supply, total cost of biomass supply chain, and mitigation measures.

Foo et al. (2013) designed mathematical models for the empty fruit bunches (EFB), palm fiber, and palm shell–based regional energy supply chain in palm oil industrial areas. These models reduce the greenhouse gas emissions of the bioenergy supply chain, provide flexibility in operations under different planning and management and effects of climate change on agricultural production. This model integrates the environmental issues, effect of climate change, supply chain network of biogas and liquid biofuel production, and other renewable sources (Lam et al., 2010; Cucek et al., 2012).

The transportation of oil-palm plantation biomass from agricultural fields to palm oil mills and then to bioenergy power plants (combined heat and power [CHP]) causes the emission of greenhouse gases that forces implementation of carbon reduction policies in the bioenergy supply chain. Memari et al. (2018) developed a mixed-integer linear programming model to obtain the palm tree biomass bioenergy supply chain planning model under carbon pricing (carbon tax) and carbon trading (cap-and-trade) policies, providing some insights on the cost increase, carbon emissions reductions, sustainability of this technology, and output of the supply chain.

2.5.5 Policy and Regulatory Issues

Policy measures and regulations in indirect incentives or direct payments to renewable energy production, especially bioenergy, affect the capital and operational performance of a biomass supply chain. Imposing a carbon tax could not promote the renewable energy production. Instead, it increased the biomass supply chain cost. Therefore, national and regional policies and regulations should be in place to increase the different types of support (monetary and non-monetary), incentives, and sustainable planning for bioenergy production (Mafakheri and Nasiri, 2014).

2.5.6 Institutional and Organizational Issues

The policies and planning of institutes and organizations have significant roles in the promotion of biomass energy and biomass supply chains (Mafakheri and Nasiri, 2014). The different parties involved with the biomass supply chain have various standards and rules, resulting in some problematic issues (Table 2.1). The community-based biomass supply is a preferred option for smooth and continuous operation of biomass supply chains (Gold and Seuring, 2011).

2.6   MODELING OF BIOMASS GASIFICATION

Biomass gasification is a thermo-chemical conversion process, and one of the most important routes for biomass-based energy generation (Baruah and Baruah, 2014). Biomass gasification generates both heat and some intermediate chemicals, such as syngas ($CO + H_2$) for commercial use to produce heat and power (CHP), drop-in diesel by catalytic Fischer–Tropsch process, electricity, synthetic natural gas (SNG), methanol, and other compounds of interest. However, they need some purification and removal of tar. The process is better than combustion for higher conversion efficiency (Liu et al., 2013). Dahlquist et al. (2013) compared gas quality produced by different gasification processes and different modeling approaches to model the gasification processes. The functioning of the gasifier is governed by few factors, such as type of fuel, the reactor configuration, and operation parameters. The advantage of computational modeling tools is to identify the optimal states of a biomass conversion reactor without conducting time-consuming and expensive experimentation. A systematic logical analysis is required to model the gasification process and efficiently disseminate the embedded information.

The gasification process, quality, and properties of produced gases and success of the gasifier are greatly influenced by some important operating parameters, such as feeding rate of biomass materials and gasifying agent, pressure, and temperature of the gasifier (Basu, 2010). The mathematical models of the gasification process that can effectively provide insights on the configuration of the reactor, flow rate of feedstock materials, and type of biomass feedstock and performance of reactor (Basu, 2010) are classified into (1) thermodynamic equilibrium, (2) kinetic, and (3) artificial neural network (ANN) routes and are discussed below.

2.6.1   THERMODYNAMIC EQUILIBRIUM MODELS

A thermodynamic equilibrium model can simulate the composition of the gas produced in the reactor (Basu, 2010; Baruah and Baruah, 2014). There are two kinds of equilibrium models, such as stoichiometric models and non-stoichiometric models. The stoichiometric models consider the equilibrium constants (Giltrap et al., 2003) and some specific chemical reactions that are used to identify the property and composition of the produced gases. However, these models do not consider some other reactions, resulting in errors that can be solved by the non-stoichiometric modeling approach (Shabbar and Janajreh, 2013). Despite this, the equilibrium models have high potential to model the gasification process in downdraft gasifiers (Table 2.2) and fluidized bed gasifiers (Table 2.3). The accuracy of the equilibrium models can be achieved by incorporating empirical correlations based on experimental studies.

TABLE 2.2
Equilibrium Model in the Study of Downdraft Gasifiers

S. No.	Author(s) (year)	Feedstock Used/Molecular Formula	Parameters Studied
1	Babu and Sheth (2006)	$CH_{3.03}O_{1.17}$	Char reactivity factor
2	Melgar et al. (2007)	Rubberwood	Air-fuel ratio and moisture content
3	Gao and Li (2008)	$CH_{3.03}O_{1.17}$	Temperature of the pyrolysis zone
4	Sharma (2008)	Douglas fir bark	Moisture content, pressure, equivalence ratio in gasifier, initial temperature in reduction zone
5	Barman et al. (2012)	$CH_{1.54}O_{0.622}N_{0.0017}$	Air-fuel ratio, mole of moisture per mole of biomass
6	Azzone et al. (2012)	Corn stalks, sunflower stalks, and rapeseed straw	Pressure, temperature, biomass humidity, and oxidant agent composition
7	Antonopoulos et al. (2012)	Olive wood, *Miscanthus*, and cardoon	Reactor temperature, feedstock, moisture content

Source: Reprinted from *Renew. Sustain. Energy Rev.*, 39, Baruah, D., and Baruah, D. C., Modeling of biomass gasification: A review, 806–815, Copyright 2014, with permission from Elsevier.

TABLE 2.3
Equilibrium Model in the Study of Fluidised Bed Gasifiers

S. No.	Author(s) (year)	Model Considerations	Feedstock Used	Parameters Studied
1	Doherty et al. (2009)	Based on Gibb's free energy minimization approach	Hemlock wood	Equivalence ratio, temperature, level of air preheating, biomass moisture, and steam injection
2	Kaushal et al. (2011)	One-dimensional steady-state model	Wood chips	Mixing of devolatilized gas, average temperature of incoming bed material, moisture content of biomass, steam-to-biomass ratio
3	Gungor (2011)	One-dimensional, isothermal and steady-state, and the fluid-dynamics are based on the two-phase theory of fluidization. Tar conversion is taken into account in the model	Biomass	Gasifier temperature, bed operational velocity, equivalence ratio, biomass particle size, and biomass-to-steam ratio
4	Loha et al. (2011)	Equilibrium model	Rice husk, sugarcane bagasse, rice straw, and groundnut shell	Gasification temperature, steam to biomass ratio
5	Hannula and Kurkela (2012)	Equilibrium model using Aspen plus simulation	Crushed wood pellets and forest residues	Heat losses, gasification pressure, steam/oxygen ratios, filtration temperature and reformer conversion levels, reforming temperature, and drying percentage
6	Xie et al. (2012)	The model uses an Eulerian method for fluid phase and a discrete particle method for solid phase, which takes particle contact force into account	Pine wood	Reactor temperature, equivalence ratio, steam-to-biomass ratio
7	Nguyen et al. (2012)	Empirical model, including biomass pyrolysis, char–gas reactions and gas-phase reaction	Pine wood chips	Gasification temperature, steam-to-fuel ratio

Source: Reprinted from *Renew. Sustain. Energy Rev.*, 39, Baruah, D., and Baruah, D. C., Modeling of biomass gasification: A review, 806–815, Copyright 2014, with permission from Elsevier.

2.6.2 KINETIC MODEL

A kinetic model is more suitable and accurate than the equilibrium model at relatively low operating temperatures for complex reactor designs. This model can simulate the gas composition, temperature, and productivity of the reactor (Basu, 2010). It considers both the reaction kinetics and reactor hydrodynamics. There are some benefits and problems associated with equilibrium or kinetic models, or a combination of both. A computational fluid dynamics (CFD) model can simulate the performance of a gasifier reactor, including the positive benefits of both models. ANN modeling can also successfully simulate the performance of gasifiers.

Gao et al. (2017) developed a kinetic model of biomass gasification, using the kinetic parameters of a micro-fluidized bed (micro-FB) in the presence of silica sand as the fluidization medium and a reaction temperature of 700°C–1000°C. Reschmeier and Karl (2016) studied the kinetic behavior of biomass gasification in the presence of CO_2 and steam. The particle size of biomass materials controlled the activation energies of the process technology. The advanced physical and chemical technologies can recover the maximum calorific value and syngas by upgrading the low-quality organic fuels into more valuable products. For example, the fluidized-bed gasification and catalytic gasification are highly recommended to produce biomass syngas (Ochoa et al., 2001).

The molecular-level kinetic model of biomass gasification was divided into two categories, a biomass composition model and the construction of the reaction network (Horton et al., 2016). The biomass composition model was divided into three sub-models, cellulose, hemicellulose, and lignin. The biomass reaction network model consists of pyrolysis, gasification, and light–gas reactions. The biomass gasification emits the lower concentration of pollutants and produces the syngas and liquid fuels, where the tar production and syngas composition are the important parameters. The pyrolysis and gasification technology of coal was developed based on a chemical percolation devolatilization (CPD) process (Grant et al., 1989; Fletcher et al., 1990, 1992) that was subsequently applied in biomass gasification. The molecular-level kinetic model of biomass gasification can detect each individual molecular species of the feedstock and products throughout the reactor and helps to understand the reaction chemistries related to biomass gasification.

2.6.3 COMPUTATIONAL FLUID DYNAMICS AND ARTIFICIAL NEURAL NETWORK MODELS

CFD models developed based on solutions of equations for conservation of mass, momentum, energy, and species can predict the temperature, composition of gas, and other parameters of the reactor. The chemistry of biomass gasification and particulate flow are used in CFD modeling of biomass gasification (Pepiot et al., 2010). A selected CFD modeling work is described in Table 2.4. Arnavat et al. (2013) developed two ANN models, one for circulating fluidized bed gasifiers (CFB) and the other for bubbling fluidized bed gasifiers (BFB) to determine the composition of gas (CO, CO_2, H_2, CH_4) and gas yield.

Two-dimensional CFD cannot accurately simulate the biomass gasification in fluidized beds. Therefore, Liu et al. (2013) and Loha et al. (2014) developed a three-dimensional CFD steady-state model by coupling the other two models to accurately simulate hydrodynamics and biomass gasification in a CFB reactor and the kinetics of homogeneous and heterogeneous reactions in the reactor. They studied the impacts of turbulence models, radiation model, water–gas shift reaction (WGSR), and equivalence ratio (ER) for a clear understanding of biomass gasification in a CFB reactor. The efficiency of fluidized bed gasification systems depends on the char conversion ratio (Bates et al., 2016). The CFB is a potential biomass gasification process due to its efficient mixing effects and heat transfer in gasifiers (Nguyen et al., 2012). CFD can accurately predict the chemical reactions occurring in the chemical process, using mainly two types of methods, the Eulerian–Lagrange approach and the Eulerian–Eulerian approach. The relatively higher calorific value fuel and higher hydrogen content can be produced from biomass gasification in fluidized bed system using air–steam mixture gasifying agents.

The biomass gasification in dual fluidized bed (DFB) reactors was modeled using the Aspen Plus simulator (Abdelouahed et al., 2012). The DFB consists of three components, biomass pyrolysis, secondary reactions, and char combustion. The model simulated the mass yields of permanent gases, water, 10 tar species, char, secondary reactions, syngas composition and flow rate. The results demonstrated that the syngas composition and flow rate are very sensitive to the gas-phase WGSR kinetic. The model simulated the mass and energy balances of the DFB gasification process. The DFB gasifies the biomass with steam or recycled gas and produce pure syngas. The heat produced from the combustion of residual char and added fuels drives the endothermic reactions in the gasifier.

TABLE 2.4

Computational Fluid Dynamics and Artificial Neural Network Models in the Study of Biomass Gasifiers

S. No.	Author(s) (year)	Type of Gasifier Studied	Feedstock Used	Model Considerations	Parameters Studied
1	Gao et al. (2012)	Air cyclone	Sawdust of walnut	Detailed CFD model of a cyclone gasifier. Models of sawdust pyrolysis and combustion of volatiles and char have been added to the standard model	Equivalence ratio, gas composition
2	Jakobs et al. (2012)	Entrained flow gasifier	Ethylene glycol	CFD model. Steady balance equations for mass, momentum, and energy are solved using a finite volume solver	Spray quality
3	Janajreh et al. (2013)	Downdraft biomass gasifier	Woody biomass	The numerical simulation is conducted on a high-resolution mesh, accounting for the solid and gaseous phases, k–e turbulence, and reacting CFD model	Gas composition, cold gas efficiency, carbon conversion efficiency, reactor temperature
4	Arnavat et al. (2013)	CFB and BFB	Woody biomass	Feed-forward ANN model	Ash, moisture, biomass composition, equivalence ratio, gasification temperature for CFB and BFB respectively, steam to dry biomass ratio (kg/kg) for BFB only
5	Sreejith et al. (2013)	Fluidised bed gasifier	Wood sawdust	Feed-forward ANN model and equilibrium correction model incorporating tar (aromatic hydrocarbons) and unconverted char	Product gas composition, heating value and thermodynamic efficiencies

Source: Reprinted from *Renew. Sustain. Energy Rev.*, 39, Baruah, D., and Baruah, D. C., Modeling of biomass gasification: A review, 806–815, Copyright 2014, with permission from Elsevier.

ANN: artificial neural network; BFB: bubbling fluidized bed gasifiers; CFB: circulating fluidized bed gasifiers; CFD: computational fluid dynamics.

2.7 PELLETIZATION AND ATTRITION OF WOOD IN BIOMASS GASIFICATION

The pelletization reduces the attrition effect of wood pellets compared to wood chips. A shrinking particle combustion model can simulate the effect of pelletization on the attrition of wood in gasification reactors (Ammendola et al., 2013). The particle size/geometry of both wood pellets and wood chips strongly controls the combustion characteristics of biomass feedstocks. The gasification-assisted attrition reduces the residence time and complete conversion of the char particles even at the maximum temperature of the reactor, 900°C. The steady model can predict the maximum allowable biomass feeding rates with respect to reactor temperature, pressure, volume, and feedstock characteristics.

The fluidized bed reactors are used to gasify the different biomass feedstocks due to their easy preparation, high rates of heat/mass transfer, and thermal inertia of the bed material. A few complex physical and chemical processes occur in the reactors. The most important processes are rapid mixing, heating, drying, and devolatilization, resulting in a mixture of light, pyrolysis products, and highly porous carbonaceous char—the latter further reacts with steam and CO_2 to produce syngas. The reactivity of char and syngas production are related the properties of the various biomass feedstocks. Reactivity refers to the rates (s−1) of gasification and combustion reactions under kinetically limited conditions.

$$C + H_2O \leftrightarrow H_2 + CO \ (+131.4 \text{ kJ/mol}) \tag{2.1}$$

$$C + CO_2 \leftrightarrow 2CO \ (+172.5 \text{ kJ/mol}) \tag{2.2}$$

$$C + 0.5O_2 \rightarrow CO \ (-393.8 \text{ kJ/mol}) \tag{2.3}$$

2.8 ENERGY PRODUCTION FROM PINE SAWDUST AND MODELING

Biomass steam explosion is one of the potential pre-treatment processes that can increase biomass storage and fuel properties by removing moisture and hemicellulose. According to the model, the temperature range of 260°C–317°C (533–590 K) and corresponding pressure of 4.7–10.8 MPa are the suitable operating conditions for steam explosion pre-treatment process to remove moisture and hemicellulose from pine sawdust (Chaula et al., 2014). The characterization and modeling demonstrated that the pine sawdust can be used as biomass feedstock to generate bioenergy via steam explosion.

2.9 CONCLUSIONS

The energy crop models and process-based models can predict the sustainable production and yield of energy crops, crop rotations, and economic and environmental issues. The mathematical models can simulate the biomass productivity and biomass yield potential, conversion of biomass into energy and biomaterials in a biorefinery system, economics of bioenergy supply chain logistics, and environmental effects of bioenergy system and lifecycle assessment models. These models should include the more accurate field-level data and estimating uncertainty to simulate the biomass yield of energy crops. More than 20 models have been demonstrated in different studies to simulate the yields of different energy crops. Different studies modeled the biomass production potential of *Eucalyptus camaldulensis*, switchgrass, miscanthus, maize, poplar, willow, and sugarcane. The perennial energy crops (e.g., switchgrass) that produce large quantities of biomass are some of the highest potential bioenergy feedstocks for sustainable and renewable energy production, because these grasses and trees can grow in poor soil substrates with low concentrations of water and nutrients and are beneficial to the environment. Three main types of mechanistic plant-growth models—radiation model, water-controlled crop model, and integrated model—were developed based on different principles or approaches. The radiation use efficiency approach is more commonly and widely used in crop models. The crop model of BioSTAR can estimate and predict the biomass productivity of energy crops and food crops at the local and regional scale, using the climate and soil-related site data. The empirical models, developed based on field data of diverse crop types, environments, geographical locations, climate, and management, can help to develop mechanistic models.

The supply of biomass feedstocks from crop residues and energy crops is not enough to meet the renewable energy target in the United Kingdom. Biomass supply chain can supply the biomass resources efficiently for biorefinery industries. The supply chain model and farm-scale economic models can explain the annual profit and uptake of energy crops compared to traditional food crops using the theory of supply chain economics. The availability of biomass resources, demand of bioenergy, effective utilization of bioenergy, and the setting up of a bioenergy industry chain can significantly improve the bioenergy sector. There are five categories of biomass supply chain modeling: harvesting of biomass, biomass pre-treatment, biomass storage, biomass transport, and biomass energy conversion. The following six types of challenges and issues exist with the biomass supply chain modeling, influencing the operations: technical, financial, social, environmental, policy/regulatory, and institutional/organizational.

The biomass gasification is a thermo-chemical conversion process, and one of the most important routes for biomass-based energy generation. Biomass gasification generates both heat and some intermediate chemicals, such as syngas ($CO + H_2$) for commercial use to produce CHP, drop-in diesel by catalytic Fischer–Tropsch process, electricity, synthetic natural gas, methanol, and other compounds. The advantage of computational modeling tools is to identify the optimal states of a biomass conversion reactor without performing time-consuming and expensive experimentation. The gasification process, quality, and properties of produced gases and success of the gasifier are greatly influenced by some important operating parameters, such as the feeding rate of biomass materials and gasifying agent, pressure, and temperature of the gasifier. The mathematical models of the gasification process are classified into (1) thermodynamic equilibrium models, (2) kinetic models, and (3) artificial neural network models. There are some benefits and problems associated with equilibrium models or kinetic models, or a combination of both. CFD models can simulate the performance of a gasifier reactor, including the positive benefits of both models. ANN modeling can also successfully simulate the performance of gasifiers.

REFERENCES

Abdelouahed, L., O. Authier, G. Mauviel, J. P. Corriou, G. Verdier, and A. Dufour. 2012. Detailed modeling of biomass gasification in dual fluidized bed reactors under Aspen Plus. *Energy Fuels* 26: 3840–3855.

Adams, P. W., G. P. Hammond, M. C. McManus, and W. G. Mezzullo. 2011. Barriers to and drivers for UK bio-energy development. *Renewable and Sustainable Energy Reviews* 15(2): 1217–1227.

Ahmadi, S. H., E. Mosallaeepour, A. A. Kamgar-Haghighi, and A. R. Sepaskhah. 2015. Modeling maize yield and soil water content with aquacrop under full and deficit irrigation managements. *Water Resources Management* 29: 2837–2853.

Alexander, P., D. Moran, M. D. A. Rounsevell, and P. Smith. 2013. Modelling the perennial energy crop market: The role of spatial diffusion. *Journal of Royal Society Interface* 10: 20130656. doi:10.1098/rsif.2013.0656.

Alexander, P., D. Moran, P. Smith, A. Hastings, S. Wang, G. Sunnenberg, A. Lovett et al. 2014. Estimating UK perennial energy crop supply using farm-scale models with spatially disaggregated data. *GCB Bioenergy* 6: 142–155. doi:10.1111/gcbb.12121.

Allen, J., M. Browne, A. Hunter, J. Boyd, and H. Palmer. 1998. Logistics management and costs of biomass fuel supply. *International Journal of Physical Distribution & Logistics Management* 28(6): 463–477.

Ammendola, P., R. Chirone, G. Ruoppolo, and F. Scala. 2013. The effect of pelletization on the attrition of wood under fluidized bed combustion and gasification conditions. *Proceedings of the Combustion Institute* 34(2): 2735–2740.

Anon. 2010. Agents of change: Conventional economic models failed to foresee the financial crisis. Could agent-based modelling do better? *The Economist*, 24 July, London, UK.

Antonopoulos, I. S., A. Karagiannidis, A. Gkouletsos, and G. Perkoulidis. 2012. Modeling of a downdraft gasifier fed by agricultural residues. *Waste Management* 32: 710–718.

Arnavat, M. P., J. C. Bruno, and A. Coronas. 2010. Review and analysis of biomass gasification models. *Renewable and Sustainable Energy Reviews* 14: 2841–2851.

Arnavat, M. P., J. A. Hernandez, J. C. Bruno, and A. Coronas. 2013. Artificial neural network models for biomass gasification in fluidized bed gasifiers. *Biomass Bioenergy* 49: 279–289.

Awudu, I., and J. Zhang. 2012. Uncertainties and sustainability concepts in biofuel supply chain management: A review. *Renewable and Sustainable Energy Reviews* 16(2): 1359–1368.

Azzone, E., M. Morini, and M. Pinelli. 2012. Development of an equilibrium model for the simulation of thermochemical gasification and application to agricultural residues. *Renewable Energy* 46: 248–254.

Babu, B. V., and P. N. Sheth. 2006. Modeling and simulation of reduction zone of downdraft biomass gasifier: Effect of char reactivity factor. *Energy Conversion Management* 47: 2602–2611.

Balat, M., and G. Ayar. 2005. Biomass energy in the world, use of biomass and potential trends. *Energy Sources* 27(10): 931–940.

Banos, R., F. Manzano-Agugliaro, F. G. Montoya, C. Gil, A. Alcayde, and J. Gomez. 2011. Optimization methods applied to renewable and sustainable energy: A review. *Renewable and Sustainable Energy Reviews* 15: 1753–1766.

Barman, N. S., S. Ghosh, and S. De. 2012. Gasification of biomass in a fixed bed downdraft gasifier: A realistic model including tar. *Bioresource Technology* 107: 505–511.

Baruah, D., and D. C. Baruah. 2014. Modeling of biomass gasification: A review. *Renewable and Sustainable Energy Reviews* 39: 806–815.

Basu, P. 2010. *Biomass Gasification and Pyrolysis: Practical Design and Theory.* Elsevier, Burlington, MA.

Bates, R. B., C. Altantzis, and A. F. Ghoniem. 2016. Modeling of biomass char gasification, combustion, and attrition kinetics in fluidized beds. *Energy Fuels* 30: 360–376.

Bauböck, R. 2014. Simulating the yields of bioenergy and food crops with the crop modeling software BioSTAR: The carbon-based growth engine and the BioSTAR ET0 method. *Environmental Sciences Europe* 26: 1.

Bauen, A. W., A. J. Dunnett, G. M. Richter, A. G. Dailey, M. J. Aylott, E. Casella, and G. Taylor. 2010. Modelling supply and demand of bioenergy from short rotation coppice and Miscanthus in the UK. *Bioresource Technology* 101: 8132–8143.

Bondeau, A., P. C. Smith, S. Zaehle, S. Schaphoff, W. Lucht, W. Cramer, D. Gerten et al. 2007. Modelling the role of agriculture for the 20th century global terrestrial carbon balance. *Global Change Biology* 13: 679–706.

Borjesson, P., and L. Gustavsson. 1996. Regional production and utilization of biomass in Sweden. *Energy* 21: 747–764. doi:10.1016/0360-5442(96)00029-1.

Busato, P., and B. Berruto. 2008. Logistics design process of biomass supply chain: simulation and optimization models. In: *Proceedings of the International Conference on Agricultural Engineering*, Crete, Greece, June 23–25.

Castillo-Villar, K. K. 2014. Metaheuristic algorithms applied to bioenergy supply chain problems: Theory, review, challenges, and future. *Energies* 7: 7640–7672. doi:10.3390/en7117640.

Castillo-Villar, K. K., H. Minor-Popocat, and E. Webb. 2016. Quantifying the impact of feedstock quality on the design of bioenergy supply chain networks. *Energies* 9: 203. doi:10.3390/en9030203.

Cembalo, L., S. Pascucci, C. Tagliafierro, and F. Caracciolo. 2014. Development and management of a bioenergy supply chain through contract farming. *International Food and Agribusiness Management Review* 17(3): 33–51.

Chaula, Z., M. Said, G. John, S. Manyele, and C. Mhilu. 2014. Modelling the suitability of pine sawdust for energy production via biomass steam explosion. *Smart Grid and Renewable Energy* 5(1): 7. doi:10.4236/sgre.2014.51001.

Chiueh, P. T., K. C. Lee, F. S. Syu, and S. H. Lo. 2012. Implications of biomass pretreatment to cost and carbon emissions: Case study of rice straw and Pennisetum in Taiwan. *Bioresource Technology* 108: 285–294.

Cundiff, J. S., N. Dias, and H. D. Sherali. 1997. A linear programming approach for designing a herbaceous biomass delivery system. *Bioresource Technology* 59(1): 47–55.

Cucek, L., J. J. Klemes, and Z. Kravanja. 2012. A review of footprint analysis tools for monitoring impacts on sustainability. *Journal of Cleaner Production* 34: 9–20.

Dahlquist, E., G. Mirmoshtaghi, E. K. Larsson, E. Thorin, J. Yan, K. Engvall, T. Liliedahl, C. Dong, X. Hu, and Q. Lu. 2013. Modelling and simulation of biomass conversion processes. In: *Modelling and Simulation (EUROSIM), 8th EUROSIM Congress.* IEEE, Cardiff, UK. doi:10.1109/EUROSIM.2013.91.

Davis, S. C., W. J. Parton, S. J. Del Grosso, C. Keough, E. Marx, P. R. Adler, and E. H. DeLucia. 2012. Impact of second-generation biofuel agriculture on greenhouse-gas emissions in the corn-growing regions of the US. *Frontiers in Ecology and the Environment* 10: 69–74.

Department of Energy and Climate Change (DECC). 2011. *UK Renewable Energy Roadmap.* London, UK.

Department for Transport. 2012. Department of Energy and Climate Change and Department for Environment Food and Rural Affairs (DTECCEFRA). UK bioenergy strategy, London, UK.

Diamantopoulou, L. K., L. S. Karaoglanoglou, and E. G. Koukios. 2011. Biomass cost index: Mapping bio-mass-to-biohydrogen feedstock costs by a new approach. *Bioresource Technology* 102(3): 2641–2650.

Di Vittorio, A. V., R. S. Anderson, J. D. White, N. L. Miller, and S W. Running. 2010. Development and optimization of an Agro-BGC ecosystem model for C4 perennial grasses. *Ecological Modelling* 221: 2038–2053.

Doherty, W., A. Reynolds, and D. Kennedy. 2009. The effect of air preheating in a biomass CFB gasifier using ASPEN Plus simulation. *Biomass Bioenergy* 33: 1158–1167.

Dunnett, A. J., C. S. Adjiman, and N. Shah. 2008. A spatially explicit whole-system model of the lignocel-lulosic bioethanol supply chain: An assessment of decentralised processing potential. *Biotechnology for Biofuels* 1: 13. doi:10.1186/1754-6834-1-13.

Eksioglu, S. D., A. Acharya, L. E. Leightley, and S. Arora. 2009. Analyzing the design and management of biomass-to-biorefinery supply chain. *Computers and Industrial Engineering* 57(4): 1342–1352.

Eksioglu, S. D., S. Li, S. Zhang, S. Sokhansanj, and D. Petrolia. 2010. Analyzing impact of inter-modal facili-ties on design and management of biofuel supply chain. *Transportation Research Record* 2191: 144–151.

Fletcher, T. H., A. R. Kerstein, R. J. Pugmire, and D. M. Grant. 1990. Chemical percolation model for devola-tilization. 2. Temperature and heating rate effects on product yields. *Energy Fuels* 4: 54–60.

Fletcher, T. H., A. R. Kerstein, R. J. Pugmire, M. S. Solum, and D. M. Grant. 1992. Chemical percolation model for devolatilization. 3. Direct use of ^{13}C NMR data to predict effects of coal type. *Energy Fuels* 6: 414–431.

Foo, D. C. Y., R. R. Tan, H. L. Lam, M. K. A. Aziz, and J. J. Klemes. 2013. Robust models for the synthesis of flexible palm oil-based regional bioenergy supply chain. *Energy* 55: 68–73.

Forsberg, G. 2000. Biomass energy transport: Analysis of bio-energy transport chains using life cycle inven-tory method. *Biomass Bioenergy* 19(1): 17–30.

Fosu, M., S. S. Buah, R. L. Kanton, and W. A. Agyare. 2012. Modeling maize response to mineral fertilizer on silty clay loam in the northern savanna zone of ghana using DSSAT model. In: *Improving Soil Fertility Recommendations in Africa using the Decision Support System for Agrotechnology Transfer (DSSAT)*. Springer, Dordrecht, the Netherlands, pp. 157–168.

Frombo, F., R. Minciardi, M. Robba, F. Rosso, and F. Sacile. 2009. Planning woody biomass logistics for energy production: A strategic decision model. *Biomass Bioenergy* 33(3): 372–383.

Gao, J., S. Koshio, M. Ishikawa, S. Yokoyama, T. Ren, C. F. Komilus, and Y. Han. 2012. Effects of dietary palm oil supplements with oxidized and non-oxidized fish oil on growth performances and fatty acid compositions of juvenile Japanese sea bass, *Lateolabrax japonicus*. *Aquaculture* (324–325): 97–103.

Gao, N., and A. Li. 2008. Modeling and simulation of combined pyrolysis and reduction zone for a downdraft biomass gasifier. *Energy Conversion and Management* 49(12): 3483–3490.

Gao, W., M. R. Farahani, M. Rezaei, A. Q. Baig, M. K. Jamil, M. Imran, and R. Rezaee-Manesh. 2017. Kinetic modeling of biomass gasification in a micro fluidized bed. *Energy Sources, Part A: Recovery, Utilization, and Environmental Effects* 39(7): 643–648. doi:10.1080/15567036.2016.1236302.

Gassman, P. W., J. R. Williams, X. Wang, A. Saleh, E. Osei, L. M. Hauck, R. C. Izaurralde, and J. Flowers. 2009. The Agricultural Policy Environmental EXtender (APEX) model: An emerging tool for landscape and watershed environmental analyses. *CARD Technical Report*, Paper 41.

Gebreslassie, B. H., Y. Yao, and F. You. 2012. Design under uncertainty of hydrocarbon biorefinery supply chains: Multiobjective stochastic programming models, decomposition algorithm, and a comparison between CVaR and downside risk. *American Institute of Chemical Engineering* 58(7): 2155–2179.

Giltrap, D. L., R. McKibbin, and G. R. G. Barnes. 2003. A steady state model of gas–char reactions in a down draft gasifier. *Sol Energy* 74: 85–91.

Gold, S., and S. Seuring. 2011. Supply chain and logistics issues of bio-energy production. *Journal of Cleaner Production* 19(1): 32–42.

Graham, R. L., B. C. English, and C. E. Noon. 2000. A Geographic Information System-based modeling sys-tem for evaluating the cost of delivered energy crop feedstock. *Biomass and Bioenergy* 18(4): 309–329.

Grant, D. M., R. J. Pugmire, T. H. Fletcher, and A. R. Kerstein. 1989. Chemical model of coal devolatilization using percolation lattice statistics. *Energy Fuels* 3: 175–186.

Gronalt, M., and P. Rauch. 2007. Designing a regional forest fuel supply network. *Biomass Bioenergy* 31(6): 393–402.

Gungor, A. 2011. Modeling the effects of the operational parameters on H2 composition in a biomass fluidized bed gasifier. *International Journal of Hydrogen Energy* 36: 6592–6600.

Gunnarsson, H., M. Ronnqvist, and J. T. Lundgren. 2004. Supply chain modelling of forest fuel. *European Journal of Operational Research* 158(1): 103–123.

Hannula, I., and E. Kurkela. 2012. A parametric modeling study for pressurised steam blown fluidised-bed gasification of wood. *Biomass Bioenergy* 38: 58–67.

Hastings, A., J. Clifton-Brown, M. Wattenbach, C. P. Mitchell, and P. Smith. 2009. The development of MISCANFOR, a new Miscanthus crop growth model: Towards more robust yield predictions under different climatic and soil conditions. *Global Change Biology Bioenergy* 1: 154–170.

Heaton, E. A., F. G. Dohleman, and S. P. Long. 2008. Meeting US biofuel goals with less land: The potential of Giant miscanthus. *Global Change Biology* 14(9): 2000–2014.

Hellmann, F., and P. H. Verburg. 2011. Spatially explicit modelling of biofuel crops in Europe. *Biomass Bioenergy* 35: 2411–2424. doi:10.1016/j.biombioe.2008.09.003.

Hoogwijk, M., A. Faaij, R. Broek Van den, G. Berndes, D. Gielen, and W. Turkenburg. 2003. Exploration of the ranges of the global potential of biomass for energy. *Biomass Bioenergy* 25(2): 119–133.

Horton, S. R., R. J. Mohr, Y. Zhang, F. P. Petrocelli, and M. T. Klein. 2016. Molecular-level kinetic modeling of biomass gasification. *Energy Fuels* 30: 1647–1661.

Huisman, W., P. Venturi, and J. Molenaar. 1997. Costs of supply chains of Miscanthus giganteus. *Industrial Crops and Products* 6(3–4): 353–366.

Iakovou, E., A. Karagiannidis, D. Vlachos, A. Toka, and A. Malamakis. 2010. Waste biomass-to-energy supply chain management: A critical synthesis. *Waste Management* 30(10): 1860–1870.

Inman-Bamber, N. G. 1991. A growth model for sugar-cane based on a simple carbon balance and the CERES-Maize water balance. *South African Journal of Plant and Soil* 8: 93–99.

Jager, H., L. M. Baskaran, C. C. Brandt, E. B. Davis, C. A. Gunderson, and S. D. Wullschleger. 2010. Empirical geographic modeling of switchgrass yields in the United States. *GCB Bioenergy* 2: 248–257. doi:10.1111/j.1757-1707.2010.01059.x.

Jain, A. K., M. Khanna, M. Erickson, and H. Huang. 2010. An integrated biogeochemical and economic analysis of bioenergy crops in the Midwestern United States. *GCB Bioenergy* 2: 217–234.

Jakobs, T., N. Djordjevic, S. Fleck, M. Mancini, R. Weber, and T. Kolb. 2012. Gasification of high viscous slurry R&D on atomization and numerical simulation. *Applied Energy* 93: 449–456.

Janajreh, I., and M. Al-Shrah. 2013. Numerical and experimental investigation of downdraft gasification of woodchips. *Energy Conversation and Management* 65: 783–792.

Jiang, R., T.-T. Wang, J. Shao, S. Guo, W. Zhu, Y.-J. Yu, S.-L. Chen, and R. Hatano. 2017. Modeling the biomass of energy crops: Descriptions, strengths and prospective. *Journal of Integrative Agriculture* 16(6): 1197–1210.

Johnson, D. M., T. L. Jenkins, and F. Zhang. 2012. Methods for optimally locating a forest biomass-to-biofuel facility. *Biofuels* 3(4): 489–503.

Kanzian, C., F. Holzleitner, K. Stampfer, and S. Ashton. 2009. Regional energy wood logistics–optimizing local fuel supply. *Silva Fennica* 43(1): 113–128.

Kaushal, P., T. Proell, and H. Hofbauer. 2011. Application of a detailed mathematical model to the gasifier unit of the dual fluidized bed gasification plant. *Biomass Bioenergy* 35: 2491–2498.

Keating, B. A., M. J. Robertson, R. C. Muchow, and N. I. Huth. 1999. Modelling sugarcane production systems I. Development and performance of the sugarcane module. *Field Crops Research* 61: 253–271.

Kenney, K., W. Smith, G. Gresham, and T. Westover. 2013. Understanding biomass feedstock variability. *Biofuels* 4: 111–127.

Khanna, M., B. Dhungana, and J. Clifton-Brown. 2008. Costs of producing miscanthus and switchgrass for bioenergy in Illinois. *Biomass and Bioenergy* 32: 482–493.

Kiniry, J. R., J. R. Williams, P. W. Gassman, and P. Debaeke. 1992. A general, process-oriented model for two competing plant species. *Transactions of the American Society of Agricultural Engineers* 35: 801–810.

Kucharik, C. J. 2003. Evaluation of a process-based AgroEcosystem Model (Agro-IBIS) across the US corn belt: Simulations of the inter-annual variability in maize yield. *Earth Interactions* 7: 1–33.

Kurian, J. K., G. R. Nair, A. Hussain, and G. S. V. Raghavan. 2013. Feedstocks, logistics and pre-treatment processes for sustainable lignocellulosic biorefineries: A comprehensive review. *Renewable and Sustainable Energy Reviews* 25: 205–219.

Lam, H. L., P. S. Varbanov, and J. J. Klemes. 2010. Optimisation of regional energy supply chains including renewables: P-graph approach. *Computers & Chemical Engineering* 34: 782–792.

Landsberg, J. J., and R. H. Waring. 1997. A generalised model of forest productivity using simplified concepts of radiation-use efficiency, carbon balance and partitioning. *Forest Ecology and Management* 95: 209–228.

Li, C., S. Frolking, and T. A. Frolking. 1992. A model of nitrous oxide evolution from soil driven by rainfall events: 1. Model structure and sensitivity. *Journal of Geophysical Research (Atmospheres)* 97: 9759–9776.

Liu, H., A. Elkamel, A. Lohi, and M. Biglari. 2013. Computational fluid dynamics modeling of biomass gasification in circulating fluidized-bed reactor using the Eulerian–Eulerian approach. *Industrial and Engineering Chemistry Research* 52: 18162–18174.

Loha, C., P. K. Chatterjee, and H. Chattopadhyay. 2011. Performance of fluidized bed steam gasification of biomass: Modeling and experiment. *Energy Conversion Management* 52: 1583–1588.

Loha, C., H. Chattopadhyay, and P K. Chatterjee. 2014. Three-dimensional kinetic modeling of fluidized bed biomass gasification. *Chemical Engineering Science* 109: 53–64.

Lu, M., and Q.-H. Zhao. 2013. Brief analysis of characteristics and further research orientation of bioenergy supply chain. *Natural Resources* 4(1): 105–109.

Mabhaudhi, T., A. T. Modi, and Y. G. Beletse. 2014. Parameterisation and evaluation of the FAO-AquaCrop model for a South African taro (Colocasia esculenta L. Schott) landrace. *Agricultural and Forest Meteorology* 192: 132–139.

MacDonald, M. 2011. *Costs of Low-Carbon Generation Technologies.* Committee on Climate Change, London, UK.

Mafakheri, F., and F. Nasiri. 2014. Modeling of biomass-to-energy supply chain operations: Applications, challenges and research directions. *Energy Policy* 67: 116–126.

McKendry, P. 2002. Energy production from biomass (Part 2): Conversion technologies. *Bioresource Technology* 83(1): 47–54.

Melgar, A., J. F. Pe´rez, H. Laget, and A. Horillo. 2007. Thermochemical equilibrium modeling of a gasifying process. *Energy Conversion Management* 48: 59–67.

Memari, A., R. Ahmad, A. R. A. Rahim, and M. R. A. Jokar. 2018. An optimization study of a palm oil-based regional bio-energy supply chain under carbon pricing and trading policies. *Clean Technologies and Environmental Policy* 20: 113–125.

Miguez, F. E., X. Zhu, S. Humphries, G. A. Bollero, and S. P. Long. 2009. A semimechanistic model predicting the growth and production of the bioenergy crop Miscanthus × giganteus: Description, parameterization and validation. *GCB Bioenergy* 1: 282–296.

Mobini, M., T. Sowlati, and S. Sokhansanj. 2011. Forest biomass supply logistics for a power plant using the discrete-event simulation approach. *Applied Energy* 88(4): 1241–1250.

Monteith, J. L. 1977. Climate and the efficiency of crop production in Britain. *Philosophical Transactions of the Royal Society of London* (*Series B - Biological Sciences*) 281: 277–294.

Murray, A. T. 1999. Spatial restrictions in harvest scheduling. *Forest Science* 45(1): 45–52.

Nair, S. S., S. Kang, X. Zhang, F. E. Miguez, R. C. Izaurralde, W. M. Post, M. C. Dietze, L. R. Lynd, and S. D. Wullschleger. 2012. Bioenergy crop models: Descriptions, data requirements, and future challenges. *GCB Bioenergy* 4: 620–633.

Ng, T. L., J. W. Eheart, X. Cai, and F. Miguez. 2010. Modeling miscanthus in the soil and water assessment tool (SWAT) to simulate its water quality effects as a bioenergy crop. *Environmental Science & Technology* 44: 7138–7144.

Nguyen, T. D. B., M. W. Seo, Y. Lim, B.-H. Song, and S.-D. Kim. 2012. CFD simulation with experiments in a dual circulating fluidized bed gasifier. *Computers and Chemical Engineering* 36: 48–56.

Nilsson, D., and P. A. Hansson. 2001. Influence of various machinery combinations, fuel proportions and storage capacities on costs for co-handling of straw and reed canary grass to district heating plants. *Biomass Bioenergy* 20(4): 247–260.

Ochoa, J., M. C. Casanello, P. R. Bonelli, and A. L. Cukerman. 2001. CO_2 gasification of Argentinean coal chars: A kinetic characterization. *Fuel Process Technology* 74: 161–176.

Papadopoulos, D. P., and P. A. Katsigiannis. 2002. Biomass energy surveying and techno-economic assessment of suitable CHP system installations. *Biomass Bioenergy* 22(2): 105–124.

Pepiot, P., C. J. Dibble, and T. D. Foust. 2010. *Computational Fluid Dynamics Modeling of Biomass Gasification and Pyrolysis.* American Petroleum Society, New York.

Perpina, C., D. Alfonso, A. Perez-Navarro, E. Penalvo, C. Vargas, and R. Cardenas. 2009. Methodology based on geographic information systems for biomass logistics and transport optimisation. *Renewable Energy* 34(3): 555–565.

Peters, R. H. 1980. Useful concepts for predictive ecology. *Synthese* 43: 257–269.

Qin, Z., Q. Zhuang, and M. Chen. 2012. Impacts of land use change due to biofuel crops on carbon balance, bioenergy production, and agricultural yield, in the conterminous United States. *GCB Bioenergy* 4: 277–288.

Reschmeier, R., and J. Karl. 2016. Experimental study of wood char gasification kinetics in fluidized beds. *Biomass Bioenergy* 85: 288–299.

Robertson, A. D., C. A. Davies, P. Smith, M. Dondini, and N. P. McNamara. 2015. Modelling the carbon cycle of Miscanthus plantations: Existing models and the potential for their improvement. *GCB Bioenergy* 7: 405–421.

Saidur, R., E. A. Abdelaziz, A. Demirbas, M. S. Hossain, and S. Mekhilef. 2011. A review on biomass as a fuel for boilers. *Renewable and Sustainable Energy Reviews* 15(5): 2262–2289.

Sampson, D. A., and R. Ceulemans. 2000. SECRETS: Simulated carbon fluxes from a mixed coniferous/deciduous Belgian forest. In: Ceulemans, R., Veroustraete, F., Gond, V., Van Rensbergen, J. B. H. F. (Eds.), *Forest Ecosystem Modeling, Upscaling and Remote Sensing.* SPB Academic Publishing, The Hague, the Netherlands, pp. 95–108.

Seay, J. R., and F. F. Badurdeen. 2014. Current trends and directions in achieving sustainability in the biofuel and bioenergy supply chain. *Current Opinion in Chemical Engineering* 6: 55–60.

Shabbar, S., and I. Janajreh. 2013. Thermodynamic equilibrium analysis of coal gasification using Gibbs energy minimization method. *Energy Conversion and Management* 65: 755–763.

Sharma, A. K. 2008. Equilibrium modeling of global reduction reactions for a downdraft (biomass) gasifier. *Energy Conversion Management* 49: 832–842.

Sherrington, C., and D. Moran. 2010. Modelling farmer uptake of perennial energy crops in the UK. *Energy Policy* 38: 3567–3578.

Sreejith, C. C., C. Muraleedharan, and P. Arun. 2013. Performance prediction of fluidised bed gasification of biomass using experimental data-based simulation models. *Biomass Conversion and Biorefinery* 3: 1–22.

Steduto, P., T. C. Hsiao, D. Raes, and E. Fereres. 2009. AquaCrop - The FAO crop model to simulate yield response to water: I. Concepts and underlying principles. *Agronomy Journal* 101: 426–437.

Stöckle, C. O., M. Donatelli, and R. Nelson. 2003. CropSyst, a cropping systems simulation model. *European Journal of Agronomy* 18: 289–307.

Stričević, R., Z. Dzeletovic, N. Djurovic, and M. Cosic. 2015. Application of the AquaCrop model to simulate the biomass of Miscanthus × giganteus under different nutrient supply conditions. *GCB Bioenergy* 7: 1203–1210.

Tatsiopoulos, I. P., and A. J. Tolis. 2003. Economic aspects of the cotton-stalk biomass logistics and comparison of supply chain methods. *Biomass Bioenergy* 24(3): 199–214.

Thornley, P., J. Rogers, and Y. Huang. 2008. Quantification of employment from biomass power plants. *Renewable Energy* 33(8): 1922–1927.

Tian, S., M. A. Youssef, G. M. Chescheir, R. W. Skaggs, J. Cacho, and J. Nettles. 2016. Development and preliminary evaluation of an integrated field scale model for perennial bioenergy grass ecosystems in lowland areas. *Environmental Modelling & Software* 84: 226–239.

Tilman, D., R. Socolow, J. A. Foley, J. Hill, E. Larson, L. Lynd, S. Pacala, J. Reilly, T. Searchinger, C. Somerville, and R. Williams. 2009. Beneficial biofuels – the food, energy, and environment trilemma. *Science* 325(5938): 270–271.

Upreti, B. R. 2004. Conflict over biomass energy development in the United Kingdom: some observations and lessons from England and Wales. *Energy Policy* 32(6): 785–800.

van den Broek, R., L. Vleeshouwers, M. Hoogwijk, A. van Wijk, and W. Turkenburg. 2001. The energy crop growth model SILVA: Description and application to eucalyptus plantations in Nicaragua. *Biomass and Bioenergy* 21: 335–349.

Velazquez-Marti, B., and E. Fernandez-Gonzalez. 2010. Mathematical algorithms to locate factories to transform biomass in bio-energy focused on logistic network construction. *Renewable Energy* 35(9): 2136–2142.

Wang, L., S. A. Agyemang, H. Amini, and A. Shahbazi. 2015. Mathematical modelling of production and biorefinery of energy crops. *Renewable and Sustainable Energy Reviews* 43: 530–544.

Williams, J. R., C. A. Jones, and P. T. Dyke. 1984. The EPIC model and its application. In: *Proceedings of the International Symposium on Minimum Data Sets for Agrotechnology Transfer.* ICRISAT Center, Patancheru, India.

Xie, J., W. Zhong, B. Jin, Y. Shao, and H. Liu. 2012. Simulation on gasification of forestry residues in fluidized beds by Eulerian Lagrangian approach. *Bioresource Technology* 121: 36–46.

You, F., and B. Wang. 2011. Life cycle optimization of biomass-to-liquid supply chains with distributed-centralized processing networks. *Industrial and Engineering Chemical Research* 50(17): 10102–10127.

Zhang, F., D. M. Johnson, and M. A. Johnson. 2012. Development of a simulation model of biomass supply chain for biofuel production. *Renewable Energy* 44: 380–391.

Zhang, W., W. Liu, Q. Xue, J. Chen, and X. Han. 2013. Evaluation of the AquaCrop model for simulating yield response of winter wheat to water on the southern Loess Plateau of China. *Water Science & Technology* 68: 821–828.

Zhu, X., X. Li, Q, Yao, and Y. Chen. 2011. Challenges and models in supporting logistics system design for dedicated-biomass-based bioenergy industry. *Bioresource Technology* 102: 1344–1351.

3 Sustainable Energy Production from Distributed Renewable Waste Resources through Major Waste-to-Energy Activities

Tao Kan, Vladimir Strezov, and Tim Evans

CONTENTS

3.1 INTRODUCTION

With increasing global population and urbanization, handling of increasing wastes, especially the municipal solid waste (MSW), has become a global concern. Landfilling, incineration, and composting are the three main routes for processing wastes. For example, in China, the MSW amount treated by landfilling, incineration, and composting were respectively 96.0, 23.2, and 1.8 million tonnes/year in 2010, compared to 64.0, 3.7, and 7.2 million tonnes/year in 2003 (Zhang et al. 2015). This is more than a 50% increase in just seven years. Landfilling is also the primary waste management method in some of the European Union (EU) countries, with the waste to energy (WtE) as the second preferred method. For example, during 2005–2006, more than 60% of MSW was landfilled in the United Kingdom. WtE plants processed less than 10% of MSW, which will increase to around 25% and contribute more than 15% of total electricity consumption by 2020 (Yassin et al. 2009). To achieve efficient and sustainable waste management, based on the US Environmental

TABLE 3.1

Hierarchy of Sustainable Waste Management

Priority	Hierarchy of Sustainable Waste Management	
	USEPA (2016)	Expanded hierarchy proposed by Themelis (2008)
Higher	Source reduction and reuse	Waste reduction
↑	Recycling and composting	Recycling
		Anaerobic composting[a]
		Aerobic composting[a]
	Energy recovery	Waste-to-energy
↓	Treatment & disposal, including physical	Modern landfill recovering and using methane
	(e.g., shredding), chemical (e.g., incineration)	Modern landfill recovering and flaring methane
Lower	and biological (e.g., anaerobic digestion,	Pre-regulation landfill
	treatments, and landfills)	

Source: Themelis, N. J. Reducing landfill methane emissions and expanded hierarchy of sustainable waste management. *Proceedings of Global Waste Management Symposium*, Rocky Mountains, CO, 2008; US EPA. Sustainable materials management: Non-hazardous materials and waste management hierarchy. Retrieved February 16, 2016, from http://www.epa.gov, 2016.

[a] Only for source-separated organics.

Protection Agency's (EPA) hierarchy of sustainable waste management, Themelis (2008) proposed the expanded hierarchy, as shown in Table 3.1.

Considering difficulties for the reuse of organic wastes, recycling or composting a large fraction of wastes, in combination with other factors, such as fossil fuel depletion and increased energy demand, WtE has become an increasingly used option. Additionally, WtE also benefits a reduced carbon footprint, decreased greenhouse gas emissions, and land conservation (Arena 2011).

The definition of sustainable development is expressed as follows: "the needs of the present generation should be met without compromising the ability of future generations to meet their own needs" (WCED 1987). In the context of sustainable energy systems, the environmental and social aspects of sustainable development should be encompassed, which means the net effect of economic energy production activities would not threaten the environmental well-being (e.g., land/water resource availability, and biodiversity), and social well-being (e.g., employment and equity) (Evangelisti et al. 2015, IRENA 2015a).

WtE is now an essential part and development trend of modern waste management. In the past decades, WtE facilities have been increasingly established. WtE refers to the energy recovery from wastes, including the distributed renewable waste resources (DRWR). DRWR means the biogenic wastes from non-fossil renewable sources (Nolan ITU Pty Ltd and TBU 2001), including the following:

1. Certain domestic refuses, such as food waste, and organic fraction of municipal solid waste (OFMSW), excluding non-biogenic organic wastes like plastic, garden waste, paper and newspaper
2. Agricultural residues, such as sugarcane bagasse, cotton straw, and animal manure
3. Certain industrial residues, such as sawdust from wood processing, residues from sugar refineries, dairy wastes, pulps, and sewage sludge (Kothari et al. 2010, Lombardi et al. 2015, O'Shea et al. 2016, Nizami et al. 2017)

The WtE technologies are mainly comprised of three categories, as shown in Figure 3.1. The flow diagram of the most prevalent WtE thermal treatment process is depicted in Figure 3.2.

A typical thermal WtE plant includes a thermal reactor (e.g., a combustor), boiler, flue gas cleaning system, and steam Rankine cycle. The hot flue gas exiting the reactor goes into the boiler integrated with the reactor to generate superheated steam, which is then introduced into a condensing turbine

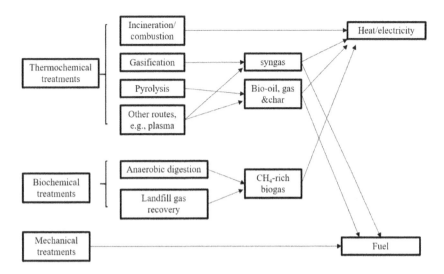

FIGURE 3.1 Technological methods for energy production from wastes. (From Consonni, S., and Vigano, F., *Waste Manage* 32: 653–666; Yue, D. J., et al. 2017. *Comput Chem Eng* 66: 36–56; Nizami, A. S., et al. 2017. *Appl Energy* 186: 189–196.)

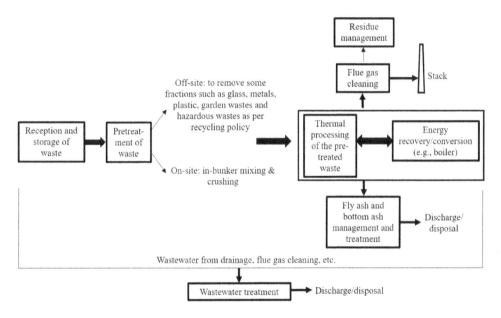

FIGURE 3.2 Flow diagram of the waste-to-energy (WtE) thermal treatment process.

for generating power (i.e., electricity) or into a back pressure or an extraction-condensing turbine combined heat and power (CHP) production (Lombardi et al. 2015). Compared to power generation only, CHP production is regarded as more beneficial to the energy recovery and environmental performances (Damgaard et al. 2010).

Energy can be recovered from wastes in a number of forms, such as heat, electricity, CH_4-rich biogas transportation fuel, syngas, bio-diesel, ethanol, methanol, and gasoline. This chapter has a strong emphasis on the energy systems in which the most prevalent energy forms of heat and electricity are the target products. This chapter aims to give an overall picture of the techno-economic performance, environmental performance and emission abatement measures, and uncertainty aspects of WtE activities.

3.2 TECHNO-ECONOMIC ASPECTS OF WASTE-TO-ENERGY ACTIVITIES

The economic performance of WtE activities is one of the primary concerns of utilizing distributed renewable waste resources. The economic performance depends on a combination of a number of aspects, including the following:

- Local heat and electricity demands and affordability decided by the population size and distribution, local income, households' living habits and activities, and industrial & business activities
- Location (i.e., the countries) of WtE activities
- Types and availability of renewable waste resources
- Transportation costs
- Degree of maturity and development of WtE technologies tailored to the practical conditions
- Capital, operational, and maintenance costs of WtE facilities, including the construction cost of relevant infrastructure, such as transportation system, water supply, wastewater discharge system, and grid connection
- Participation of industrial and other entities, as well as WtE energy customers
- Financial and policy support from governments at different levels

In the following sections, some of these important factors are briefly introduced to give an overview of the economic aspect of energy generation from distributed renewable waste resources.

3.2.1 ENERGY COST IN DIFFERENT COUNTRIES

Globally, the cost of electricity from biomass processing is about 30–240 (typically 50–100) USD/MWh, compared to the lower cost of large hydropower at 30–150 (typically 30–50) cents/MWh (IRENA 2013, 2015b). However, the cost of renewable wastes expressed as USD/GJ is dependent on the feedstock type, collection, and transportation, as shown in Table 3.2.

3.2.2 COMPARISON OF DIFFERENT WASTE TO ENERGY TECHNOLOGIES

Incineration, gasification, pyrolysis, and anaerobic digestion (AD) are the major WtE technologies, and their brief description, equipment types, advantages, disadvantages, and status are listed in Table 3.3.

TABLE 3.2
Typical Prices of Renewable Wastes for Waste-to-Energy Plants

Waste Types	Price (USD/GJ)	Comments
OFMSW/sewage sludge	≤0	Transportation cost excluded
Wood wastes	0.5–2.5	Collection and
Landfill gas	0.9–2.8[a]	transportation costs included
Forest residues	1.3–2.6	
Agricultural wastes	1.7–4.3	

Sources: US EPA, Combined heat and power: Catalog of technologies. Retrieved November 11, 2015, from http://www.epa.gov, 2007; IRENA, Renewable power generation costs in 2012: An overview, Retrieved June 6, 2016, from https://www.irena.org/, 2013.

[a] Including gas collection and flare.

OFMSW: organic fraction of municipal solid waste.

TABLE 3.3

Technological Aspects and Current Status of Different Waste-to-Energy (WtE) Technologies

WtE Technology	Technology Description	Equipment Type	Advantages	Disadvantages/Comments	Current Status
Incineration	It is currently the most dominant technology for waste thermal treatment and energy production. Surplus oxygen is supplied to obtain complete oxidation of waste materials with $\lambda = 1.5$.[a] Incineration converts the chemical energy in wastes to heat-producing waste by-products, such as flue gas, fly ash, and bottom ash. The flue gas consists of CO_2, steam and O_2	Moving grate combustor	Withstands 1250°C; maximum capacity of 120 MW/line on LHV basis	In EU, the grate combustor is widely utilized, with a dominant share of approximately 90% of the incineration plants	Incineration with ORC: early commercial; incineration with steam circle: commercial
		Fixed-grate combustor Rotary kiln	Accepts all types of waste; withstands 1400°C; maximum capacity of 30 MW/line on LHV basis	/	
		Fluidized bed	Suitable for materials with homogenous particle distribution	Requirement of material pretreatment; relatively low temperature of 800°C–900°C	
Gasification	Material treated in a partial oxidation atmosphere (oxidants: air, steam, CO_2, and/or O) with $\lambda = \sim 0.5$ at 550°C–1300°C. The aim of gasification is to produce syngas with high heating value	Fixed bed (updraft and down-draft), fluidized bed (bubbling/circulating fluidized bed), entrained bed, vertical shaft, moving grate furnace, rotary kiln and plasma reactor	The produced syngas is easier to process than flue gas from waste incineration; Reducing the generation of some pollutants, such as PCDD/Fs and NOx; Options for further utilization of syngas, e.g., synthesis and combustion in internally-fired devices to enhance energy efficiency	Preparation of homogeneous materials is required; Concerns about the explosive nature of the syngas; Gasification process is more complicated, resulting in higher cost, operation and maintenance complexity; further utilization of syngas is costly	Gasification with fuel cell: between R&D and demonstration; BICGT and BIGCC: generally at demonstration scale; Gasification with engine: early commercial; Gasification with steam: near commercial stage

(Continued)

TABLE 3.3 (Continued)
Technological Aspects and Current Status of Different Waste-to-Energy (WtE) Technologies

WtE Technology	Technology Description	Equipment Type	Advantages	Disadvantages/Comments	Current Status
Pyrolysis	Wastes are heated at 450°C–800°C in non-oxidizing atmosphere. Bio-oil, combustible gas and solid bio-char are the main products	Fixed bed, fluidized bed (bubbling/circulating fluidized bed), ablative reactor, vacuum pyrolysis reactor, rotating cone reactor, auger reactor, microwave reactor, plasma reactor	Acceptance of a wide range of feedstock; Bio-oils can be further upgraded to produce various energy types; Almost all materials can be recovered with nearly no emissions to environment	Preparation of homogeneous materials are required. High water content, corrosivity and complexity of bio-oils make upgrading of bio-oils very difficult	Generally at early commercial stage. First commercial plant was set up in 1984 in Germany, and there were 11 plants in Japan by 2009
Anaerobic digestion (AD)	The process can be briefly described as: wastes → monomer sugars + amino acids + higher fatty acids → fatty acids + ethanol + H_2 + CO_2 + NH_3 + H_2S → acetic acid + H_2 + CO_2 + formic acid + methanol → CH_4 + CO_2	Various digesters, such as fixed film, anaerobic fluidized bed, complete mix, plug flow and suspended media	Combined treatment of different types of feedstock; co-generation of biogas and fibrous fertilisers; stable process	Lower production rate than thermal methods; odor generation due to sulphur-bearing compounds	Two-stage AD: between early commercial and commercial; one-stage AD and landfill gas recovery: commercial

Source: IEA Bioenergy, Accomplishments from IEA Bioenergy task 36: Integrating energy recovery into solid waste management systems (2007–2009), Retrieved June 6, 2016, from www.ieabioenergy.com, 2010; Consonni, S., and Vigano, F., *Waste Manage.*, 32, 653–666, 2012; ISWA, Waste-to-energy. State-of-the-art-report. Statistics, 6th ed., Retrieved February 22, 2016, from http://www.iswa.org/media/publications/knowledge-base, 2012; Astrup, T. F. et al., *Waste Manag.*, 37, 104–115, 2015; IRENA, 2015a; Lombardi, L. et al., *Waste Manage.*, 37, 26–44, 2015; Pan, S. Y., et al., *J. Clean. Prod.*, 108, 409–421, 2015.

[a] Ratio.

BIGCC: biomass internal gasification combined cycle; BICGT: biomass internal combustion gas turbine; LHV: lower heating value; ORC: organic Rankine cycle; PCDD/Fs: polychlorinated dibenzodioxines and furans.

3.2.3 ECONOMIC PERFORMANCE OF DIFFERENT TECHNOLOGIES

The economic analysis of two commercially available and widely used WtE technologies, incineration and anaerobic digestion, is presented in this section.

3.2.3.1 Incineration

The electricity production efficiency for incineration systems should be around 0.3–0.7 MWh$_e$/t waste, with the preferable range of >0.55 MWh$_e$/t waste. This efficiency can be improved by increasing the steam pressure and temperature before the turbine or decreasing the pressure behind (Tabasova et al. 2012). Generally, the larger the plant processing capacity, the lower the capital and operating costs (per tonne/year) with all other conditions (e.g., geographical area, feedstock, and WtE technology) being the same or equivalent. For example, for Danish incineration plants, the capital and operating costs in 2004 were, respectively, 650 and 48.8 (7.5% of 650) euro/tpa at the plant size of 40 ktpa, with lower costs of 560 and 36.5 (6.5% of 560) euro/tpa at a larger size of 230 ktpa (Murphy and McKeogh 2004).

Typically, electricity of 0.4–0.7 MWh$_e$ or heat of 1.3–2.6 MWh$_{th}$ can be derived from one tonne of treated MSW. The energy production efficiencies (MWh$_{th}$ or MWh$_e$ per tonne of treated MSW) of selected waste incineration plants in European countries, based on the French plants, are exhibited in Figure 3.3. Generally, the heat recovery efficiency is higher than the electricity generation due to the distinct efficiency factors for external heat production and electrical conversion (i.e., 1 MWh is equivalent to 0.38 MWh$_e$ or 0.91 MWh$_{th}$) (European IPPC Bureau 2006). The difference between the "production" and "export" values in Figure 3.3 is attributed to the consumption of heat or electricity by the plants. The energy efficiency of the European MSW incineration plants is also expressed in percentage (%, based on energy from both wastes and used fuels), as shown in Figure 3.4. In the more advanced energy production mode of CHP, the production and export efficiencies are approximately 60% and 50%, respectively.

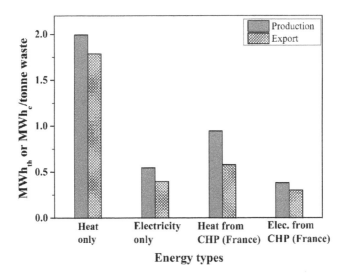

FIGURE 3.3 Average energy production efficiencies (MWh$_{th}$ or MWh$_e$/tonne waste) of selected MSW incineration plants in European countries. (From European IPPC Bureau. European IPPC Bureau: Reference document of the best available techniques for waste incineration, 2006.)

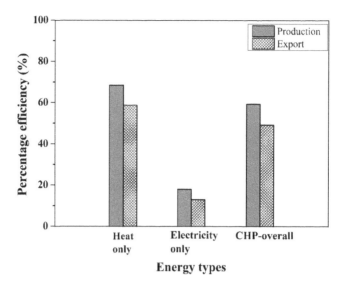

FIGURE 3.4 Average percentage efficiencies (%) of selected municipal solid waste (MSW) incineration plants in the European countries. (From European IPPC Bureau. European IPPC Bureau: Reference document of the best available techniques for waste incineration, 2006.)

3.2.3.2 Anaerobic Digestion for Bio-Gas Production

3.2.3.2.1 Bio-Gas to Combined Heat and Power

For CHP production from AD processing of wastes, the produced bio-gas is combusted in a bio-gas engine to generate electricity. A minority of the produced heat (e.g., approximately 10%) will be used to pre-heat the wastes, and a part of the produced electricity (e.g., approximately 35%) will be consumed in some processes (e.g., pre-treating waste and driving pumps) (Murphy and McKeogh 2004). Figure 3.5 shows the efficiency of energy export at different plant capacities (DRANCO process from Organic Waste Systems, Year 2001). The efficiency of heat export is about 0.27 MWh_{th}/tonne waste at the plant capacity of 5000 tpa, which is not obviously affected by the plant capacity in the

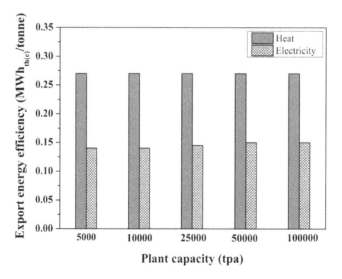

FIGURE 3.5 Efficiency of energy export at different plant capacities for DRANCO process from organic waste systems in 2001. (From Six, W., Project engineer with organic waste systems, personal communication, 2001.)

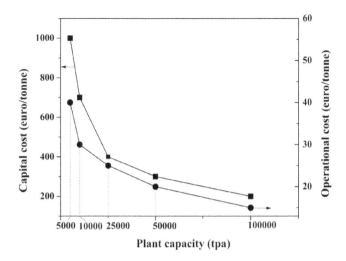

FIGURE 3.6 Capital and operating costs with the plant capacity of anaerobic digestion of wastes for combined heat and power (CHP) production (DRANCO process from Organic Waste Systems, Year 2001). (From Six, W., Project engineer with organic waste systems, personal communication, 2001; Murphy, J.D. and McKeogh, E., 2004. *Renew. Energy*, 29, 1043–1057, 2004.)

investigated range. The efficiency of heat export is about 0.14 MWh_e/tonne waste at 5000 tpa and slightly increases with increasing the capacity from 10,000 to 50,000 tpa.

The capital cost (euro/tpa) and operational cost (euro/tonne) with the plant capacity are shown in Figure 3.6. As evident, the increase in the plant capacity from 5,000 to 100,000 tpa results in the decrease in the capital cost from 1000 to 200 euro/tpa and the operating cost from 40 to 15 euro/tonne.

3.2.3.2.2 *Bio-Gas as Transportation Fuel*

When the end bio-gas product is used as transportation fuel, the cost of the CHP plant (1.0–2.5 million euro/MW_e depending on the CHP output) can be saved, but additional scrubbers (7,000–8,000 euro/m^3 bio-gas/hour) are needed to clean the crude bio-gas to meet the requirements of vehicle engines (Murphy and McKeogh 2004). After processing, a biogas with methane content of 95%–98% can be obtained with great reduction in CO_2, H_2S, and moisture. The efficiency of using methane-rich bio-gas in vehicles has been proven to be comparable to that of using petroleum, which leads to its extensive application as vehicle fuel in many International Energy Agency (IEA) Task 36 participating countries, such as Germany, Finland, and Denmark (Murphy and McKeogh 2004, IEA Bioenergy 2014). IEA Task 36 refers to the three-year program Integrating Energy Recovery into Solid Waste Management Systems, approved by the Executive Committee of IEA Bioenergy in 2006.

3.2.4 SUMMARY OF CAPITAL, OPERATIONAL, AND MAINTENANCE COSTS

The capital costs for WtE facilities generally include costs for equipment, infrastructure and logistics, civil works, grid connection, land use, planning, pre-financing and loan interests, and others (IRENA 2013). The operating costs mainly cover the expenditures for feedstock, electricity and water consumption, personnel salaries, depreciation costs, and emission (ash, wastewater, and gas) processing (Zhao et al. 2016). Table 3.4 shows the typical ranges of capital, operating, and maintenance and levelized costs, as well as the proportion of feedstock costs, for different WtE technologies.

The revenues are generally comprised of three main parts (i.e., heat and electricity sale, gate fees, and value-added tax return). Gate fees, as a subsidy, are the payment from governments and are based on the quantities of waste that WtE plants process. It was reported that the gate fees

TABLE 3.4

Typical Cost Ranges (Year 2010) for Different Electricity Generation Technologies from Biomass Wastes in OECD (Organization for Economic Co-operation and Development) Countries

Technologies (Electricity Generation from Biomass Wastes)	Capital Cost (×10⁶ USD/ MWᵃ)	Operating and Maintenance Costs		Levelized Cost of Electricity (USD/MWh)	Proportion of Feedstock Cost in Levelized Cost of Electricity (%)
		Fixed (% of Capital Cost Per Annual)	Variable (USD/MWh)		
Stoker combustion	1.8–4.2	3.2–6	3800–4700	0.06–0.21	8–49
BCB/CFB combustion	2.2–4.4				
Gasification	2.1–5.6	3–6	3700	0.06–0.24	9–44
CHP	3.5–6.8	Not provided	Not provided	0.07–0.29	13–85
Anaerobic digestion	2.5–6.2	2.1–7	4200	0.06–0.15	Not provided
Landfill gas	2.05–2.6	11–20	Not applicable	0.06–0.15	12–25
Co-firing with coal	0.3–0.85	2.5–3.5	Not provided	0.04–0.13	82–87

Source: IRENA, Renewable power generation costs in 2012: An overview, Retrieved June 6, 2016, from https://www.irena .org/, 2013; IRENA, 2015b.

ᵃ Megawatts.

for WtE plants via gasification and pyrolysis processes were in the range of 35–140 euro/tonne waste in 2009, and the new large-scale incineration plants were around 50–80 euro/tonne in United Kingdom (Yassin et al. 2009). Murphy and McKeogh (2004) estimated the gate fees for incineration, gasification, bio-gas to CHP, and biogas to transport fuel were, respectively 55, 40, 20 and 11 euro/tonne waste in 2004 in Ireland, based on the assumption that energy was supplied to a city with 1,000,000 population, electricity price at euro 70/MWh, petrol price at euro 0.89/L, and 25% tax on CH_4-rich transport fuel.

3.3 EMISSIONS FROM WASTE-TO-ENERGY ACTIVITIES

The final levels of emissions to the air, water, and soil depend on a number of parameters, such as the waste type and chemical composition, the applied WtE technologies, reactor type, process parameters, pollution abatement technologies of the gas, water and solid effluents, and emission regulations in different countries/regions.

3.3.1 Environmental Aspects of Waste-to-Energy Activities

For the most commonly used technology of WtE incineration, the flue gas from waste incineration generally contains particulate matter (PM), elements of S, Cl, F, and N, and components of N_2, CH_4, CO, NO_x, and SO_2, volatile organic compounds (VOCs), vaporised heavy metals (e.g., Pb, Hg, and Cd), polychlorinated dibenzodioxins and furans (PCDD/Fs), polycyclic aromatic hydrocarbons (PAH), HCl, HF, NH_3, and others (Damgaard et al. 2010, Chen et al. 2017). Total organic carbon (TOC) is also used as a parameter that refers to the total carbon amount of the organic compounds. Table 3.5 lists typical component concentrations (in descending order) in the crude flue gas from MSW incineration after the boiler and before the flue-gas cleaning systems.

TABLE 3.5

Typical Concentrations of Pollutants in the Crude Flue Gas after the Boiler and before the Flue-Gas Cleaning Systems

Flue Gas Components	H_2O	CO_2	Dust	HCl	SO_x	NO_x (excluding N_2O)	N_2O	HF	CO	TOC	Cd + Tl	Hg	PCDD/Fs
Values (unit: mg/Nm³ª)	10%–20%	5%–10%	1000–5000	500–2000	200–1000	250–500	<40	5–20	5–50	1–10	<3	0.05–0.5	0.5–10 ngTEQ/Nm³

Source: European IPPC Bureau, European IPPC Bureau: Reference document of the best available techniques for waste incineration, 2006.

ª Different units for H_2O, CO_2 and PCDD/Fs.

The wet-gas cleaning systems are responsible for the most noteworthy amounts of wastewater generated during cleaning of flue gas. As an example, the European Association for Co-Processing (EUCOPRO) regulated the permit limit levels of some water quality indices for WtE plants of the member companies to ensure they do not pose negative impacts on the quality of natural groundwater. The permit limit values of indices are shown in Table 3.6.

TABLE 3.6

EUCOPRO Permit Limit Levels of Water Quality Indices for Water Discharge from WtE Plants to Environment

Water Quality Indexes		Permit Limit Values (mg/L)
pH		5.5–9.5
Temperature		Max. 30°C–45°C
Total suspended solids		30–60
Chemical oxygen demand		50–300
Hydrocarbons		2–10
Biochemical oxygen demand		30–40
Nitrogen-kjeldahl		n.a.–40
Nitrogen (total)		10–50
Phosphates (total)		1–10
Heavy metals	Cr (VI)	0.01–0.1
	Cr (total)	0.02–0.5
	Fe	10–15
	Ni	0.05–0.5
	Cu	0.03–0.5
	Zn	0.3–2
	Cd	0.05–0.2
	Sn	0.01–2
	Hg	0.05–0.15
	Pb	0.05–0.5

Source: Eucopro, Hazardous waste preparation for energy recovery, Retrieved June 16, 2016, from www.eucopro.org, 2003; EU IPPC, Integrated Pollution Prevention and Control: Reference document on best available techniques for the waste treatments industries, 2006.

The solid wastes mainly originate from the sludge from the wastewater treatment, bottom ash, and fly ash.

3.3.2 Gas Emission Standards and Abatement Technologies

The presence of toxic compounds makes the cleaning of flue gas a basic requirement to meet the emission limit regulations in different countries (Tabasova et al. 2012). As the global leading countries/regions, Europe (EU EEA) and the United States (US EPA) set strict emission limit standards for the pollutants emitted to the atmosphere, especially for air pollution control (APC). Table 3.7 lists the air emission standards in the US, EU, and China.

As shown in Table 3.7, EU countries and the US are implementing more stringent air emission regulations than China.

The physical and chemical measures for gas emission abatement are listed in Table 3.8. In practical applications, these measures are combined to obtain the optimum gas cleaning performance. A large variety of factors need to be taken into account when selecting suitable measures and equipment.

3.3.3 Emission Factors for Different Waste-to-Energy Technologies

The AD processing of source-separated renewable food waste and agricultural waste for bio-gas production mainly involves the mechanical pre-treatment of materials, followed by two- or multi-stage digestion. Ammonia (NH_3) is the major concern of pollution emisssion from anaerobic digestion of renewable wastes. The relevant emission factors of different AD facilities and practices are shown in Table 3.9. It is indicated that the NH_3 emission factors drop in the range of 5.6–230.0 g per tonne of solid or liquid substance before/after digestion. Table 3.10 gives the typical emission factors (uncontrolled, in unit of g/tonne waste) of incineration of domestic/municipal wastes and sewage sludge from the EU EEA and US EPA.

TABLE 3.7
Air Emission Standards in the US, EU and China

Emissions (mg/Nm³)	Emission Standards		
	USEPA	EU EEA	China
Particulate matter (PM)	11	10	80
Nitrogen oxides (NO_x)	264	200	400
Sulphur dioxide (SO_2)	63	50	260
Hydrogen chloride (HCl)	29	10	75
Carbon monoxide (CO)	45	50	150
Mercury (Hg)	0.06	0.05	0.2
Cd	/	0.05	0.1
Total organic carbon (TOC)	N/A	10	
Dioxins (TEQ, ng/Nm³)	0.14	0.10	1

Source: Themelis, N. J., Reducing landfill methane emissions and expanded hierarchy of sustainable waste management, *Proceedings of Global Waste Management Symposium*, Rocky Mountains, CO, 2008.

TABLE 3.8

Formation Mechanisms of Gas Emissions from Thermal WtE Plants and the Abatement Measures

Emissions (mg/Nm³)	Chemical Nature/Mechanisms of Formation	Main Dependent Factors	Effects/Comments	Abatement Measures	
				Primary	Secondary
Particulate matter (PM)	Inorganic matter and tars depositing on particulates in gas flow	Material properties; combustor configuration and operation	Fly ash and tar cause fouling and corrosion to the pipelines; one of the most important air quality indicators	Reducing SO$_x$ and NO$_x$ generation as precursors for PM$_{2.5}$; Ensuring sealing of equipment	Cyclones, multi-cyclones, fabric filters; dry electrostatic precipitators (ESP); wet ESP; wet scrubbers
Nitrogen oxides (NO$_x$)	Oxidation of nitrogen content in wastes (contributing 70%–80% of NO$_x$) and atmosphere gas due to complex mechanisms of fuel-NO$_x$, thermal-NO$_x$ and/or prompt NO$_x$ formation routes[a]	Material properties; combustor configuration and operation	Cause acid rains; reduction of NO$_x$ benefits the increase in energy efficiency of WtE plants	Air-staged combustion; gas re-burning or fuel-staged combustion	Selective Non-Catalytic Reduction (SNCR), e.g., reduction of NO$_x$ by NH$_3$ or urea; Selective Catalytic Reduction (SCR); Very-low NO$_x$ process (VLN)
SO$_x$, HCl and HF	Besides NO$_x$, SO$_2$ and HCl are the other acidic gases together with general presence of SO$_3$, HF, HBr	Material properties	Cause acid rain; no relations between acid gas contents and combustion conditions	Pre-treatment of wastes to reduce the contained sulphur and halogen elements, e.g., desulfurization using catalysts	SO$_x$, HCl and HF are removed by dry neutralization with Ca(OH)$_2$ or NaHCO$_3$, semi-dry neutralization with Ca(OH)$_2$ and wet scrubbing
Carbon monoxide (CO)	Incomplete combustion in some zones	Combustor configuration, operation, and material distribution in combustor	CO content is an indicator of efficiency, instability and non-uniformity of combustion process	Combustor design and operation modification	Further oxidation (e.g., combustion)
VOCs	Include a variety of organic compounds with low boiling points (usually <100°C)		Some types of VOCs cause ozone generation		Absorption by some types of activated carbon

(Continued)

TABLE 3.8 (Continued)
Formation Mechanisms of Gas Emissions from Thermal WtE Plants and the Abatement Measures

Emissions (mg/Nm^3)	Chemical Nature/Mechanisms of Formation	Main Dependent Factors	Effects/Comments	Abatement Measures Primary	Abatement Measures Secondary
Trace metals	Present with PM and as vapours (e.g., Hg)	Material properties; operating conditions	Human health impacts; some are human carcinogens (known or probable)	Pre-treatment of wastes; change of operating conditions	Trace metals removal by absorption by activated carbon, especially for Hg, and wet scrubbing
Polycyclic aromatic hydrocarbons (PAH)	Formation during and after combustion; formation of		High risk of carcinogenicity	Process and equipment modifications	Absorption by some types of activated carbon
PCDD/Fs (TEQ, ng/ Nm^3)	PCDD/Fs may take place during any combustion reactions in the presence of C, O and Cl	Material properties and plant operation		Reducing chlorine in raw materials; Optimising process parameters such as reducing the residence time of flue gas in temperature zone of 200°C–450°C	Absorption by activated carbon; absorption by carbon-impregnated materials/ carbon slurries in wet scrubbers; catalytic reduction in presence of selective catalysts (e.g., TiO$_2$); catalytic filter bags; re-burn of adsorbents

Source: USEPA, 1995a; Dean, A. M. and Bozzelli, J. W., Chapter 2. Combustion chemistry of nitrogen, in *Gas-Phase Combustion Chemistry*, W. C. Gardiner (Ed.), Springer Science+Business Media, New York, 2000; European IPPC Bureau, European IPPC Bureau: Reference document of the best available techniques for waste incineration, 2006; Gohlke, O. et al., *Waste Manage.*, 30, 1348–1354, 2010; Poggio, A., and Grieco, E., *Waste Manage.*, 30, 1355–1361, 2010; Yang, Y. et al., *Energy Fuels*, 31, 5491–5497, 2017.

TABLE 3.9
Emission Factors of Anaerobic Digestion of Renewable Wastes

Facility Type	Practices	Emissions	Emission Factors	
			Value	Unit
Industrial[a]	Solid storage (before digestion)	NH_3	5.6	g/tonne biogenic material
Industrial[a]	Intermediate storage of liquid digestate (after digestion)	NH_3	80	g/tonne liquid digestate
Industrial[a]	Solid storage (after digestion)	NH_3	104	g/tonne solid fermentation substance (fresh substance)
Agricultural[b]	Solid storage (before digestion)	NH_3	230	g/tonne biogenic wastes (fresh substance)
Agricultural[b]	Final storage of liquid digestate (after digestion)	NH_3	80	g/tonne liquid digestate
Agricultural[b]	Solid storage (after digestion)	NH_3	104	g/tonne solid (fresh substance) digestate

Source: EEA, Biological treatment of waste – anaerobic digestion at biogas facilities, 2016.

[a] Feedstocks are from food processing industries, restaurants, municipalities without dung or manure.

[b] Feedstocks are from farms using dung or manure (~80%) with the remaining as agricultural waste, food waste, or glycerine.

TABLE 3.10
Typical Emission Factors (Uncontrolled, in Unit of g/t Waste) of Incineration of Domestic/Municipal Wastes and Sewage Sludge

Pollutant Emissions	Domestic/ Municipal Wastes	Municipal Solid Waste		Wood	Sewage Sludge	
	EU EEA	US EPA		US EPA	US EPA	
	/	Mass Burn and Modular Excess Air Combustors	Modular Starved-Air Combustors	Trench Combustors	Multiple Hearth Sewage Sludge Combustors	Fluidized Bed Combustors
CO_2	/	985,000	985,000			
NO_x	1800	1830[a]/1130[b]/1230[c]/1240[d]	1580	2000	2500	880
SO_2	1700	1730	1610	50	1400	150
HCl	/	3200	1080	/	/	
CO	700	232[a]/383[b]/685[c]/ND[d]	1500	ND	1550	1100
VOCs (non-methane)	20	/				
PM_{10}	13,700	12,600[e]	1720[e]	6500[e]	5200[e]	2300[e]
$PM_{2.5}$	9200					
Pb	104	107	ND		50	20
Cd	3.4	5.45	1.2		16	2.2

(Continued)

TABLE 3.10 (*Continued*)

Typical Emission Factors (Uncontrolled, in Unit of g/t Waste) of Incineration of Domestic/Municipal Wastes and Sewage Sludge

Pollutant Emissions	Domestic/Municipal Wastes EU EEA /	Municipal Solid Waste US EPA Mass Burn and Modular Excess Air Combustors	Modular Starved-Air Combustors	Wood US EPA Trench Combustors	Sewage Sludge US EPA Multiple Hearth Sewage Sludge Combustors	Fluidized Bed Combustors
Hg	2.8	2.8	2.8		/	/
As	2.14	2.14	0.33		4.7	2.2
Cr	0.18	4.49	1.65		14	/
Cu	0.09				40	/
Ni	0.12	3.93	2.76		8	17.8
Zn	0.9	/			66	/
Al					240	/
S					3600	
PCB (Polychlorinated biphenyl)	0.0053	/				
Dixons and furans	3.5 mg I-TEQ/t waste	0.0835 g/t[a]; ND[b]; 0.0075 g/t[c]; ND[d]	0.0015 g/t		0.0097 g/t	1.1 µg/t

Source: USEPA 1995b, European Environmental Agency, EMEP/EEA air pollutant emission inventory guidebook – 2013, Retrieved June 16, 2016.

[a] Mass burn waterwall combustors.
[b] Mass burn rotary waterwall combustors.
[c] Mass burn refractory wall combustors.
[d] Modular excess air combustors.
[e] Total PM.
ND: Not detected.

3.4   UNCERTAINTY ASPECT OF WASTE-TO-ENERGY ACTIVITIES

Sustainable energy production from distributed renewable wastes requires planning from initial market research to sustainable and steady energy production and supply for the end users. During the processes, there are some uncertainties requiring attention.

3.4.1   Sustainable Supply of Materials

Power generation from renewable wastes can be beneficial for addressing the energy security problem that relates to the availability of energy sources (Iychettira et al. 2017). However, as one of the most outstanding barriers for energy production from renewable waste in the distributed mode, the challenge for sustainable availability of materials is due to the nature of different waste types

(IRENA 2015a). For the agricultural residues and some process wastes, such as cotton straw, sug-arcane bagasse, and rice husks, their available amounts vary by the inherent growing period and seasonal change and are vulnerable to the meteorological conditions. Thus, the material storage and regular production are vitally important to the sustainable supply for the continual energy produc-tion (especially heat and electricity) from WtE facilities. In terms of food waste and farm waste for heat, electricity, and transportation fuel (CH_4-rich bio-gas) generation through anaerobic digestion, the material availability is relatively stable. The steady availability also applies to the collection of OFMSW.

In remote rural areas and islands, the development of WtE systems using local waste resources is in high demand, as the transportation of fossil fuels is costly (Antizar-Ladislao and Turrion-Gomez 2010). The cost of manpower for collecting the dispersed waste resources is also a major concern. Thus, maintaining the availability of renewable waste is a high priority, and the acceptance of diver-sified waste types by WtE facilities can alleviate this issue to some degree.

3.4.2 TECHNOLOGICAL ASPECT OF WASTE-TO-ENERGY SYSTEMS

The economic and environmental performance of WtE plants is highly dependent on the WtE tech-nology, which is the result of comprehensive and solid considerations based on local conditions. For example, livestock manure from a large farm near an urban area can be co-digested with OFMSW to produce fertilizers, CHP, and/or gaseous transportation fuel (Kothari et al. 2010). However, accessibility to appropriate and advanced WtE technologies and engineering know-how are critical for economically feasible performance of the technologies. This is partly responsible for the worse economic and environmental performances of the WtE plants in the developing countries than those in the developed countries, such as the EU, US, and Japan (Antizar-Ladislao and Turrion-Gomez 2010).

Another uncertainty involves the technological development of WtE systems. WtE technologies (especially advanced gasification and pyrolysis) are still in development and not fully commercial-ized. Energy production efficiency in reactors (e.g., fixed beds and fluidized beds) and conversion efficiency to final products, such as electricity, in energy conversion systems (e.g., the Organic Rankine Cycle) require improvement. There is a growing trend of electricity generation from WtE plants, which enhances the need for modifying existing devices and engineering novel technologies to upgrade the current average electricity efficiency of approximately 20% to a higher level of 30% or more (Tabasova et al. 2012). Besides, WtE technologies are also restricted by the costly pollution control measures that are subjected to the increasingly stricter emission regulations (Kothari et al. 2010). The current efforts for development of advanced WtE technologies include the tri-generation of heat, power, and cooling (Chevalier and Meunier 2005, Rentizelas et al. 2009) and the creation of new software (Tabasova et al. 2012) for better WtE modelling. In addition to the technological uncertainties for WtE systems, the integration of WtE electricity (if electricity is the product) into an existing utility grid is also an important consideration that affects the reliability (e.g., voltage stabil-ity) of WtE systems (Antizar-Ladislao and Turrion-Gomez 2010, Tabasova et al. 2012).

3.4.3 POLICY INSTRUMENTS AND INCENTIVES

The additional considerations for designing the WtE systems are the funding arrangements for the capital, operating and maintenance costs, as well as the fiscal and policy support from governments at different levels. Currently, the energy prices (e.g., electricity price) from WtE plants are generally higher than market commercial prices, which force governments to create incentives to encourage WtE activities. For well-established markets in which the entry of WtE activities will face signifi-cant obstacles, governments are required to play a key role to enable market channels for energy products from WtE. It is not only local government's, but also the state/country's, responsibility to drive the development of WtE. Generally, governments support WtE developments through various means of policy support, financial support through subsidies and industrial grants, tax reduction/ returns, and concession loans. For example, currently in China at the national level for every MWh_e electricity generated by WtE plants, a feed-in tariff of around USD 30–40 is provided as subsidy to offset the electricity purchase price nationwide, along with an additional subsidy of around USD 9–14/t waste for WtE incineration (Themelis 2008, Li et al. 2015, Zhang et al. 2015).

However, government support also has drawbacks, by increasing the WtE beneficiary dependence, which, to some extent, impedes the progress of WtE. Thus, flexible and adjustable policy instruments are essential to promote the transformation of WtE technologies from quantitative to qualitative. Some measures, such as building standards, issuing licenses to qualified participants, and continuous monitoring, can be applied to ensure the sustainability of WtE activities (IRENA 2015a).

3.4.4 SOCIAL AND POLITICAL FACTORS

Social and political aspects of WtE activities are also of great concern. The local communities around the WtE plants generally welcome the increased job opportunities. However, the public opposition to the installation of WtE plants may also occur due to the increasing awareness and concerns of living environment protection, which may influence the social stability (Zhang et al. 2015). The introduction of public and stakeholders' participation in the decision-making and implementation of WtE projects, the supervision by the third party, and adequate compensation to affected residents may be useful measures to ensure social acceptance for development of WtE (Mohammed et al. 2013a, 2013b).

3.4.5 LACK OF ACCESSIBLE DATA

With the rapid development of gasification and pyrolysis of renewable wastes, the establishment of an up-to-date database on the economic and environmental performance of these advanced WtE technologies is an urgent task. For example, the emission factors for the commercialized technologies of waste incineration and anaerobic digestion are accessible for the public from the EU EEA or US EPA. However, systematic data on WtE by gasification and pyrolysis are not available at this stage.

During the past decade, life-cycle assessment (LCA) has been extensively applied to assess the impacts (benefits and disadvantages) of WtE processes on the environment. The procedure evaluates the entire life-cycle of a particular energy product, from the very beginning of gaining raw material, through the whole production process to the end of life and the product's "liquidation" (Tabasova et al. 2012), which is highly dependent on the availability and sufficiency of data sources (Lee et al. 2007). As the most used method to assess the environmental performance of WtE activities, LCA is also facing some challenges.

According to Astrup et al. (2015), around 80% of the LCA case studies on WtE technologies since 1995 focused on the waste incineration processes, approximately 50% of which involved a moving grate design. This indicates that the other, more advanced, thermal WtE technologies, such as gasification (8% of case studies) and pyrolysis (5% of case studies), require further research attention.

Consonni and Vigano (2012) also stated that the biggest challenge to hinder the commercialization of waste gasification (and other technologies other than incineration) is the limited data and experience from practical operations of waste gasification plants. Besides, the published information on APC technologies is insufficient. Sustainable processing of residues (bottom ash, fly ash, acid gases, PCDD/F, NO_x, and sludge will need to be further explored (Astrup et al. 2015). The sensitivity analysis and evaluation of scenario uncertainties in the LCA studies are also to be further addressed.

3.5 CONCLUSIONS

Energy production from distributed renewable waste resources can economically, environmentally, and socially benefit community development. This chapter aimed to give an overall picture of the state-of-art of WtE activities. Incineration and anaerobic digestion are the most mature technologies, with extensive commercial applications. More advanced waste processing technologies, mainly including gasification and pyrolysis, are also gaining more popularity. The prices of renewable wastes for WtE production typically vary in the range of below zero to USD 4.3/GJ depending on the waste type, collection, and transportation costs. Typically, 0.4–0.7 MWh_e electricity or 1.3–2.6 MWh_{th} heat can be obtained from incineration of one tonne of treated municipal solid waste. Pollutant emission standards of EU, US, and China were compared. Formation mechanisms of gas emissions from thermal WtE plants and the corresponding primary and secondary abatement measures were also summarized. In addition, uncertainty factors of WtE activities (including sustainable supply of materials, technological aspects of WtE systems, policy instruments and incentives, social and political factors, and lack of accessible data) were underlined at the end of this chapter. On the way to complete acceptance of WtE by markets, it is essential that financial support by local governments is guaranteed, with other uncertainty factors under satisfactory control.

REFERENCES

Antizar-Ladislao, B. and Turrion-Gomez, J. L. (2010). "Decentralized energy from waste systems." *Energies* 3(2): 194–205.

Arena, U. (2011). "Gasification: An alternative solution for waste treatment with energy recovery." *Waste Management* 31(3): 405–406.

Astrup, T. F., Tonini, D., Turconi, R. and Boldrin, A. (2015). "Life cycle assessment of thermal Waste-to-Energy technologies: Review and recommendations." *Waste Management* 37: 104–115.

Chen, X., Liaw, S. B. and Wu, H. (2017). "Important role of volatile–char interactions in enhancing PM1 emission during the combustion of volatiles from biosolid." *Combustion and Flame* 182: 90–101.

Chevalier, C. and Meunier, F. (2005). "Environmental assessment of biogas co- or tri-generation units by life cycle analysis methodology." *Applied Thermal Engineering* 25(17–18): 3025–3041.

Consonni, S. and Vigano, F. (2012). "Waste gasification vs. conventional Waste-To-Energy: A comparative evaluation of two commercial technologies." *Waste Management* 32(4): 653–666.

Damgaard, A., Riber, C., Fruergaard, T., Hulgaard, T. and Christensen, T. H. (2010). "Life-cycle-assessment of the historical development of air pollution control and energy recovery in waste incineration." *Waste Management* 30(7): 1244–1250.

Dean, A. M. and Bozzelli, J. W. (2000). Chapter 2. Combustion chemistry of nitrogen. In *Gas-Phase Combustion Chemistry*, W. C. Gardiner (Ed.). New York, Springer Science+Business Media.

EEA. (2016). "Biological treatment of waste – anaerobic digestion at biogas facilities." Retrieved June 16, 2016.

EU IPPC. (2006). "Integrated Pollution Prevention and Control: Reference document on best available techniques for the waste treatments industries." Retrieved June 16, 2016.

Eucopro. (2003). "Hazardous waste preparation for energy recovery." Retrieved June 16, 2016, from www.eucopro.org.

European Environmental Agency. (2013). "EMEP/EEA air pollutant emission inventory guidebook – 2013." Retrieved June 16, 2016.

European IPPC Bureau. (2006). "European IPPC Bureau: Reference document of the best available techniques for waste incineration."

Evangelisti, S., Lettieri, P., Clift, R. and Borello, D. (2015). "Distributed generation by energy from waste technology: A life cycle perspective." *Process Safety and Environmental Protection* 93: 161–172.

Gohlke, O., Weber, T., Seguin, P. and Laborel, Y. (2010). "A new process for NOx reduction in combustion systems for the generation of energy from waste." *Waste Management* 30(7): 1348–1354.

IEA Bioenergy. (2010). "Accomplishments from IEA Bioenergy task 36: Integrating energy recovery into solid waste management systems (2007–2009)." Retrieved June 6, 2016, from www.ieabioenergy.com/.

IEA Bioenergy. (2014). "IEA Bioenergy task 37 country reports summary." Retrieved June 6, 2016, from www.ieabioenergy.com/.

IRENA. (2013). "Renewable power generation costs in 2012: An overview." Retrieved June 6, 2016, from https://www.irena.org/.

IRENA. (2015a). "Biomass for heat and power: Technology brief." Retrieved June 6, 2016, from https://www.irena.org/.

IRENA. (2015b). "Renewable power generation costs in 2014." Retrieved June 6, 2016, from https://www.irena.org/.

ISWA. (2012). "Waste-to-energy. State-of-the-art-report. Statistics, 6th ed." Retrieved February 22, 2016, from http://www.iswa.org/media/publications/knowledge-base/.

Iychettira, K. K., Hakvoort, R. A., Linares, P. and de Jeu, R. (2017). "Towards a comprehensive policy for electricity from renewable energy: Designing for social welfare." *Applied Energy* 187: 228–242.

Kothari, R., Tyagi, V. V. and Pathak, A. (2010). "Waste-to-energy: A way from renewable energy sources to sustainable development." *Renewable & Sustainable Energy Reviews* 14(9): 3164–3170.

Lee, S. H., Choi, K. I., Osako, M. and Dong, J. I. (2007). "Evaluation of environmental burdens caused by changes of food waste management systems in Seoul, Korea." *Science of the Total Environment* 387(1–3): 42–53.

Li, Y., Zhao, X. G., Li, Y. B. and Li, X. Y. (2015). "Waste incineration industry and development policies in China." *Waste Management* 46: 234–241.

Lombardi, L., Carnevale, E. and Corti, A. (2015). "A review of technologies and performances of thermal treatment systems for energy recovery from waste." *Waste Management* 37: 26–44.

Mohammed, Y. S., Mokhtar, A. S., Bashir, N. and Saidur, R. (2013a). "An overview of agricultural biomass for decentralized rural energy in Ghana." *Renewable & Sustainable Energy Reviews* 20: 15–25.

Mohammed, Y. S., Mustafa, M. W., Bashir, N. and Mokhtar, A. S. (2013b). "Renewable energy resources for distributed power generation in Nigeria: A review of the potential." *Renewable & Sustainable Energy Reviews* 22: 257–268.

Murphy, J. D. and McKeogh, E. (2004). "Technical, economic and environmental analysis of energy production from municipal solid waste." *Renewable Energy* 29(7): 1043–1057.

Nizami, A. S., Shahzad, K., Rehan, M., Ouda, O. K. M., Khan, M. Z., Ismail, I. M. I., Almeelbi, T., Basahi, J. M. and Demirbas, A. (2017). "Developing waste biorefinery in Makkah: A way forward to convert urban waste into renewable energy." *Applied Energy* 186: 189–196.

Nolan ITU Pty Ltd and TBU. (2001). "Guideline for determining the renewable components in waste for electricity generation. Prepared for Australian Government."

O'Shea, R., Kilgallon, I., Wall, D. and Murphy, J. D. (2016). "Quantification and location of a renewable gas industry based on digestion of wastes in Ireland." *Applied Energy* 175: 229–239.

Pan, S. Y., Du, M. A., Huang, I. T., Liu, I. H., Chang, E. E. and Chiang, P. C. (2015). "Strategies on implementation of waste-to-energy (WTE) supply chain for circular economy system: A review." *Journal of Cleaner Production* 108: 409–421.

Poggio, A. and Grieco, E. (2010). "Influence of flue gas cleaning system on the energetic efficiency and on the economic performance of a WTE plant." *Waste Management* 30(7): 1355–1361.

Rentizelas, A. A., Tatsiopoulos, I. P. and Tolis, A. (2009). "An optimization model for multi-biomass trigeneration energy supply." *Biomass & Bioenergy* 33(2): 223–233.

Six, W. (2001). "Project engineer with organic waste systems, personal communication."

Tabasova, A., Kropac, J., Kermes, V., Nemet, A. and Stehlik, P. (2012). "Waste-to-energy technologies: Impact on environment." *Energy* 44(1): 146–155.

Themelis, N. J. (2008). "Reducing landfill methane emissions and expanded hierarchy of sustainable waste management." *Proceedings of Global Waste Management Symposium*, Rocky Mountains, CO.

USEPA. (1995a). "AP 42, Compilation of air pollutant emission factors, fifth edition." Retrieved November 15, 2015, from http://www.epa.gov.

USEPA. (1995b). "Compilation of air pollutant emission factors AP-42." Retrieved November 11, 2015, from http://www.epa.gov.

USEPA. (2007). "Combined heat and power: Catalog of technologies." Retrieved November 11, 2015, from http://www.epa.gov.

USEPA. (2016). "Sustainable materials management: Non-hazardous materials and waste management hierarchy." Retrieved February 16, 2016, from http://www.epa.gov.

WCED. (1987). *Our Common Future: The World Commission on Environment and Development*. Oxford, UK, Oxford University Press.

Yang, Y., Dai, F., Li, C., Xiang, S., Yaseen, M. and Zhang, S. (2017). "Kinetic evaluation of hydrodesulfurization and hydrodenitrogenation reactions via a lumped model." *Energy & Fuels* 31(5): 5491–5497.

Yassin, L., Lettieri, P., Simons, S. J. R. and Germana, A. (2009). "Techno-economic performance of energy-from-waste fluidized bed combustion and gasification processes in the UK context." *Chemical Engineering Journal* 146(3): 315–327.

Yue, D. J., You, F. Q. and Snyder, S. W. (2014). "Biomass-to-bioenergy and biofuel supply chain optimization: Overview, key issues and challenges." *Computers & Chemical Engineering* 66: 36–56.

Zhang, D. L., Huang, G. Q., Xu, Y. M. and Gong, Q. H. (2015). "Waste-to-energy in China: Key challenges and opportunities." *Energies* 8(12): 14182–14196.

Zhao, X. G., Jiang, G. W., Li, A. and Wang, L. (2016). "Economic analysis of waste-to-energy industry in China." *Waste Management* 48: 604–618.

4 Technical and Economic Assessment of Biogas and Liquid Energy Systems from Sewage Sludge and Industrial Waste

Lifecycle Assessment and Sustainability

Hossain M. Anawar and Vladimir Strezov

CONTENTS

4.1   INTRODUCTION

Given the increasing population, demand of energy, climate change, and environmental concerns, new technologies are being developed to produce renewable energy and partially replace fossil fuel (Balat and Balat, 2009; Jeihanipour and Bashiri, 2015). Municipal and agricultural waste-generated, thermochemical-based biodiesel has high potential to reduce greenhouse gas emissions and present an economic process to produce liquid and gaseous biofuels through fermentation and anaerobic digestion (AD). This process also present sustainable waste management and resource recovery benefits (Zah, 2010). With the increasing population, the production rate of municipal solid waste (MSW) is estimated to increase from 1.3 billion tonnes per year in 2012 to 2.2 billion tonnes per year by 2025 (Hoornweg and Bhada-Tata, 2012). The amount of sewage sludge is increasing rapidly in the European Union (EU) and other parts of the world, making a significant worldwide challenge for its management (Ahmed and Lan, 2012; Kelessidis and Stasinakis, 2012; Luo et al., 2014). MSWs are incinerated and landfilled in many countries, creating environmental problems, land crises, and economic disadvantages. However, these problems can be alleviated when renewable energy and valuable materials are produced from the MSW, whether alone or in combination with other organic wastes, including crop residue, biomass, industrial waste, and forestry by-products. In addition to the solid waste, the MSW leachate usually poses major environmental issues, causing severe contamination and eutrophication of the receiving aquatic and land ecosystems. Therefore, the leachate should be treated properly before being released to the environment or digested.

A number of renewable energy sources, such as methanol, ethanol, hydrogen, biogas, and biodiesel have been studied for use in vehicles. They are either blended with petroleum or used in their pure form. Biogas is an economic renewable energy produced from waste materials with high organic contents (Sorathia et al., 2012). The microorganisms produce biogas and biosolids by breaking down organic materials in the digestion process under anaerobic conditions (Nghiem et al., 2014). The AD process can treat and reduce the volume of sewage sludge, and produce methane, heat, electricity, and biofuels, as well as generate biosolids/compost with high contents of nutrients (Carrère et al., 2010). AD processes are gaining high priority to produce biogas and stabilize the waste due to increasing concerns of climate change, energy crises, and environmental management of waste (Khanal et al., 2008). The existing AD infrastructure can be applied at wastewater treatment plants to achieve the above goals. This chapter will conduct the technical and economic assessment of biogas and liquid energy systems from sewage sludge and industrial waste, with a focus on lifecycle assessment and sustainability.

4.2 BIOGAS PROCESSES FOR SUSTAINABLE DEVELOPMENT

There is an increasing need for efficient utilization of resources in the world (Marchaim, 1992). Sustainable system development is necessary for the recovery and utilization of wastes, considering the emerging environmental agenda. AD plays a significant role in an integrated resource recovery system and sustainability in waste management. Further technological development of this process is focusing on applications in developing countries for more economical uses of bioenergy processes. The role of microbial activities is essential for efficient degradation of organic wastes and methane production. The World Conservation Strategy has several objectives, out of which the main objective is to protect healthy ecological processes, fertile land, and clean water to sustain the existence of living beings in the world. The production of biogas from organic agricultural wastes, such as waste biomass, wood, dung, and crop wastes in the rural and regional communities, is one of the procedures that can provide benefits at the farm level.

4.3 PRINCIPLES, DIFFERENT PROCESSES, AND POTENTIAL OF ANAEROBIC DIGESTION OF WASTE-ACTIVATED SLUDGE

The AD of sewage sludge plays an important role to convert organic matter (OM) into biogas (Appels et al., 2008). This process kills the pathogens of the sludge and reduces odor problems, thus reducing the cost of wastewater treatment plants (WWTPs) and environmental problems. Hydrolysis controls the rate of the complex digestion process. Several bacterial groups are inhibited in the AD of sewage sludge due to toxic effects of some chemical species in the solution. The various pretreatments, such as mechanical, thermal, chemical, and biological interventions of the feedstock, can augment the hydrolysis rate, digestion of feedstock, and production of biogas. The (co-) incineration of sludge, where permits can be obtained, can solve the problem of land application due to policy restrictions for acceptable limits of composition (European Commission, 1986; Werther and Ogada, 1999; Dewil et al., 2007; Van de Velden et al., 2008).

The pre-treatment process in the water purification part of a WWTP commonly removes about 50%–60% of the suspended solids and 30%–40% of the biological oxygen demand (BOD) (Qasim, 1999; Metcalf and Eddy, 2003). After pre-treatment, aerobic microorganisms in the biological step degrade the remaining (or nearly total) BOD and suspended solids. The process simultaneously removes nitrogen (N) and phosphorus (P). The subsequent process releases the effluent with the acceptable limit of contamination leaving the bottom sludge. The sludge is transferred to the sludge treatment units of the WWTP. After pre-treatment, the sludge generated from the primary and secondary processes undergoes further treatment involving various steps. In the thickening step, the gravity, flotation, or belt filtration reduces the volume of sludge, and the separated water is added to the influent of the WWTP. After this step, the sludge undergoes specific biochemical treatment, and AD then converts the organic matter into biogas.

Airtight tanks are used in the AD of sludge, where generally all forms of organic matter are digested, except for stable woody materials, because the bacteria cannot biodegrade lignin. The biogas produced in the AD process is a high-value, beneficial renewable energy source. Although AD has many advantages, it has also some disadvantages. These limitations include (1) partial degradation of organic matter, (2) slow reaction rate, high volume of sludge, and high cost of the digesters, (3) some unexpected barriers to the process, (4) production of low-quality supernatant, (5) production of harmful and unexpected gases, such as carbon dioxide (CO_2), hydrogen sulphide (H_2S), and excess moisture, (6) possible presence of volatile siloxanes in the biogas, and (7) high contents of heavy metals and refractory "organics" in the residual sludge. A process flowchart of the sludge-processing steps is shown in Figure 4.1.

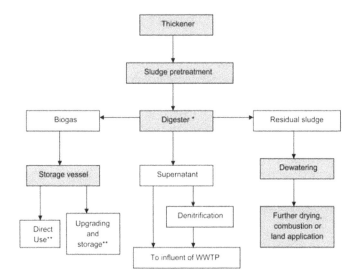

FIGURE 4.1 Process flowchart of the sludge processing steps. (Reprinted from Appels, L. et al. 2008. Principles and potential of the anaerobic digestion of waste-activated sludge. *Prog Energy Combust Sci* 34: 755–781. Copyright 2008, with permission from Elsevier.)

4.4 TYPES OF ANAEROBIC DIGESTION PROCESSES PRODUCING BIOFUEL FROM WASTE

The anaerobic digestion processes, such as up-flow anaerobic sludge blanket (UASB) reactor (Peng et al., 2008; Ye et al., 2011) and anaerobic moving-bed biofilm reactor (MBBR) (Chen et al., 2008a) have shown high efficiency in chemical oxygen demand (COD) removal and average methane yield, while treating leachates with high COD concentrations (Luo et al., 2014). The anaerobic fluidized bed reactor (AFBR) and anaerobic membrane bioreactor (AnMBR) also demonstrate similar positive results (Turan et al., 2005; Zayen et al., 2010).

Additional reactors currently available to anaerobically treat waste include granular sludge bed (EGSB) reactor or internal circulation (IC) reactor. The bench-scale EGSB reactor has high efficiency in COD removal (Dang et al., 2013). An integrated internal and external circulation (IIEC) reactor, adapted from EGSB and IC construction, showed higher efficiency of anaerobic degradation. The achievement of these advanced reactors depends on the reactive granular characteristics of sludge and bacterial activities (Liu et al., 2006; Abreu et al., 2007).

4.5 TREATMENT OF ORGANIC WASTES BY ANAEROBIC DIGESTION AND DECOMPOSITION

Anaerobic digestion has a significant role in waste management and biogas production (Ostrem, 2004). The biogas, composed of up to 65% methane, can be combusted in a cogeneration unit generating green energy. The biosolid or solid digestate is applied to agricultural soil and other lands as a fertilizer or an organic soil amendment, providing a waste management strategy more than 20 countries. The following strategies are suggested to overcome the drawbacks of AD development: new technological development, proper pricing for waste management, strict environmental regulations, and knowledge distribution. Many biotic and abiotic factors, which are broadly classified into aerobic and anaerobic relating to the presence or absence of oxygen, control the action/reaction in the decomposition technology (Christy et al., 2013). Carbohydrate, protein, lipid, and fats can proceed in anaerobic or aerobic conditions. Although aerobic degradation of organic materials is quicker than anaerobic processes, the latter option is more profitable.

4.6 MODEL-BASED OPTIMIZATION OF BIOGAS PRODUCTION

Feedstock characteristics, gas generation efficiency, compost quality, and odor emissions are the main problems of the AD plants (Arsova, 2010). The financial investment, maintenance costs, and income generation from biogas and compost products are the main challenges of this technology. The anaerobic acetogenesis of organic matter can also produce biogas. Instead of methanol use, the use of volatile fatty acids from food waste can increase the denitrification rate in WWTPs, decrease the cost of denitrification and the entire operating costs of the AD, and decrease the carbon footprint.

Popov (2010) described modelling of anaerobic co-digestion process in a WWTP and adapted the AD mathematical model 1 (ADM1) to the reactor system. The theoretical estimation using the measured data demonstrated 0.6142 m^3 of methane production per kg COD removal, showing a 29.7% difference between the theoretical estimation and real field production. The simulation study demonstrated that the simulated biogas production was lower than the measured value. The higher biodegradability of food waste is responsible for the increasing amount of biogas production. The increased biogas production occurs due to high amount of polysaccharides and carbohydrates in the food waste. Therefore, food waste co-digestion with raw sludge can substantially augment the waste management and biogas production in a WWTP (Arsova, 2010).

4.7 MUNICIPAL SEWAGE SLUDGE CO-DIGESTION WITH INDUSTRIAL ORGANIC WASTES

4.7.1 RESTAURANT GREASE TRAP WASTE/FATTY RESIDUES

Grease trap waste (GTW) consists of spent fat, oil, and grease, in addition to solids and debris from food waste (Zhu et al., 2011). The GTW deposits and coats inside the pipes and pumps of the sewer collection systems, resulting in the destruction of the system and pump failure (Chu and Ng, 2000). GTW contains high concentrations of organic carbon. GTW can generate about 145 L methane/L GTW, which is more than 15 times higher than that of municipal waste sludge (MWS) (8.9 L methane/L MWS). Less than 4% (vol/vol) addition of GTW and co-digestion with MWS can substantially increase the methane production in the AD process (Zhu et al., 2011). However, use of GTW alone without MWS in the AD plant encounters some barriers, such as nutrient imbalance, rapid acidification, and deposition of inhibitory compounds (Nakhla et al., 2003). Razaviarani (2014) indicated that addition of GTW/restaurant fatty wastes should be within the limit that provides up to a maximum loading of 23% volatile solids (VS) or 58% COD. This rate of GTW addition to MWS can produce the biogas amount 67% greater than that of the control. However, addition of GTW higher than this limit can significantly reduce the biogas production, due to process inhibition by long-chain fatty acid accumulation.

4.7.2 BIODIESEL GLYCERIN WASTE

Biodiesel glycerin waste (BGW) is an organic by-product generated from biodiesel production (Athanasoulia et al., 2014) that can affect the environment and is expensive to recover. However, the co-digestion of BGW with sewage sludge in AD plants can increase biogas production and be stored for future use. The maximum safe limit of BGW co-digestion was found at 23% and 35% of the total 1.04 kg VS/(m^3·d) and 2.38 kg COD/(m^3·d) loadings, respectively. The rate of BGW addition to MSW exceeding the above limit can reduce the efficiency and stability of the AD process. The biogas and methane production rates in the AD plant were 1.65 and 1.83 times higher than the control with only MWS, respectively. The co-digestion of BGW will not only produce the renewable energy, but also increase the environmental safety and economic profit. Biomethane potential test and BioWin simulations indicated that the intermittent use of BGW (0.63% v/v or 3%v/v) could increase the biogas production (Nghiem et al., 2014), where the lower rate of BGW addition (0.63% v/v) showed the higher efficiency for biogas generation.

Crude glycerol or g-phase, oil cakes, or oil meals can be co-digested with MWS after careful consideration of inhibitory barriers (Kolesarova et al., 2011). The high salinity of the feedstock might inhibit the growth of the methanogenic microorganisms in case of anaerobic digestion of crude glycerol. The low content of nutrient sources from ammonium can increase the microbial growth, while its high concentrations exert an inhibition effect in the digestion plant.

4.7.3 DOMESTIC REFUSE (SWILL)

Zupancic et al. (2008) studied the mixture of municipal sludge and organic waste from domestic refuse with an average organic loading rate of 0.8 kg m^{-3} d^{-1} of volatile suspended solids. A slight increase of organic loading rate by 25% to 1.0 kg m^{-3} d^{-1} augmented the biogas quantity by 80%, electricity generation by 130%, and heat production by 55%.

4.7.4 PULP AND PAPER INDUSTRY WASTE

The secondary sludge from pulp and paper mills' wastewater treatment contains a high concentration of organic matter (Hagelqvist, 2013) that can produce methane of pure quality. Their co-digestion with municipal sewage sludge increases the rate of methane production. On-site setting of AD plants can provide multiple financial benefits to mills by decreasing the supply chain cost of waste, operation costs of AD, availability of heat from mills, and income generation from biogas business. The wastewater of this industry contains sulphites, sulphates, lignin, resin acids, fatty acids, and terpenes that can inhibit the AD process.

4.7.5 PRE-TREATMENTS

Primary sewage sludge can produce more biomethane than waste-activated sludge (WAS), such as 300–350 L CH$_4$ per kg of OM feed for primary sludge, 260–290 L CH$_4$ kg^{-1} OM feed for mixed sludge, and 140–210 L CH$_4$ kg^{-1} OM feed for WAS (Kepp and Solheim, 2000). Among WAS, sludge obtained from extended aeration processes is less biodegradable than that from high-load-activated sludge processes (Carrere et al., 2008). Thermal, biological, chemical, or mechanical pre-treatments can increase the potential biogas production from sewage sludge (Weemaes and Verstraete, 1998; Elliott and Mahmood, 2007; Appels et al., 2008). These pre-treatments can be applied separately or combined to obtain the maximum benefits. Mechanical treatments, such as thermal and ultrasonic treatments, are the most applied, especially at full-scale plants (Chauzy et al., 2007; Neis et al., 2007).

4.8 UPDATES ON TECHNOLOGICAL DEVELOPMENT FOR ENERGY RECOVERY FROM WASTEWATERS

Different technologies have been developed to treat the wastewater and recover the renewable energy, such as biogas, bioethanol, and heat (Stafford et al., 2013). Based on the current data analysis, the estimates showed that 3,200–9,000 MWth of energy might be recovered from wastewater generated from livestock and industrial and domestic wastewater in South Africa. Besides energy recovery, the additional economic and environmental benefits may include reduction in pollution, water reclamation, reduced carbon footprint, and nutrient and material recovery. Various technological options involving energy recovery from wastewater include three steps, in which the inputs are converted into intermediates and then intermediates are converted into energy outputs.

4.9 APPROPRIATE TECHNOLOGIES TO RECOVER ENERGY FROM WASTEWATER

The technologies currently applied to obtain renewable energy from wastewater are shown in Table 4.1. The principles of these technologies, case studies, and more complex, integrated applications are reviewed below.

TABLE 4.1

Comparison of Appropriate Energy from Wastewater Technologies

Technology	Wastewater Characteristics	Advantages	Disadvantages	Comments
Fermentations for biomass and secondary products	• Nutrients (C, N) • Non-toxic effluent for microbial growth • Dissolved or suspend organic	• Can produce high value secondary metabolites as by-products • Can remove toxic and recalcitrant chemicals	• Chemicals such as phenol are inhibitory • pH, salinity, aeration need to be adjusted for growth of the microbe	• More biomass produced from aerobic compared to anaerobic fermentations • Biomass production for use as a feedstock for bioethanol production and for gasification
Anaerobic digestion	• Works best at warmer (30°C–60°C) temperatures pH: 5.5–8.5 • Good design to control digestion and collection of gas • Dissolved or suspend organic	• Suitable with most substrates • Can achieve 90% conversion • Help contain odor • Produces biogas for heating, electricity generation and steam • Produces less biomass than aerobic fermentation	• H_2S oxidized to SO_2 when combined with water vapor can form sulphuric acid, which is corrosive • High capital investment	• Produces biogas fuel rich in methane/hydrogen (60%) and carbon dioxide (40%) • The bioliquid and sludge can be used as fertilizer and compost for soils, as feed for biodiesel production, or can be gasified
Combustion gasification	• Biomass • Low water content and suspended organic matter	• Heat energy • Destruction/conversion of all hazardous material • Mature technology available • 95% fuel-to-feed efficiency	• Electricity costs are higher than for a coal-fired power station • Could produce hazardous off-gases	• Organic compounds are converted to syngas for use as power, chemicals, Fischer–Tropsch liquids and gaseous fuels, fertiliser and steam • Metals can be recovered • Ash and tar wastes
Algal growth for biodiesel production	• Phosphate and nitrogen sources • Non-toxic effluent for growth • Dissolved organic—COD removal with heterotrophic growth	• Low energy requirements—use energy of sunlight for algal growth • Can result in CO_2 sequestration • Can utilize dilute wastewater streams	• Algal ponding area can represent considerable land area • Photobioreactors have large capital costs • No suspended solids • Evaporation	• Algal oils that are converted to biodiesel fuel by transesterification by-product filter cake rich in proteins and carbohydrates from algae • Valuable secondary products

(Continued)

TABLE 4.1 (*Continued*)
Comparison of Appropriate Energy from Wastewater Technologies

Technology	Wastewater Characteristics	Advantages	Disadvantages	Comments
Bioethanol production	• Carbon and nitrogen sources • Non-toxic effluent for microbial growth • Carbohydrate (sugar) rich • Dissolved organics (or suspended with emerging technology)	• Established technology producing fuel suitable for a variety of combustion engines	• Cost of carbohydrate-rich raw materials • Large volumes of bioreactors needed • Non-dilute waste waters required	• Ethanol fuel, carbon dioxide, and biodunder
Microbial fuel	• Non-variable wastewater sources • Non-toxic effluent for microbial growth	• Can be used at less than 20°C • Suitable for use with low concentration of organics in waste water • Efficient (direct conversion to electricity)	• Capital intensive • Still in development • Variable COD reduction depending on waste waters	• Direct conversion of waste to electricity • Off-gas mainly carbon dioxide • Some microbial sludge formed
Heat recovery	• Waste waters with temperature above ambient	• Direct heat recovery	• Heat above ambient and the need for heat energy	• Heat for household heating, steam generation, reduces electricity requirements

Source: Stafford, W. et al., *J. Energy South. Afr.*, 24, 15–26, 2013.
COD: chemical oxygen demand.

4.9.1 Biogas Production by Anaerobic Digestion

The OM AD in an oxygen-free condition produces biogas. Biogas consists of a mixture of gases, typically containing 50%–70% methane. The *Rhodobacter* or *Enterobacter* microbial species should dominate to produce hydrogen as the major product. The hydrogen fermentation may become more attractive in the fuel cell technology. The non-methane components should be removed from biogas for applications in vehicles and generators. Proper maintenance is required to ensure optimal operation of the biodigesters.

4.9.2 Fermentation to Bioethanol and Gas

Bioethanol is a renewable liquid fuel that can be used alone or in combination with natural petroleum. The yeast *Saccharomyces sp.* and bacteria *Zymomonas sp.* drive the fermentation to produce bioethanol from sugars under anaerobic conditions (Shuler and Kargi, 2002). Although chemical or biological pre-treatment or novel microorganisms are used to assist fermentation of a broader range of organic substrates in waste streams, there are some challenges. Vieitez and Ghosh (1999) developed a two-phase fermentation method to increase the microbial activities, methane production, and storage of methane.

4.9.3 Microbial Fuel Cells

Microbial fuel cells generate electricity by microbial-mediated oxidation of the organic matter in different types of wastewater. The electrons move to the anode and then travel via a circuit to the cathode, to combine with protons and oxygen to form water. This technology has high potential for electricity generation and wastewater treatment simultaneously. However, the low electricity output and high operation and maintenance costs do not encourage widespread application. Therefore, further research is continuing to improve this technology.

4.10 EXAMPLES OF APPLICATION OF TECHNOLOGIES FOR ENERGY FROM WASTEWATER

There are plenty of examples worldwide to recover energy from wastewater and produce the valuable materials.

4.10.1 Home-Based Biogas Plant

Some countries, such as China, India, Nepal, Vietnam, and Sri Lanka, implemented a large-scale installation of home-based biogas plants. China alone installed millions of biogas plants to supply energy to households in rural areas within a few years, and the technological development continued. However, implementation of biogas plants has been minimal in African countries due to the requirement of initial high investment and operating costs.

4.10.2 Biogas Production Using Agricultural Waste

Poultry and livestock slaughterhouses generate large amounts of wastewater in some villages of China that is used to produce biogas to meet the demand of more than 300 households and 7,200 tonnes of organic fertilizer each year (ISIS, 2006). The wastewater-generated biogas supplies high-quality fuel for public transport buses in Linköping, Sweden (IEA Biogas, undated). In Ireland, the AD of farms and food processing wastewater produces renewable energy, such as electricity and heat, and supplies the energy to small farming communities. An estimated 150,000 kWh of electricity and 500,000 kWh of heat energy are generated per year in this plant (Stafford et al., 2013). The biogas energy generates 680 kW of heat and electricity in a combined heat and power facility in the agricultural town of Hamlar, Germany (INNOVAS and DGE GmbH, undated).

4.10.3 BIOGAS IN HORDALAND, BALTIC BIOGAS BUS

For the development of an environmentally friendly and climate-neutral transport sector, Bjørlykke and Rojas (2012) investigated the biogas production from waste materials in Hordaland, Norway, with an emphasis on identifying new raw material sources, such as raw materials from aquaculture and forestry. The aquaculture industry sludge consists of both feces and fish waste, which serve as raw materials for biogas production. Wood is the greatest resource, including tree branches and tops, overgrowth along roads and in the cultural landscape, or forest on building lots and the wood exported to grinding. These unused resources constitute a biogas potential large enough to run Skyss' entire bus fleet in Bergen, Norway. Norway will lead an active and comprehensive policy if these raw materials are utilized with the aim of realizing existing resources and running climate-neutral heavy transport, which would give a substantial commitment to the objectives in the climate policy.

4.10.4 BIOETHANOL

The VTT Technical Research Centre of Finland produced ethanol by fermentation of food processing wastewater, developing a technology enabling production even at a small scale. It is estimated to have potential to meet 2% of the total volume of petrol sold in Finland and is currently being commercialized by St1 Biofuels Oy (Stafford et al., 2013).

4.11 RENEWABLE ENERGY FROM WASTEWATER USING INTEGRATED TECHNOLOGIES

The integration of technologies and waste streams can provide the renewable energy supply and sustainable and cost-effective wastewater treatment in several systems. One of these systems is integrated algal biodiesel production. The recovery of algal biodiesel and wastewater treatment can be performed simultaneously. This integrated process would be more lucrative if animal feed, valuable biomaterials, and energy products were recovered. For example, in Aquaflow Bionomic (New Zealand), the growth of algae and treatment of waste streams are conducted in the same oxidation ponds, and then the crude oil is harvested (Stafford et al., 2013). Further developments are required to make the projects more sustainable and economic.

These integrated systems significantly contribute to the overall cost of microalgae biomass production, wastewater treatment, renewable energy production, and recovery of biobased materials. According to some recent lifecycle analysis studies, these systems can produce cost-effective biodiesel from microalgae and add value to treating the wastewater itself. A biorefinery design can integrate the municipal wastewater management with the oleaginous microalgae culture.

4.12 RENEWABLE ENERGY FROM WASTEWATER AND LIFECYCLE ASSESSMENT

The lifecycle assessment can indicate the potential of energy production from wastewater by analyzing the costs, benefits, and technologies appropriate for a specific wastewater. The types and yields of energy products can be provided by a pre-feasibility study. The wastewater solid materials are divided into three categories, including streams of dissolved solids, suspended solids, and sludge. The following important considerations should be included in a feasibility study:

1. The value of the produced energy types (heat, electricity, combined heat and power, or fuel) and the demand in the market decide the most appropriate technologies and their limitations.
2. The ease of separation of the energy fuel product from water is often the key to feasibility. For example, it is easy to separate biogas from the wastewater by phase separation, whereas energy-intensive distillation is required for bioethanol.

3. The dissolved solids, suspended solids, and sludges can be treated by anaerobic biogas technology. However, the latter require pre-treatment to maximize degradation.
4. The solid waste with limited moisture content, such as dewatered and solar-dried (or previously stockpiled) sewage sludge, can be treated by combustion/gasification in limited cases.
5. The fruit and sugar-cane processing wastewater can be fermented for bioethanol in the agricultural sector.
6. The biodiesel production from algal biomass is not yet feasible as a stand-alone technology and requires integrated technologies.

4.13 BIOGAS PRODUCTION FROM FOOD WASTE AND EFFECT OF TRACE ELEMENTS

Microorganisms need an adequate supply of nutrients for growth in biogas digesters (Feng et al., 2010). Certain trace elements are necessary in addition to the macronutrients (C, N, P, etc.) to strongly impact the biogas production. Cobalt (Co), nickel (Ni), iron (Fe), zinc (Zn), molybdenum (Mo), and/or tungsten (W) are the main trace elements for the activity of enzymes in methanogenic systems. The performance of biogas production can be improved by additions of trace elements resulting in rapid decomposition of organic matter and lower levels of volatile fatty acids (Pobeheim et al., 2010). However, combinations of several trace elements have shown both synergetic and antagonistic effects. The mixture of Co and Ni enhanced the degradation of acetate more than each did individually (Murray and van den Berg, 1981). Furthermore, Mo has augmented the performance of reactor only in combination with Co and Ni. Fe and Co demonstrated the synergetic effects, while Co and Ni showed the antagonistic effects on the total methanogenic activity in methane production from sulphate-rich wastewater (Patidar and Tare, 2006).

4.14 MICROBIAL POPULATION IN DIFFERENT BIOGAS REACTORS: EFFECT OF SUBSTRATE AND OPERATIONAL PARAMETERS

The microbial community has a significant role in conversion of organic materials to methane and carbon dioxide, plus small amounts of hydrogen sulphide in anaerobic digestion through four main reactions: hydrolysis, acidogenesis, acetogenesis, and methanogenesis. A few enzymes depolymerize carbohydrates, lipids, and proteins into soluble compounds in the hydrolysis process of anaerobic degradation. The acetogenic bacteria produce acetate from hydrogen and carbon sources in the acetogenesis reaction (Daniel et al., 1993). Methanogens belong to the Archaeal phylum Euryarchaeota (Anderson et al., 2009), and methane is produced in the last step of the anaerobic process. The methane-producing microorganisms, which usually dominate in biogas reactors, are the acetoclastic methanogens (Zinder, 1993). The primary substrate for methane production by the hydrogenotrophic methanogens is carbon dioxide and hydrogen, and this group consists of several methanogenic orders: Methanobacteriales, Methanococcales, and Methanomicrobiales (Garcia et al., 2000; Liu and Whitman, 2008).

The syntrophic acetate oxidation reaction and subsequent methane production occurs in digestor with a high content of ammonia and fatty acids. The bacteria involved in this process are *Clostridium ultunense*, *Tepidanaerobacter acetatoxydans*, and *Syntrophaceticus schinkii*, followed by hydrogenotrophic methanogen—for example, Methanomicrobiales and Methanobacteriales (Schnurer et al., 1999; Westerholm et al., 2011). The acetogenic bacteria and the methanogenic Archaea differ largely in respect to nutrition and environmental situations (Chen et al., 2008b). Furthermore, the methanogens grow more slowly than the acidogenic bacteria (Pandey et al., 2011), leading to accumulation of intermediate degradation products. The imbalance between these two groups of microorganisms is a common reason for biogas reactor instability (Demirel and Yenigun, 2002). Solli et al. (2014)

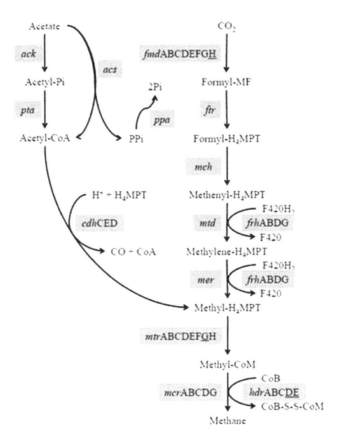

FIGURE 4.2 The methanogenesis pathway. Enzymes are shown in blue boxes. Subunits missing in all our datasets (R1, R2, R3, R4, and IN) after search against the KO database at MG-RAST are underlined. Abbreviations used in the figure are Acetyl-Pi: acetyl phosphate; ack: acetate kinase; acs: acetyl-CoA synthetase; cdh: acetyl-CoA decarbonylase/synthase; CO: carbon monoxide; CoA: coenzyme A; CoB: coenzyme B; CoB-S-S-CoM: coenzyme M 7-mercaptoheptanoylthreonine-phosphate heterodisulfide; F420: coenzyme F420; fmd: formylmethanofuran dehydrogenase; Formyl-H$_4$MPT: 5-formyl-5,6,7,8-tetrahydromethanopterin; Formyl-MF: formylmethanofuran; frh: coenzyme F420 hydrogenase; ftr: formylmethanofuran-tetrahydromethanopterin N-formyltransferase; H$_4$MPT: 5,6,7,8-tetrahydromethanopterin; hdr: heterodisulfide reductase; mch: methenyl-tetrahydromethanopterin cyclohydrolase; mcr: methyl-coenzyme M reductase; mer: 5,10-methylenetetrahy dromethanopterin reductase; KO: KEGG orthology; Methenyl-H$_4$MPT: 5,10-methenyl-5,6,7,8-tetrahydro-methanopterin; Methyl-CoM: methylcoenzyme M; Methylene-H$_4$MPT: 5,10-methylenetetrahydromethanopterin; Methyl-H$_4$MPT: 5-methyl-5,6,7,8-tetrahydromethanopterin; mtd: methylenetetrahydromethanopterin dehydrogenase; mtr: tetrahydromethanopterin S-methyltransferase; ppa: inorganic diphosphatase; pta: phosphate acetyltransferase. (Reproduced from Solli L. et al. 2014. *Biotechnol Biofuels* 7: 146.)

found that the bacteria *Clostridium* and *Syntrophomonas* were highly abundant, and the dominating methanogen was the hydrogenotrophic *Methanoculleus* in biogas reactors digesting manure and fish waste silage. Figure 4.2 shows the results from analysis of functional enzymes involved in methane production.

The hydrogen sulphide has corrosive properties, causing damage on equipment and reducing the microbial growth. Therefore, the hydrogen sulphide should be removed in industrial-scale pro-duction (Appels et al., 2008) through precipitation of sulphides with ferric or ferrous iron in the digester, aeration of the gas to obtain elemental sulphur, or biological treatment with bacteria such as *Thiobacillus* strains (Moestedt et al., 2013).

4.15 LIGNOCELLULOSIC WASTES AND PRE-TREATMENT FOR ETHANOL AND BIOGAS PRODUCTION

The solid waste and wastewater from different industry and municipal sources have lignocelluloses as a major composition. Hydrolysis of waste materials facilitates the digestion to biogas (methane) or fermentation to ethanol. Without pretreatment, hydrolysis of lignocelluloses is poor due to high stability of the materials to enzymatic or bacterial attacks attributed to crystallinity, accessible surface area, and protection by lignin and hemicellulose (Taherzadeh and Karimi, 2008). There are a couple of methods for pre-treatment of lignocellulosic materials that include "physical pre-treatment," "physico-chemical pre-treatment," "chemical pre-treatment," and "biological pre-treatment" (Wyman, 1996; Berlin et al., 2006; Karimi et al., 2006). The pre-treatment methods have diverse advantages and benefits to enhance ethanol and/or biogas production (Figure 4.3), which include milling, irradiation, microwave, steam explosion, ammonia fiber explosion (AFEX), supercritical carbon dioxide and its explosion, alkaline hydrolysis, liquid hot-water pretreatment, organosolv processes, wet oxidation, ozonolysis, dilute- and concentrated-acid hydrolyses, and biological pre-treatments. The efficient utilization of the hemicelluloses may reduce the cost of ethanol or biogas production. However, some methods are costly and not economical and sustainable (Eggeman and Elander, 2005), while biological pre-treatments are extremely slow (Sun and Cheng, 2002).

4.16 CONCLUSIONS

Sustainable system development is necessary for proper use and exploitation of municipal and industrial wastes, considering the emerging environmental problems, public interest, and survival of human, animal and plant populations. AD plays a significant role in an integrated resource recovery system and sustainability in waste management. This process produces biogas, reduces the volume of waste for final disposal or recycling, destroys the pathogens, and reduces odor problems. Hydrolysis is an important rate-controlling step in the digestion of complex substrates. Various pretreatments to the feedstock, such as mechanical, thermal, chemical, and biological interventions, can be applied to enhance the hydrolysis of feedstock and augment the digestion and production of biogas. Many biotic and abiotic factors, mainly aerobic and anaerobic, control the action/reaction in the decomposition technology.

There are several anaerobic reactor processes, such as up-flow anaerobic sludge blanket reactor, anaerobic moving-bed biofilm reactor, anaerobic fluidized bed reactor, anaerobic membrane bioreactor, expanded granular sludge bed reactor or internal circulation reactor, bench-scale EGSB reactor, and integrated internal and external circulation reactor. The characteristics of feedstock, compost quality, and odor emissions are the main problems of the anaerobic digestion plants. The financial investment, expense of operation, and income generation from biogas and biobased fertilizer are the biggest challenges of this technology.

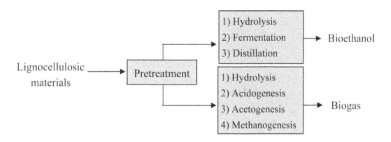

FIGURE 4.3 Pretreatment of lignocellulosic materials prior to bioethanol and biogas production. (Reproduced from Taherzadeh M. J. and Karimi K. 2008. *Int J Mol Sci* 9: 1621–1651.)

Different technologies have been developed to recover the renewable energy from wastewater. Various technological options involving energy recovery from wastewater include three steps: (1) input streams, (2) process intermediates, and (3) energy outputs. The technology has been developed for ethanol production by fermentation of industrial wastewater. The integration of technologies and waste streams, such as integrated algal biodiesel and advanced integrated wastewater pond systems, can provide sustainable renewable energy supply and wastewater management. The lifecycle assessment can indicate the potential of energy recovery from wastewater, by which the costs and benefits of technologies can be analyzed for a specific wastewater.

Based on pyrosequencing study, the *Methanosaeta* and *Methanomicrobium* were the dominant acetoclastic and hydrogenotrophic methanogen genera, respectively, during stable reactor operation. The microbial community has a significant role in conversion of organic materials to biogas, carbon dioxide, plus small amounts of hydrogen sulphide in anaerobic digestion through four main reactions: hydrolysis, acidogenesis, acetogenesis, and methanogenesis. Certain trace elements are necessary in addition to the nutrients (C, N, P, etc.) to strongly impact the biogas production. Several bacterial groups are inhibited in the anaerobic reactors that are sensitive to process parameters.

The waste streams from different sources, such as forestry, agriculture, and municipalities, have lignocelluloses as a major component. Without pre-treatment, hydrolysis of lignocelluloses is generally poor due to high stability of the materials to enzymatic or bacterial attacks.

REFERENCES

Abreu, A. A., J. C. Costa, P. Araya-Kroff, E. C. Ferreira, and M. M. Alves. 2007. Quantitative image analysis as a diagnostic tool for identifying structural changes during a revival process of anaerobic granular sludge. *Water Research* 41: 1473–1480.

Ahmed, F. N., and C. Q. Lan. 2012. Treatment of landfill leachate using membrane bioreactors: A review. *Desalination* 287: 41–54.

Anderson, I., L. E. Ulrich, B. Lupa, D. Susanti, I. Porat, S. D. Hooper, A. Lykidis et al. 2009. Genomic characterization of methanomicrobiales reveals three classes of methanogens. *PLoS One* 4: 6.

Appels, L., J. Baeyens, J. Degreve, and R. Dewil. 2008. Principles and potential of the anaerobic digestion of waste-activated sludge. *Progress in Energy and Combustion Science* 34: 755–781.

Arsova, L. 2010. Anaerobic digestion of food waste: Current status, problems and an alternative product. M.S. Degree thesis in Earth Resources Engineering, Department of Earth and Environmental Engineering, Columbia University, New York, May 2010.

Athanasoulia, E., P. Melidis, and A. Aivasidis. 2014. Co-digestion of sewage sludge and crude glycerol from biodiesel production. *Renewable Energy* 62: 73–78.

Balat, M., and M. Balat. 2009. Political, economic and environmental impacts of biomass-based hydrogen. *International Journal of Hydrogen Energy* 34(9): 3589–3603.

Berlin, A., M. Balakshin, N. Gilkes, J. Kadla, V. Maximenko, S. Kubo, and J. Saddler. 2006. Inhibition of cellulase, xylanase and beta-glucosidase activities by softwood lignin preparations. *Journal of Biotechnology* 125: 198–209.

Bjørlykke, S., and N. Rojas. 2012. Biogas in Hordaland. *Baltic Biogas Bus*, WP4.3, Bergen, August 20, 2012. Stockholm, Sweden.

Carrere, H., C. Bougrier, D. Castets, and J. P. Delgenès. 2008. Impact of initial biodegradability on sludge anaerobic digestion enhancement by thermal pretreatment. *Journal of Environmental Science and Health Part A* 43: 1551–1555.

Carrère, H., Y. Rafrafi, A. Battimelli, M. Torrijos, J.-P. Delgenès, and G. Ruysschaert. 2010. Methane potential of waste activated sludge and fatty residues: Impact of codigestion and alkaline pretreatments. *The Open Environmental Engineering Journal* 3: 71–76.

Chauzy, J., D. Cretenot, A. Bausseon, and S. Deleris. 2007. Anaerobic digestion enhanced by thermal hydrolysis: First reference BIOTHELYS® at Saumur, France, in Facing sludge diversities: Challenges, risks and opportunities. *Water Practice and Technology* 3(1).

Chen, S., D. Z. Sun, and J. S. Chung. 2008a. Simultaneous removal of COD and ammonium from landfill leachate using an anaerobic–aerobic moving-bed biofilm reactor system. *Waste Management* 28: 339–346.

Chen, Y., J. J. Cheng, and K. S. Creamer. 2008b. Inhibition of anaerobic digestion process: A review. *Bioresource Technology* 99: 4044–4064.

Christy, P. M., L. R. Gopinath, and D. Divya. 2013. A review on decomposition as a technology for sustainable energy management. *International Journal of Plant, Animal and Environmental Sciences* 3: 44–50.

Chu, W., and F. L. Ng. 2000. Upgrading the conventional grease trap using a tube settler. *Environment International* 26: 17–22.

Dang, Y., J. Ye, Y. Mu, B. Qiu, and D. Sun. 2013. Effective anaerobic treatment of fresh leachate from MSW incineration plant and dynamic characteristics of microbial community in granular sludge. *Applied Microbiology and Biotechnology* 97: 10563–10574.

Daniel, S. L., and H. L. Drake. 1993. Oxalate- and glyoxylate-dependent growth and acetogenesis by *Clostridium thermoaceticum*. *Applied and Environmental Microbiology* 59: 3062–3069.

Demirel, B., and O. Yenigun. 2002. Two-phase anaerobic digestion process: A review. *Journal of Chemical Technology and Biotechnology* 77: 743–755.

Dewil, R., J. Baeyens, and L. Appels. 2007. Enhancing the use of waste activated sludge as bio-fuel through selectively reducing its heavy metal content. *Journal of Hazardous Materials* 144: 703–707.

Eggeman, T., and R. T. Elander. 2005. Process and economic analysis of pretreatment technologies. *Bioresource Technology* 96: 2019–2025.

Elliott, A., and T. Mahmood. 2007. Pretreatment technologies for advancing anaerobic digestion of pulp and paper biotreatment residues. *Water Research* 41: 4273–4286.

European Commission. 1986. Council Directive 86/278/EEC on the protection of the environment, and in particular soil, when sewage sludge is used in agriculture. Brussels, Belgium.

Feng, X. M., A. Karlsson, B. H. Svensson, and S. Bertilsson. 2010. Impact of trace element addition on biogas production from food Industrial waste—linking process to microbial communities. *FEMS Microbiological Ecology* 74: 226–240.

Garcia, J. L., B. K. C. Patel, and B. Ollivier. 2000. Taxonomic, phylogenetic and ecological diversity of methanogenic Archaea. *Anaerobe* 6: 105–226.

Hagelqvist, A. 2013. Batchwise mesophilic anaerobic co-digestion of secondary sludge from pulp and paper industry and municipal sewage sludge. *Waste Management* 33: 820–824.

Hoornweg, D., and P. Bhada-Tata. 2012. *What a Waste: A Global Review of Solid Waste Management*. World Bank, Washington, DC.

IEA Biogas. (undated). 100% Biogas for urban transport in Linkoping, Sweden. Available online: http://www.iea-biogas.net/files/daten-redaktion/download/linkoping_final.pdf, accessed June 2018.

INNOVAS, DGE GmbH. (undated). Industrial Biogas plant with wastewater treatment. Available online: http://www.dge-wittenberg.de/english/produkte/alternativenergie/news11_eng.pdf, accessed January 2012.

Institute of Science in Society (ISIS). (2006). Biogas China. Available online: http://www.i-sis.org.uk/BiogasChina.php, accessed June 2018.

Jeihanipour, A., and R. Bashiri. 2015. Perspective of biofuels from wastes. In K. Karimi (Ed.), *Lignocellulose-Based Bioproducts, Biofuel and Biorefinery Technologies 1*. Springer International Publishing, Cham, Switzerland. doi:10.1007/978-3-319-14033-9_2.

Karimi, K., S. Kheradmandinia, and M. J. Taherzadeh. 2006. Conversion of rice straw to sugars by dilute acid hydrolysis. *Biomass Bioenergy* 30: 247–253.

Kelessidis, A., and A. S. Stasinakis. 2012. Comparative study of the methods used for treatment and final disposal of sewage sludge in European countries. *Waste Management* 32(6): 1186–1195.

Kepp, U., and O. E. Solheim. 2000. Thermo dynamical assessment of the digestion process. In *CIWEM/Aqua Enviro 5th European Biosolids and Organic Residiuals Conference*. Cedar Court, Wakefield, UK. Available at: http://www.cambi.no/photoalbum/view2/P3NpemU9b3JnJmlkPTIyMDAxOCZ0eXBlPTE, accessed March 2010.

Khanal, S. K., M. Rasmussen, P. Shrestha, H. J. Van Leeuwen, C. Visvanathan, and H. Liu. 2008. Bioenergy and biofuel production from wastes/residues of emerging biofuel industries. *Water Environment Research* 80(10): 1625–1647.

Kolesarova, N., M. Hutnan, I. Bodik, and V. Spalkova. 2011. Utilization of biodiesel by-products for biogas production. *Journal of Biomedicine and Biotechnology* 2011(ID-126798): 15.

Liu, Y. H., Y. L. He, S. C. Yang, and Y. Z. Li. 2006. The settling characteristics and mean settling velocity of granular sludge in upflow anaerobic sludge blanket (UASB)-like reactors. *Biotechnology Letters* 28: 1673–1678.

Liu, Y., and W. B. Whitman. 2008. Metabolic, phylogenetic, and ecological diversity of the methanogenic Archaea. *Annals of the New York Academy of Sciences* 1125: 171–189.

Luo, J., J. Zhou, G. Qian, and J. Liu. 2014. Effective anaerobic biodegradation of municipal solid waste fresh leachate using a novel pilot-scale reactor: Comparison under different seeding granular sludge. *Bioresource Technology* 165: 152–157.

Marchaim, U. 1992. *Biogas Processes for Sustainable Development.* Migal Galilee Technological Centre, Kiryat Shmona, Israel.

Metcalf, and Eddy. 2003. *Wastewater Engineering: Treatment and Reuse.* 4th ed. McGraw-Hill, New York.

Moestedt, J., S. N. Påledal, and A. Schnürer. 2013. Effect of substrate and operational parameters on sulphate-reducing bacteria in industrial anaerobic biogas digesters. *Bioresource Technology* 132: 327–332.

Murray, W. D., and L. van den Berg. 1981. Effects of nickel, cobalt, and molybdenum on performance of methanogenic fixed-film reactors. *Applied and Environmental Microbiology* 42: 502–505.

Nakhla, G., M. Al-Sabawi, A. Bassi, and V. Liu. 2003. Anaerobic treatability of high oil and grease rendering wastewater. *Journal of Hazardous Materials* 102: 243–255.

Neis, U., K. Nickel, and A. Lunden. 2007. Improving anaerobic and aerobic degradation by ultrasonic disintegration of biomass. In *Facing Sludge Diversities: Challenges, Risks and Opportunities*, Antalya, Turkey, pp. 125–132.

Nghiem, L. D., T. T. Nguyen, P. Manassa, S. K. Fitzgerald, M. Dawson, and S. Vierboom. 2014. Co digestion of sewage sludge and crude glycerol for on-demand biogas production. *International Biodeterioration and Biodegradation* 95: 160–166.

Ostrem, K. 2004. Greening waste: Anaerobic digestion for treating the organic fraction of municipal solid wastes. M.S. thesis in Earth Resources Engineering, Department of Earth and Environmental Engineering, Columbia University, New York, May 2004.

Pandey, P. K., P. M. Ndegwa, M. L. Soupir, R. J. Alldrege, and M. J. Pitts. 2011. Efficacies of inocula on the startup of anaerobic reactors treating dairy manure under stirred and unstirred conditions. *Biomass Bioenergy* 35: 2705–2720.

Patidar, S. K., and V. Tare. 2006. Effect of nutrients on biomass activity in degradation of sulfate laden organics. *Process Biochemistry* 41: 489–495.

Peng, Y. Z., S. J. Zhang, W. Zeng, S. W. Zheng, T. Mino, and H. Satoh. 2008. Organic removal by denitritation and methanogenesis and nitrogen removal by nitritation from landfill leachate. *Water Research* 42: 883–892.

Pobeheim, H., B. Munk, J. Johansson, and G. M. Guebitz. 2010. Influence of trace elements on methane formation from a synthetic model substrate for maize silage. *Bioresource Technology* 101: 836–839.

Popov, J. 2010. Model based optimization of biogas production at SNJ plant. Master of Science thesis in Environmental Technology, department of Mathematics and Natural Science, Faculty of Science and Technology at University of Stavanger, Spring semester, Stavanger, Norway.

Qasim, S. R. 1999. *Wastewater Treatment Plants: Planning. Design and Operation.* 2nd ed. Boca Raton, FL: CRC Press.

Razaviarani, V. 2014. Anaerobic co-digestion of municipal sewage sludge with selected commercial and industrial organic wastes. PhD thesis in Environmental Engineering, Department of Civil and Environmental Engineering, University of Alberta, Edmonton, Canada.

Schnurer, A., G. Zellner, and B. H. Svensson. 1999. Mesophilic syntrophic acetate oxidation during methane formation in biogas reactors. *FEMS Microbiology Ecology* 29: 249–261.

Shuler, M. L., and F. Kargi. 2002. *Bioprocess Engineering. Basic Concepts.* 2nd ed. Prentice Hall, Upper Saddle River, NJ.

Solli, L., O. E. Håvelsrud, S. J. Horn, and A. G. Rike. 2014. A metagenomic study of the microbial communities in four parallel biogas reactors. *Biotechnology for Biofuels* 7: 146.

Sorathia, H. S., P. P. Rathod, and A. S. Sorathiya. 2012. Bio-gas generation and factors affecting the bio-gas generation – a review study. *International Journal of Advanced Engineering Technology* III: 72–78.

Stafford, W., B. Cohen, S. Pather-Elias, H. von Blottnitz, R. van Hille, S. T. L. Harrison, and S. G. Burton. 2013. Technologies for recovery of energy from wastewaters: Applicability and potential in South Africa. *Journal of Energy in Southern Africa* 24(1): 15–26.

Sun, Y., and J. Cheng. 2002. Hydrolysis of lignocellulosic materials for ethanol production: A review. *Bioresource Technology* 83: 1–11.

Taherzadeh, M. J., and K. Karimi. 2008. Pretreatment of lignocellulosic wastes to improve ethanol and biogas production: A review. *International Journal of Molecular Sciences* 9: 1621–1651.

Turan, M., H. Gulsen, and M. S. Celik. 2005. Treatment of landfill leachate by a combined anaerobic fluidized bed and zeolite column system. *Journal of Environmental Engineering* 131: 815–819.

Van de Velden, M., R. Dewil, J. Baeyens, L. Josson, and P. Lanssens. 2008. The distribution of heavy metals during fluidized bed combustion of sludge (FBSC). *Journal of Hazardous Materials* 151: 96–102.

Vieitez, E. R., and S. Ghosh. 1999. Biogasification of solid wastes by two-phase anaerobic fermentation. *Biomass and Bioenergy* 16: 299–309.

Weemaes, M., and W. Verstraete. 1998. Evaluation of current wet sludge disintegration techniques. *Journal Chemistry Technology Biotechnology* 73: 83–92.

Werther, J., and T. Ogada. 1999. Sewage sludge combustion. *Progress in Energy and Combustion* 25: 55–116.

Westerholm, M., J. Dolfing, A. Sherry, N. D. Gray, and A. Schnürer. 2011. Quantification of syntrophic acatete oxidizing microbial communities in biogas processes. *Environmental Microbiology Report* 3: 500–505.

Wyman, C. E. 1996. *Handbook on Bioethanol: Production and Utilization.* Taylor & Francis Group, Washington, DC.

Ye, J. X., Y. J. Mu, X. Cheng, and D. Z. Sun. 2011. Treatment of fresh leachate with high strength organics and calcium from municipal solid waste incineration plant using UASB reactor. *Bioresource Technology* 102: 5498–5503.

Zah, R. 2010. *Future Perspectives of 2nd Generation Biofuels,* vol. 55. vdf Hochschulverlag AG, Zurich, Switzerland.

Zayen, A., S. Mnif, F. Aloui, F. Fki, S. Loukil, M. Bouaziz, and S. Sayadi. 2010. Anaerobic membrane bio-reactor for the treatment of leachates from Jebel Chakir discharge in Tunisia. *Journal of Hazardous Materials* 177: 918–923.

Zhu, Z., M. K. Hsueh, and Q. He. 2011. Enhancing biomethanation of municipal waste sludge with grease trap waste as a co-substrate. *Renewable Energy* 36: 1802–1807.

Zinder, S. H. 1993. Physiological ecology of methanogenesis. In J. G. Ferry (Ed.), *Methanogenesis. Ecology, Physiology, Biochemistry and Genetics,* vol. 1. 1st ed. Chapman and Hall, New York, pp. 128–206.

Zupancic, G. D., N. Uranjek-Zevart, and M. Ros. 2008. Full-scale anaerobic co-digestion of organic waste and municipal sludge. *Biomass and Bioenergy* 32: 162–167.

5 Mutual Effects of Climate Change and Energy Crops and Their Controls on Biomass and Bioenergy Production

Hossain M. Anawar and Vladimir Strezov

CONTENTS

5.1 INTRODUCTION

Renewable bioenergy production is attracting attention for future energy supply (Caspeta et al., 2013) due to the demands of increasing population, fossil fuel depletion of reserves, and environmental impact of greenhouse gas (GHG) emissions (Mendu et al., 2012; Gelfand et al., 2013; Wang et al., 2014). The large amounts of GHG emissions into the atmosphere derived from fossil fuel combustion, deforestation and other human activities cause climate change resulting in considerable impacts on ecosystems and human societies (Parry et al., 2007; Popp et al., 2014a). The production of bioenergy has potential impacts on energy and food security, anthropogenic climate change mitigation, deforestation, water use patterns, maintenance of biodiversity and GHG emission changes through direct or indirect land-use change (LUC) and through a life cycle analysis (LCA) (Fargione et al., 2008; Martindale and Trewavas, 2008; Searchinger et al., 2008; Hill et al., 2006, 2009; Tilman et al., 2006, 2009; Georgescu et al., 2011). Biomass can provide energy in different forms, which may include heat, electricity, and gaseous, solid, and liquid fuels. Bioenergy, especially liquid biofuel, is produced from bioenergy crops and biomass (Villamil et al., 2012) that provides a number of significant benefits for sustainable development, enhancement of national energy security, and opportunities for rural economic development (McLaughlin et al., 2002; Fargione et al., 2008; Charles, 2009; Tulbure et al., 2012).

Although the production of ethanol biofuel from starch-based feedstocks, such as corn (*Zea mays* L.) grain in the United States (Landis et al., 2008) has created conflict with the food security, the second-generation biofuels produced from the lignocellulose of plant biomass have mitigated the conflict of food/feed and fuel (Antizar-Ladislao and Turrion-Gomez, 2008). This development notwithstanding, 136 billion L of liquid biofuels are recommended by the Energy Independence and Security Act of 2007 to be produced in the United States by 2022, with 57 billion L of corn-based ethanol and 80 billion L of advanced biofuels per year derived from non-corn starch products (Miresmailli et al., 2013). Recently, many wild plant species, such as the American native plant switchgrass (*Panicum virgatum*), have been introduced in different regions and countries, including China and Europe, and cultivated for bioenergy production (Alexopouloua et al., 2008; Xiong et al., 2008). Some introduced species have high primary productivity and are beneficial for bioenergy production in some regions; however, they are detrimental to the regional ecosystems (Barney and Ditomaso, 2008). For example, the giant reed (*Arundo donax*) has the highest biofuel potential in Europe (Herrera and Dudley, 2003; Lewandowskia et al., 2003), but it is one of the world's top 100 worst invaders (Buddenhagen et al., 2009; Gordon et al., 2011; Flory et al., 2012).

The bioenergy crops planted widely can have adverse environmental impacts (Miresmailli et al., 2013). Most plants release biogenic volatile organic compounds (BVOCs) as signals and cues for pollination (Balao et al., 2011), direct and indirect defense (Gershenzon and Dudareva, 2007), intra-specific communication (Baldwin et al., 2006), and protection against environmental stressors, such as ozone (Loreto and Velikova, 2001) and elevated temperature (Singsaas et al., 1997). Furthermore, BVOCs can influence the chemical and physical constituents of the atmosphere (Loreto et al., 2008; Penuelas and Staudt, 2010) due to higher reactivity with ozone (O_3), hydroxyl radical (OH), and nitrate radical (NO_3) (Loreto et al., 2008; Goldstein et al., 2009), and formation of photochemical oxidants (Finlayson-Pitts and Pitts, 1997) and tropospheric ozone (Chameides et al., 1988). BVOCs can also contribute to accumulation of methane and other GHGs (De Carlo et al., 2004). Atmospheric chemical reactions contribute to changes of the highly volatile compounds into less-volatile compounds (Goldstein and Galbally, 2007), which then contribute to formation of secondary organic aerosol particles that can have impacts on human health (Goldstein et al., 2009). The effect of global climate change can be mitigated by reduction in deforestation, promoting afforestation and reforestation, and maintaining forests with higher carbon density (Torvelainen et al., 2014; Torssonen et al., 2016).

5.2 EFFECTS OF CLIMATE CHANGE ON TEMPORAL STABILITY OF PLANT COMMUNITY BIOMASS PRODUCTION

Since 1880, the average global temperature has increased by 0.065°C per decade, resulting in significant changes in precipitation patterns (IPCC, 2013). The adverse effects of rapid climate change are significantly changing the earth's ecosystems functioning (Garcia et al., 2014; Seddon et al., 2016) and influencing community structure (Wang et al., 2012; Wipf et al., 2013; Elmendorf et al., 2015) and species interactions (Yang et al., 2011; Baldwin et al., 2014; Rudgers et al., 2014), resulting in changes in community biomass stability (Fussmann et al., 2014; Shi et al., 2016; Yang et al., 2016). The stable ecosystems can provide sustainable ecosystem functioning and services to humanity (Oliver et al., 2015). Therefore, it is important to understand the drivers of ecological stability when many ecosystems are experiencing significant anthropogenic change (Hooper et al., 2005; Grman et al., 2010; Hautier et al., 2014, 2015).

Different studies have investigated the effect of climate change on ecological stability, its underlying mechanisms, and the temporal stability of plant community biomass in an alpine grassland located on the Tibetan Plateau (Ma et al., 2017). Global warming lowers biomass temporal stability through a reduction in the degree of species asynchrony. However, the precipitation alteration does not influence stability. Furthermore, plant species diversity does not influence the biomass temporal stability. Ma et al. (2017) state that the changes in climate may change the stability of ecological communities, which can, in turn, potentially affect their potential to achieve ecosystem services for humanity.

5.3 PARTIAL REPLACEMENT OF CORN WITH SECOND-GENERATION ENERGY CROP GIANT REED UNDER CLIMATE CHANGE

Agricultural production of corn and other crops is affected by climate change effects, such as rising temperatures, unfavorable rainfall distribution, and increasing frequency of extreme weather events (Porter and Semenov, 2005; Kang et al., 2009) in many regions. The potential impacts of climate change on the productivity of agricultural systems can be estimated by the process-based biophysical models that can support local stakeholders and policy makers in defining effective and site-specific adaptation strategies (White et al., 2011; Fernandes et al., 2012). The water crisis and competition with urban water supply can increase the cost of corn-based cropping systems. These factors are driving farmers to adopt alternative business models focusing on the agrofuel sector.

The giant reed (*Arundo donax* L.) can provide a potential source of renewable energy in the Mediterranean countries (Cappelli et al., 2015). Therefore, the giant reed–based supply chains and technologies have been investigated to explore their potential for (1) biogas production through anaerobic digestion, (2) combustion for production of electricity and thermal energy, and (3) production of ethanol and biodiesel from anaerobic fermentation. Some compounds of interest for the chemical and pharmaceutical industries (i.e., cellulose, alkaloids) can be extracted from this crop plant that can be used for phytoremediation because of their ability to absorb nitrates, phosphates, and other pollutants. A perennial invasive grass, giant reed can grow in different soil types (Perdue, 1958) and marginal land with low agronomic inputs (Christou et al., 2000; Williams et al., 2009), but can produce attainable energy per hectare higher than those obtained with corn (Schievano et al., 2012); these factors motivated farmers to grow this crop partially instead of corn. The amount of methane produced from anaerobic digestion of giant reed ranges from 7.17 $dam^3\,ha^{-1}\,y^{-1}$ to 11.28 $dam^3\,ha^{-1}\,y^{-1}$ (all gas volumes measured at standard conditions of 298 K and 101.3 kPa) (Schievano et al., 2012; Ragaglini et al., 2014).

However, the effects of future climate on this crop are to be clearly understood. Cappelli et al. (2015) studied the climate change impact on production of giant reed in the Lombardy Plain (northern Italy) and included information on biogas plants, land use, crop management and distribution, and weather conditions for current and projected future climates (Table 5.1). The results suggested increased biomass production rates from giant reed under the predicted future climate conditions (+20% in 2020 and +30% in 2050). For this region, the giant reed by far outweighs the other potential bioenergy crops in terms of the low production costs, payback time, and potential biogas yield on a per hectare–based assessment.

TABLE 5.1

Costs and Productivity of Giant Reed and Alternative Pilot Cropping Systems Targeted for Biogas Production in Lombardy

	Cropping System	15-Years Lifetime Production Costs (€ ha⁻¹)	Biogas Production Rate on Dry Matter (m³ t⁻¹)	Payback Time (Years)	Potential Biogas Yield (m³ ha⁻¹ year⁻¹)	Potential Biomethane Yield (m³ ha⁻¹ year⁻¹)
1	Giant reed	10,500	450	6.4–8.5	21,555–28,350	10,539–13,875
2	Corn	31,590	694	11.3	14,761	7,133
3	Rye-corn	49,275	577	18.2	15,002	7,342
4	Triticale-sorghum	43,545	540	9.0	16,740	8,193
5	Triticale-corn	50,190	631	10.1	21,454	10,500
6	Grass-corn	46,290	584	12.0	18,104	8,861

Source: Reprinted from *Global Change Biol.*, 7, Schievano, A. et al., Biogas from dedicated energy crops in Northern Italy: Electric energy generation costs, 899–908, Copyright 2015, with permission from Elsevier.

5.4 EFFECT OF CLIMATE CHANGE AND GENETIC CONTROLS ON YIELDS OF A MODEL BIOENERGY SPECIES, SWITCHGRASS

Switchgrass (*Panicum virgatum* L.), a warm-season perennial grass and herbaceous species native to North America, has high potential as a bioenergy feedstock (McLaughlin and Walsh, 1998; Parrish and Fike, 2005; Tulbure et al., 2012). This plant species has high yields, a wide geographic distribution in the world, tolerance and sound growth over a wide range of climatic and edaphic conditions, and high nutrient and water use efficiency (McLaughlin and Walsh, 1998; Parrish and Fike, 2005), accompanied by a range of environmental benefits, such as soil erosion prevention, increased carbon sequestration, reduced water runoff, and provision of habitat for wildlife (Parrish and Fike, 2005). There are two genetically and phenotypically distinct cytotypes of switchgrass across the world. These are cytotypes grown in (1) wetter and southern latitudes lowland, and (2) drier, higher latitudes upland (Sanderson et al., 1996; Casler et al., 2004).

Based on the general additive models (GAMs) model and a 39-year climate dataset, Tulbure et al. (2012) demonstrated the spatio-temporal variability in lowland and upland switchgrass yield due to climate variables alone and environmental variables. The variability in switchgrass yield was dominantly controlled by fertilizer application, genetics, precipitation, and management practices. The switchgrass yield demonstrated different relationships with climate variables for upland and lowland cultivars. The prime cropland fields, such as in the Corn Belt of the United States, are the best suited for a constant and high switchgrass biomass yield with the lowest variability. The agricultural yield is not productive in regions with less-suitable climates, where much lignocellulosic feedstock production can be done. However, interannual variability in yields should be expected and incorporated into operational planning.

According to the random forest (RF) algorithm results (Breiman, 2001), the determining parameter for increased switchgrass yields was the nitrogen fertilizer. The switchgrass yields are predicted to show high variabilities due to changes in climate conditions. Genetic variables become important when the two switchgrass cytotypes were analyzed together, with both cytotypes showing decrease in switchgrass biomass yields with increase in average growing season temperature, albeit the change was different among the two cytotypes (Tulbure et al., 2012). The negative relationship between switchgrass yields and temperature was mostly found to be due to the higher moisture stress under elevated temperature conditions.

5.5 EFFECTS OF CLIMATE CHANGE ON PRODUCTION AND UTILIZATION OF ENERGY BIOMASS

Many studies have explored the potential climate impacts of forest-based bioenergy by applying the metric of radiative forcing (i.e., the change in the net energy balance of the earth system), affected by changes in concentration of CO_2 in the atmosphere (Kirkinen et al., 2008; Repo et al., 2011, 2012, 2015; Sathre and Gustavsson, 2011, 2012; Kilpeläinen et al., 2012; Cherubini et al., 2013; Gustavsson et al., 2015). The LCA can compare the net CO_2 exchange between the forest-based system and the fossil system; thus, it can demonstrate how replacement of fossil fuels with forest-based energy can contribute to the climate change mitigation potential (Cherubini and Strømman, 2011).

Using a forest ecosystem model simulation in an LCA, Torssonen et al. (2016) conducted a study on the effect of climate change and management of forests on the radiate forcing impacts on production and use of biomass energy in Finland. The study found that substitution of coal with biomass for energy production reduces radiative forcing, resulting in cooling climate impacts compared to the fossil system as found by Sathre and Gustavsson (2011, 2012) and Gustavsson et al. (2015). Sustainable management of nitrogen fertilization could improve climate and economic impacts of biomass energy utilization. Sustainable management of harvesting residues should also be practiced, as this is recommended for sites that are sufficiently or moderately fertile to ensure the future carbon sequestration potential of the forests (Äijälä et al., 2014). In some cases, nitrogen fertilization may be required to compensate for losses in nutrients (Kuusinen and Ilvesniemi, 2008). Nitrogen fertilization increases the net ecosystem CO_2 exchange by up to 7% when compared to conditions with no fertilization. Under the climate change conditions, the climate change mitigation efficiency with biomass energy was found to be lower compared to that under the current climate.

Bioenergy crops have potential to reduce CO_2 emissions in energy supply and contribute to the European energy policy, provided that the bioenergy supply is sustainable and resilient to climate change, with no impacts on agriculture at both global and regional scales (Cosentino et al., 2012). Cosentino et al. (2012) modeled the potential distribution of selected bioenergy crops in agronomic and environmental constraints for the European Union (EU) for the years 2020 and 2030. The studies on the effects of climate change, technological development, and site-specific models show that in northern Europe, the bioenergy crop yields might increase due to climate change and technological development, while in southeastern Europe, the negative effect of climate change will be mitigated by technological development. The requirements for bioenergy supply, as claimed in the European directive 2009/28/EC, may not be fulfilled by the estimated total biomass production in Europe, on the basis of future yields and surplus land made available for energy crops (Cosentino et al., 2012).

5.6 DIRECT CLIMATE EFFECTS OF PERENNIAL BIOENERGY CROPS IN THE UNITED STATES

The widespread cultivation of perennial grasses, such as switchgrass (*Panicum virgatum* L.) or miscanthus (*Miscanthus × giganteus*) is an important strategy for bioenergy production. Georgescu et al. (2011) used the Weather Research and Forecasting Model (WRF) to determine the effect of climate on conversion of agricultural areas in the central United States to perennial crops. Studies on side-by-side co-cultivation of miscanthus and maize by Dohleman and Long (2009) showed the one-month shifting impact of modification of surface vegetation properties (albedo, leaf area index, and vegetation fraction) by which vegetation characteristics are advanced in the spring season by one month relative to the default and delayed by one month in autumn.

The hypothetical conversion of annual to perennial bioenergy crops across the central United States provides biogeophysical effects and a significant local to regional cooling with considerable implications for the reservoir of stored soil water (Georgescu et al., 2011), increase in transpiration, and higher albedo. The higher albedo under the studied condition can reduce radiative forcing equivalent to carbon emission reduction of 78 t C/ha or six times above the annual biogeochemical effects from offsetting fossil fuel. An important aspect of climate impacts of biofuels is the bio-geophysical effects, even at the global scale. This simulated cooling effect can partially offset the warming due to increasing GHGs over the next few decades locally. The bioenergy-related LUC and their costs and benefits must include potential impacts on the surface energy and water balance to comprehensively address important concerns for local, regional, and global climate change.

5.7 IMPACTS OF CLIMATE CHANGE ON DISTRIBUTION OF MAJOR NATIVE NON-FOOD BIOENERGY PLANTS

The non-food bioenergy crops can be grown on the marginal lands that are in relatively poor natural condition and are not currently used for agricultural production (Zhuang et al., 2011; Wang et al., 2014). For the last few years, several bioenergy plants have been introduced and grown in China with some introduced species, including *Ricinus communis* and *Helianthus tuberosus*, showing risk of biological invasion (Axmacher and Sang, 2013). It is recommended that native non-food bioenergy plants are favor-able over introduced plant species. The impact of climate on cultivation of non-food bioenergy crops is an important variable for investigation. Species distribution models (SDMs) are typically applied to determine the growing areas under current and future climate condition. There are different types of SDM models developed to determine the species distribution from presence-only species record, which include Genetic Algorithm for Rule-Set Production (GARP), Ecological Niche Factor Analysis (ENFA), Maximum Entropy (MaxEnt) and BIOCLIM. Out of these models, MaxEnt usually produces satisfac-tory predictions of species distribution (Phillips et al., 2006; Elith et al., 2011; Garcia et al., 2013; Booth et al., 2014). The various applications of this model include spatial distribution of plant species (Yang et al., 2013), the spread of invasive species (O'Donnell et al., 2012), and the effects of climate change on plant distributions (Khanum et al., 2013; Warren et al., 2013) and identification of the current and future suitable areas for growing bioenergy crops (Evans et al., 2010; Trabucco et al., 2010).

Wang et al. (2014) applied the MaxEnt species distribution model to investigate the impacts of cli-mate change on the distribution of nine non-food bioenergy plants native to China (viz., *Pistacia chi-nensis, Cornus wilsoniana, Xanthoceras sorbifolia, Vernicia fordii, Sapium sebiferum, Miscanthus sinensis, M. floridulus, M. sacchariflorus*, and *Arundo donax*). According to the MaxEnt modeling results, the most important bioclimatic variables for most of the nine plants were "precipitation of the warmest quarter" and "annual mean temperature." There will be little effect on the total distri-bution area of each plant, although global warming in coming decades may result in a decrease in the extent of suitable habitat in the tropics. Therefore, it will be possible to grow these plants on the marginal lands of the regions east of the Mongolian Plateau and the Tibetan Plateau in the future.

5.8 UNCERTAINTY OF FOSSIL ENERGY RESOURCES AND IMPLICATIONS ON BIOENERGY USE AND CARBON EMISSIONS

The quantity and type of the global fossil fuel resources and their consumption determine the evolu-tion of the future global energy system, corresponding fossil fuel carbon emissions and its impact on global climate change (Nakicenovic et al., 2000; Calvin et al., 2016). To meet the climate policy goals, several modeling studies have investigated the potential role, resource size, and land-use emissions impact due to replacement of some portion of fossil energy use with bioenergy (Chum et al., 2011; Havlik et al., 2011; Popp et al., 2012). Calvin et al. (2016) explored the links and implica-tions of the potential impact of the availability of fossil fuels on agriculture, land use, ecosystems, and carbon emissions from LUC due to the interaction with biomass energy. The limited availability of oil resources and higher price of liquid fuels will encourage the production of bioenergy crops and liquid fuels, particularly in a business-as-usual scenario. The increased production of bioenergy crops will convert unmanaged ecosystems to produce bioenergy, and higher rates of terrestrial carbon emissions from land use.

5.9 LANDSCAPE PATTERNS OF BIOENERGY IN A CHANGING CLIMATE AND COMPETITION FOR CROP LAND

Rural landscapes are encountering multiple challenges due to climate change, shifting development pressure, and loss of agricultural land (Graves et al., 2016). Although the cultivation of perennial bioenergy crops on agricultural soils can contribute to conservation of rural landscapes and produce energy crops at the same time, there are still concerns regarding the food-energy-environment trilemma. The assessment of bioenergy potential in complex landscapes is complicated by heterogeneity of climate, soils, and land use, creating challenges to evaluating future tradeoffs. Graves et al. (2016) reported that the production of perennial bioenergy can contribute to diversion of agricultural land conversion to development. They provided assessment of potential bioenergy crop growth in a southern Appalachian Mountain region until 2100 using a process-based crop model and three crops (switchgrass *Panicum virgatum*, giant miscanthus *Miscanthus* × *giganteus*, and hybrid poplar *Populus* × sp.) under current climate and climate-change scenarios. The average annual perennial grass yield decreased from −4% to −39%, while the production of hybrid poplar increased from +8% to +20%. These results indicate that a switch to woody crops would maximize bioenergy crop production under both climate-change scenarios. In all future climate scenarios, the distribution of bioenergy crops varied significantly from that in the current climate to maximize landscape yield. The maximum yield crop required allocation of most of the agricultural landscape to miscanthus (70%), followed by switchgrass (17%) and hybrid poplar (13%) under current climate scenarios. Maximizing bioenergy production under both climate scenarios required allocating more land to switchgrass and poplar and less land to miscanthus by 2050. The allocation of land to poplar increased to more than 90% by 2100, while land allocation to miscanthus decreased to close to zero.

The maximum landscape yield in the twenty-first century increased by up to 90,000 Mg/yr (6%) due to increased poplar production. The bioenergy crop production areas (>18 Mg·ha^{-1}·yr^{-1}) consistently overlapped with urban development likelihood and the food crop production. When the bioenergy production is constrained to marginal (non-crop) lands, the landscape yield is shown to decrease by 27%. The losses of up to 670,000 Mg/yr (40%) can be incurred when the land with high development probability is removed from crop production. The study of Graves et al. (2016) indicated that tradeoffs among bioenergy production, crop production, and exurban expansion in mountain areas can vary spatially over time with climate change. Bioenergy crops have potential to counter the losses of agricultural land due to development in case the bioenergy market develops to its potential.

5.10 EFFECT OF CLIMATE CHANGE ON UNCERTAINTY OF AGRICULTURE AND FORESTRY-BASED BIOMASS CONTRIBUTIONS TO RENEWABLE ENERGY RESOURCES

The amount of available biomass for production of bioenergy largely depends on the land-use policies and the effect of climate change on the crop yields (Gutsch et al., 2015). Climate change will have substantial impact on the potential supply of biomass for production of bioenergy. Gutsch et al. (2015) provided a study on how the projected climate change for Germany will impact the bioenergy crop production under the three climate change scenarios of maximum, average, and minimum temperature change and for four biomass feedstocks, consisting of forest, short-rotation coppices, cereal straw, and energy maize. Overall, Germany's annual biomass potential, estimated at 1,500 PJ, would vary between −5% and +8%, depending on the assumed climate-change scenario. This change, however, would not impact the country's bioenergy targets for 2020 (1,287 PJ) or 2050 (1,534 PJ), considering the predicted potential energy would vary between 1,425 and 1,620 PJ. The major impacted crops from climate change would be energy maize and winter wheat. Climate change is expected to have mainly positive effects on forest yields, while there are both positive and negative effects on the yields of short-rotation coppices (Gutsch et al., 2015).

5.11 LAND-USE TRANSITION FOR BIOENERGY AND CLIMATE STABILIZATION: MODEL COMPARISON OF DRIVERS, IMPACTS, AND INTERACTIONS WITH OTHER LAND USE–BASED MITIGATION OPTIONS

Popp et al. (2014b) analyzed the drivers and impacts of bioenergy production on the global land system using three integrated assessment models (GCAM, IMAGE, and ReMIND/MAgPIE). Their results indicate that dedicated bioenergy crops and biomass residues will contribute significantly to the energy system. However, the results of different models vary strongly in terms of deployment rates, land-cover allocation, land-use constraints, energy crop yields, non-bioenergy land mitigation options modeled, feedstock composition, and land-use and GHG implications. The bioenergy cropland will make between 10% and 18% of the total cropland by 2100, depending on the model used, and will promote enlargement of croplands at the expense of carbon-richer ecosystems under a scenario with no climate change mitigation. The predicted emissions of GHGs in 2100 from LUC and agricultural intensification will increase from 14 to 113 Gt CO_2-eq.

5.12 IMPACTS OF HERBACEOUS BIOENERGY CROPS ON BIOGENIC VOLATILE ORGANIC COMPOUNDS AND CONSEQUENCES FOR GLOBAL CLIMATE CHANGE

There is potential that vast areas of land will be occupied by the bioenergy crops due to their low energy density. The new bioenergy crops will change the quantity and composition of plant-derived BVOCs emitted to the atmosphere. Miresmailli et al. (2013) determined the types and amount of different BVOCs emitted to the atmosphere by different herbaceous perennial plants, which included miscanthus (*Miscanthus × giganteus*), switchgrass (*Panicum virgatum*), and a restored prairie assemblage of 28 species. Although the prairie assemblages have relatively low rates of biomass production, they have a number of advantages as a bioenergy feedstock, which include low requirements for nitrogen and water, and high potential to rebuild biodiversity in agricultural lands (Tilman et al., 2006). The concentration of BVOCs determined in this study was found to differ between the plant canopies. The BVOC emissions were found to be higher among all plants for the upper canopy levels, with miscanthus producing lower emissions, compared to the other grass species. The chemical composition of BVOCs was significantly different between the studied vegetation types. Miscanthus produced BVOCs depleted in terpenoids compared to the other plant communities. The prairie assemblage produced high-carbon-flux BVOC emissions, when compared to miscanthus and switchgrass. The study concluded that the widespread adoption of herbaceous bioenergy crops could substantially change the atmospheric composition of BVOCs when the findings are extrapolated to the landscape scale level, which would influence the global warming potential and other environmental impacts, including formation of atmospheric particles, photochemical smog, and interactions between plants and arthropods. When planted to the landscape level, miscanthus was suggested to have lower predicted environmental impacts from BVOC emissions compared to the other plants.

5.13 CLIMATE CHANGE MITIGATION BY GLOBAL CROPLANDS AND BIOENERGY CROPS FOR ENHANCED WEATHERING

The large amounts of carbon dioxide (CO_2) and nitrous oxide (N_2O) are emitted to the atmosphere due to conventional row crop agriculture for both food and fuel. The increased production on agricultural land increases the potential for soil carbon loss and soil acidification due to fertilizer use. The global climate change can be mitigated and nutrient availability to plants can be increased by enhanced weathering (EW) in agricultural soils; this can be accomplished by applying crushed silicate rock as a soil amendment (Kantola et al., 2017). EW in the land producing food and fuel has high potential to sequester carbon (C) and reduces N_2O loss through pH buffering, while benefitting

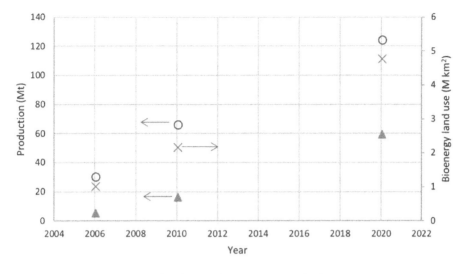

FIGURE 5.1 Projections of global biodiesel/bioethanol production and 1G bioenergy crop land use. (Modified from Kantola, I.B. et al., *Biol. Lett.*, 13, 20160714, 2017.)

crop production and global climate and reducing fossil fuel combustion. However, there exist some uncertainties in the long-term effects and global implications of large-scale efforts to directly manipulate Earth's atmospheric CO_2 composition. The natural chemical weathering of silicate rocks controls and sequesters the atmospheric CO_2 on geologic timescales that can be accelerated by applying crushed, fast-weathering basalt or Ca- and Mg-rich silicate rocks to the land surface as EW (Schuiling and Krijgsman, 2006; Moosdorf et al., 2011; Hartmann et al., 2013; Taylor et al., 2016) while reducing N loss, counteracting soil acidification, and supplying nutrients through the by-products of the weathering processes. The 10–15 M km^2 of global cropland (FAO, 2012) offers a host of environments for deployment of EW substrates, with a potential return of 200–800 kg sequestered CO_2 t^{-1} basalt (Renforth, 2012).

The growing annual and perennial bioenergy crops in the agricultural and marginal lands is increasing due to growing interest in biofuels and reduction in fossil fuel consumption (Anderson-Teixeira et al., 2009, 2012; Davis et al., 2012; Popp et al., 2014b). Over the past 20 years, the bioenergy production from 1G bioenergy crops augmented from near zero in 1990 to 85 million tonnes of bioethanol and biodiesel in 2010 (Figure 5.1). This increasing trend continues, following the models of Brazil, the EU, and the United States, with subsidies and mandates for fossil fuel reductions (Kantola et al., 2017). Perennial crops are more effective than annuals at weathering, due to their longer growing seasons than annuals and extensive root systems that support large biotic communities (Hinsinger et al., 2001; Anderson-Teixeira et al., 2012, 2013).

5.14 CONCLUSIONS

Biomass and bioenergy crops can produce heat, electricity, and gaseous, solid, and liquid fuels that increase national energy security, rural economic development, and sustainable development. The climate change affects the agricultural production of corn and other crops and is driving farmers to replace corn with second-generation energy crops, such as giant reed, under climate change. Climate change has significant effects on the potential supply of biomass for energy production. The climate scenarios lead to decreasing yields of energy maize and winter wheat in northern European countries, such as Germany. Forest yields are mainly positively impacted by climate change, while the short-rotation crops can exhibit both positive and negative yield effects by climate change. Many wild plant species, including switchgrass (*Panicum virgatum* L.), are cultivated for bioenergy production. Switchgrass has high potential as a bioenergy feedstock, a wide geographic distribution, tolerance and growth over a wide range of climatic and edaphic conditions, and high nutrient and water use efficiency. Switchgrass shows different relationships with climate variables for upland and lowland cultivars. The giant reed can provide a potential source of renewable energy in Mediterranean countries under future climate projections. The bioenergy-related LUC and their costs and benefits should account for the water balance and surface energy to address the concerns of climate change.

Energy production using bioenergy crops can reduce CO_2 emission and radiative forcing, resulting in a cooling climate impact compared to the fossil system, provided that the bioenergy supply is sustainable and resilient to climate change. The global climate change can be mitigated and nutrient availability to plants can be increased by EW in agricultural soils through applying crushed silicate rock as a soil amendment, while the perennial crops are more effective than annuals at weathering. The effect of global climate change can be mitigated by reduction in deforestation and degradation of forests, promoting afforestation and reforestation and maintaining higher carbon density in forests.

REFERENCES

Äijälä O, Koistinen A, Sved J, Vanhatalo K, Vaisanen P (Eds.) (2014) *Recommendations for Forest Management in Finland*. Forestry Development Centre Tapio, Metsakustannus Oy, Helsinki, Finland. (In Finnish)

Alexopouloua E et al. (2008) Biomass yields for upland and lowland switchgrass varieties grown in the Mediterranean region. *Biomass Bioenerg* 32: 926–933.

Anderson-Teixeira KJ, Davis SC, Master MD, DeLucia EH (2009) Changes in soil organic carbon under biofuel crops. *Glob Change Biol Bioenergy* 1: 75–96. doi:10.1111/j.1757-1707.2008.01001.x.

Anderson-Teixeira KJ, Duval BD, Long SP, DeLucia EH (2012) Biofuels on the landscape: Is 'land sharing' preferable to 'land sparing'? *Ecol Appl* 22: 2035–2048. doi:10.1890/12-0711.1.

Anderson-Teixeira KJ, Masters MD, Black CK, Zeri M, Hussain MZ, Bernacchi CJ, DeLucia EH (2013) Altered belowground carbon cycling following land-use change to perennial biofuel crops. *Ecosystems* 16: 508–520. doi:10.1007/s10021-012-9628-x.

Antizar-Ladislao B, Turrion-Gomez JL (2008) Second-generation biofuels and local bioenergy systems. *Biofuel Bioprod Bior (Biofpr)* 2: 455–469.

Axmacher JC, Sang W (2013) Plant invasions in China - challenges and chances. *PLoS One* 8: e64173.

Balao F, Herrera J, Talavera S, Dotterl S (2011) Spatial and temporal patterns of floral scent emission in Dianthus inoxianus and electroantennographic responses of its hawkmoth pollinator. *Phytochemistry* 72: 601–609.

Baldwin AH, Jensen K, Schonfeldt M (2014) Warming increases plant biomass and reduces diversity across continents, latitudes, and species migration scenarios in experimental wetland communities. *Glob Change Biol* 20: 835–850.

Baldwin IT, Halitschke R, Paschold A, Von Dahl CC, Preston CA (2006) Volatile signalling in plant-plant interactions: "Talking trees" in the genomics era. *Science* 311: 812–815.

Barney JN, Ditomaso JM (2008) Nonnative species and bioenergy: Are we cultivating the next invader? *BioScience* 58: 64–70.

Booth TH, Nix HA, Busby JR, Hutchinson MF (2014) BIOCLIM: The first species distribution modelling package, its early applications and relevance to most current MAXENT studies. *Divers Distrib* 20: 1–9.

Breiman L (2001) Random forests. *Machine Learning* 45: 5–32.

Buddenhagen CE, Chimera C, Clifford P (2009) Assessing biofuel crop invasiveness: a case study. *PLoS One* 4: e5261.

Calvin K, Wise M, Luckow P, Kyle P, Clarke L, Edmonds J (2016) Implications of uncertain future fossil energy resources on bioenergy use and terrestrial carbon emissions. *Clim Change* 136: 57–68.

Cappelli G, Yamac SS, Stella T, Francone C, Paleari L, Negri M, Confalonieri R (2015) Are advantages from the partial replacement of corn with second-generation energy crops undermined by climate change? A case study for *giant reed* in northern Italy. *Biomass Bioenerg* 80: 85–93.

Casler MD, Vogel KP, Taliaferro CM, Wynia RL (2004) Latitudinal adaptation of switchgrass populations. *Crop Science* 44: 293–303.

Caspeta L, Buijs NAA, Nielsen J (2013) The role of biofuels in the future energy supply. *Energ Environ Sci* 6: 1077–1082.

Chameides WL, Lindsay RW, Richardson J, Kiang CS (1988) The role of biogenic hydrocarbons in urban photochemical smog: Atlanta as a case study. *Science* 241: 1473–1475.

Charles D (2009) BIOFUELS corn-based ethanol flunks key test. *Science* 324: 587–587.

Cherubini F, Guest G, Strømman AH (2013) Bioenergy from forestry and changes in atmospheric CO2: Reconciling single stand and landscape level approaches. *J Environ Manage* 129: 292–301.

Cherubini F, Strømman AH (2011) Life cycle assessment of bioenergy systems: State of the art and future challenges. *Biosource Technol* 102: 437–451.

Christou M et al. (2000) Screening of *Arundo donax* L. population in South Europe. In *Proceedings 1st World Conference on Biomass for Energy and Industry*, Sevilla, Spain, June 5–9, pp. 2048–2051.

Chum H et al. (2011) Bioenergy. In *IPCC Special Report on Renewable Energy Sources and Climate Change Mitigation*, Edenhofer O, Pichs-Madruga R, Sokona Y, Seyboth K (Eds.), pp. 209–332. Cambridge, UK: Cambridge University Press.

Cosentino SL, Testa G, Scordia D, Alexopoulou E (2012) Future yields assessment of bioenergy crops in relation to climate change and technological development in Europe. *Ital J Agron* 7: 22.

Davis SC, Parton WJ, Del Grosso SJ, Keough C, Marx E, Adler PR, DeLucia EH (2012) Impact of second generation agriculture on greenhouse-gas emissions in the corn-growing regions of the US. *Front Ecol Environ* 10: 69–74. doi:10.1890/110003.

De Carlo P et al. (2004) Missing OH reactivity in a forest: Evidence for unknown reactive biogenic VOCs. *Science* 304: 722–725.

Directive 2009/28/EC (2009) On the promotion of the use of energy from renewable sources and amending and subsequently repealing Directives 2001/77/EC and 2003/30/EC.2009.04.23. *Off J Eur Union* L140: 16–62.

Dohleman FG, Long SP (2009) More productive than maize in the Midwest: How does Miscanthus do it? *Plant Physiol* 150: 2104–2115.

Elith J, Phillips SJ, Hastie T, Dudik M, Chee YE et al. (2011) A statistical explanation of MaxEnt for ecologists. *Divers Distrib* 17: 43–57.

Elmendorf SC et al. (2015) Experiment, monitoring, and gradient methods used to infer climate change effects on plant communities yield consistent patterns. *Proc Natl Acad Sci USA* 112: 448–452.

Evans J, Fletcher Jr RJ, Alavalapati J (2010) Using species distribution models to identify suitable areas for biofuel feedstock production. *GCB Bioenergy* 2: 63–78.

FAO (2012) World agriculture: Towards 2030/2050: The 2012 Revision. ESA Working Paper No. 12-03. Rome, Italy. http://www.fao.org/docrep/016/ap106e/ap106e.pdf.

Fargione J, Jason H, Tilman D, Polasky S, Hawthorne P (2008) Land clearing and the biofuel carbon debt. *Science* 319: 1235–1238.

Fernandes ECM, Soliman A, Confalonieri R, Donatelli M, Tubiello F (2012) Climate change and agriculture in Latin America, 2020–2050. Projected impacts and response to adaptation strategies. Washington, DC: World Bank, 88 p., Report No. 69265.

Finlayson-Pitts BJ, Pitts Jr JN (1997) Tropospheric air pollution: Ozone, airborne toxics, polycyclic aromatic hydrocarbons, and particles. *Science* 276: 1045–1052.

Flory SL, Lorentz KA, Gordon DR, Sollenberger LE (2012) Experimental approaches for evaluating the invasion risk of biofuel crops. *Environ Res Lett* 7: 045904.

Fussmann KE, Schwarzmüller F, Brose U, Jousset A, Rall BC (2014) Ecological stability in response to warming. *Nat Clim Change* 4: 206–210.

Garcia A, Ortega-Huerta MA, Martinez-Meyer E (2013) Potential distributional changes and conservation priorities of endemic amphibians in western Mexico as a result of climate change. *Environ Conserv* 41: 1–12.

Garcia RA, Cabeza M, Rahbek C, Araujo MB (2014) Multiple dimensions of climate change and their implications for biodiversity. *Science* 344: 1247579.

Gelfand I et al. (2013) Sustainable bioenergy production from marginal lands in the US Midwest. *Nature* 493: 514–517.

Georgescu M, Lobell DB, Field CB (2011) Direct climate effects of perennial bioenergy crops in the United States. *PNAS* 108(11): 4307–4312.

Gershenzon J, Dudareva N (2007) The function of terpene natural products in the natural world. *Nat Chem Biol* 3: 408–414.

Goldstein AH, Galbally IE (2007) Known and unknown organic constituents in the earth's atmosphere. *Environ Sci Technol* 41: 1514–1521.

Goldstein AH, Koven CD, Heald CL, Fung IY (2009) Biogenic carbon and anthropogenic pollutants combine to form a cooling haze over the southeastern United States. *Proc Natl Acad Sci USA* 106: 8835–8840.

Gordon DR, Tancig KJ, Onderdonk DA, Gantz CA (2011) Assessing the invasive potential of biofuel species proposed for Florida and the United States using the Australian Weed Risk Assessment. *Biomass Bioenerg* 35: 74–79.

Graves RA, Pearson SM, Turner MG (2016) Landscape patterns of bioenergy in a changing climate: implications for crop allocation and land-use competition. *Ecol Appl* 26(2): 515–529.

Grman E, Lau JA, Schoolmaster DR, Gross KL (2010) Mechanisms contributing to stability in ecosystem function depend on the environmental context. *Ecol Lett* 13: 1400–1410.

Gustavsson L, Haus S, Ortiz CA, Sathre R, Truong NL (2015) Climate effects of bioenergy from forest residues in comparison to fossil energy. *Appl Energ* 138: 36–50.

Gutsch M, Lasch-Born P, Lüttger AB, Suckow F, Murawski A, Pilz T (2015) Uncertainty of biomass contributions from agriculture and forestry to renewable energy resources under climate change. *Meteorol Z* 24(2): 213–223. doi:10.1127/metz/2015/0532.

Hartmann J, West AJ, Renforth P, Kohler P, De La Rocha CL, Wolf-Gladrow DA, Durr HH, Scheffran J (2013) Enhanced chemical weathering as a geoengineering strategy to reduce atmospheric carbon dioxide, supply nutrients, and mitigate ocean acidification. *Rev Geophys* 52: 113–149. doi:10.1002/rog.20004.

Hautier Y et al. (2014) Eutrophication weakens stabilizing effects of diversity in natural grasslands. *Nature* 508: 521–525.

Hautier Y et al. (2015) Anthropogenic environmental changes affect ecosystem stability via biodiversity. *Science* 348: 336–340.

Havlik P et al. (2011) Global land-use implications of first and second generation biofuel targets. *Energy Policy* 39: 5690–5702.

Herrera AM, Dudley TL (2003) Reduction of riparian arthropod abundance and diversity as a consequence of giant reed (Arundo donax) invasion. *Biol Invasions* 5: 167–177.

Hill J et al. (2009) Climate change and health costs of air emissions from biofuels and gasoline. *Proc Natl Acad Sci USA* 106: 2077–2082.

Hill J, Nelson E, Tilman D, Polasky S, Douglas T (2006) Environmental, economic, and energetic costs and benefits of biodiesel and ethanol biofuels. *Proc Natl Acad Sci USA* 103: 11206–11210.

Hinsinger P, Barrios ONF, Benedetti MF, Noack Y, Callot G (2001) Plant-induced weathering of a basaltic rock: Experimental evidence. *Geochim Cosmochim Acta* 65: 137–152. doi:10.1016/S0016-7037(00)00524-X.

Hooper DU et al. (2005) Effects of biodiversity on ecosystem functioning: A consensus of current knowledge. *Ecol Monogr* 75: 3–35.

IPCC (2013) Climate change 2013: The physical science basis. In *Contribution of Working Group I to the Fifth Assessment Report of the Intergovernmental Panel on Climate Change*. Cambridge, UK: Cambridge University Press, 2013.

Kang Y, Khan S, Ma X (2009) Climate change impacts on crop yield, crop water productivity and food security e a review. *Prog Nat Sci* 19(12): 1665–1674.

Kantola IB, Masters MD, Beerling DJ, Long SP, DeLucia EH (2017) Potential of global croplands and bioenergy crops for climate change mitigation through deployment for enhanced weathering. *Biol Lett* 13: 20160714.

Khanum R, Mumtaz AS, Kumar S (2013) Predicting impacts of climate change on medicinal asclepiads of Pakistan using Maxent modeling. *Acta Oecol* 49: 23–31.

Kilpeläinen A, Kellomaki S, Strandman H (2012) Net atmospheric impacts of forest bioenergy production and utilization in Finnish boreal conditions. *GCB Bioenergy* 4: 811–817.

Kirkinen J, Palosuo T, Holmgren K, Savolainen I (2008) Greenhouse impact due to the use of combustible fuels: Life cycle viewpoint and relative radiative forcing commitment. *Environ Manage* 42: 458–469.

Kuusinen M, Ilvesniemi H (Eds.) (2008) *Environmental Effects on Energy wood Harvesting, Study Report*. Forestry Development Centre Tapio and Finnish Forest Research Institute publications, pp. 74. (in Finnish).

Landis D, Gardiner M, van der Werf W, Swinton S (2008) Increasing corn for biofuel production reduces biocontrol services in agricultural landscapes. *Proc Natl Acad Sci USA* 105: 20552–20557.

Lewandowskia I, Scurlockb JMO, Lindvallc E, Christoud M (2003) The development and current status of perennial rhizomatous grasses as energy crops in the US and Europe. *Biomass Bioenerg* 25: 335–361.

Loreto F, Kesselmeier J, Schnitzler JP (2008) Volatile organic compounds in the biosphere-atmosphere system: a preface. *Plant Biol (Stuttgart, Germany)* 10: 2–7.

Loreto F, Velikova V (2001) Isoprene produced by leaves protects the photosynthetic apparatus against ozone damage, quenches ozone products, and reduces lipid peroxidation of cellular membranes. *Plant Physiol* 127: 1781–1787.

Ma Z, Liu H, Mi Z, Zhang Z, Wang Y, Xu W, Jiang L, He J-S (2017) Climate warming reduces the temporal stability of plant community biomass production. *Nat Commun* 8: 15378.

Martindale W, Trewavas A (2008) Fuelling the 9 billion. *Nat Biotechnol* 26:1068–1070.

McLaughlin S, Walsh M (1998) Evaluating environmental consequences of producing herbaceous crops for bioenergy. *Biomass Bioenerg* 14: 317–324.

McLaughlin SB, Ugarte DGDL, Garten CT, Lynd LR, Sanderson MA, Tolbert VR, Wolf DD (2002) High-value renewable energy from prairie grasses. *Environ Sci Technol* 36: 2122–2129.

Mendu V et al. (2012) Global bioenergy potential from high-lignin agricultural residue. *Proc Natl Acad Sci USA* 109: 4014–4419.

Miresmailli S, Zeri M, Zangeri AR, Bernacchi CJ, Berenbaum MR, Delucia EH (2013) Impacts of herbaceous bioenergy crops on atmospheric volatile organic composition and potential consequences for global climate change. *GCB Bioenergy* 5: 375–383.

Moosdorf N, Hartmann J, Lauerwald R, Hagedorn B, Kempe S (2011) Atmospheric CO_2 consumption by chemical weathering in North America. *Geochim Cosmochim Acta* 75: 7829–7854. doi:10.1016/j.gca.2011.10.007.

Nakicenovic N et al. (2000) *IPCC Special Report on Emissions Scenarios*. Cambridge, UK: Cambridge University Press.

O'Donnell J et al. (2012) Invasion hotspots for non-native plants in Australia under current and future climates. *Global Change Biol* 18: 617–629.

Oliver TH et al. (2015) Declining resilience of ecosystem functions under biodiversity loss. *Nat Commun* 6: 10122.

Parrish D, Fike J (2005) The biology and agronomy of switchgrass for biofuels. *Crit Rev Plant Sci* 24: 423–459.

Parry ML, Canziani OF, Palutikof JP, van der Linden PJ, Hanson CE (Eds.) (2007) Climate change 2007: Impacts adaptation and vulnerability. In *Contribution of Working Group II to the Fourth Assessment Report of the Intergovernmental Panel on Climate Change*. Cambridge, UK: Cambridge University Press.

Penuelas J, Staudt M (2010) BVOCs and global change. *Trends Plant Sci* 15: 133–144.

Perdue RE (1958) Arundo donax e source of musical reeds and industrial cellulose. *Econ Bot* 12(4): 368–404.

Phillips SJ, Anderson RP, Schapire RE (2006) Maxent entropy modeling of species geographic distributions. *Ecol Model* 190: 231–259.

Popp A et al. (2014b) Land-use transition for bioenergy and climate stabilization: Model comparison of drivers, impacts and interactions with other land use-based mitigation options. *Clim Change* 123: 495–509.

Popp A, Krause M, Dietrich JP, Lotze-Campen H, Leimbach M, Beringer T, Bauer N (2012) Additional CO_2 Emissions from land use change - forest conservation as a precondition for sustainable production of second generation bioenergy. *Ecol Econ* 74: 64–70.

Popp J, Lakner Z, Harangi-Rakos M, Fari M (2014a) The effect of bioenergy expansion: Food, energy, and environment. *Renew Sustain Energy Rev* 32: 559–578. doi:10.1016/j.rser.2014.01.056.

Porter R, Semenov MA (2005) Crop responses to climatic variability. *Philos Trans R Soc Lond B Biol Sci* 360(1463): 2021e35.

Ragaglini G, Dragoni F, Simone M, Bonari E (2014) Suitability of giant reed (*Arundo donax* L.) for anaerobic digestion: Effect of harvest time and frequency on the biomethane yield potential. *Bioresour Technol* 152: 107–115.

Renforth P (2012) The potential of enhanced weathering in the UK. *Int J Greenh Gas Control* 10: 229–243. doi:10.1016/j.ijggc.2012.06.011.

Repo A, Kankanen R, Tuovinen J-P, Antikainen R, Tuomi M, Vanhala P, Liski J (2012) Forest bioenergy climate impact can be improved by allocating forest residue removal. *GCB Bioenergy* 4: 202–212.

Repo A, Tuomi M, Liski J (2011) Indirect carbon dioxide emissions from producing bioenergy from forest harvest residues. *GCB Bioenergy* 3: 107–115.

Repo A, Tuovinen J-P, Liski J (2015) Can we produce carbon and climate neutral forest bioenergy? *GCB Bioenergy* 7: 253–262.

Rudgers JA et al. (2014) Responses of high-altitude graminoids and soil fungi to 20 years of experimental warming. *Ecology* 95: 1918–1928.

Sanderson M et al. (1996) Tischler Switchgrass as a sustainable bioenergy crop. *Bioresour Technol* 56: 83–93.

Sathre R, Gustavsson L (2011) Time-dependent climate benefits of using forest residues to substitute fossil fuels. *Biomass Bioenerg* 35: 2506–2516.

Sathre R, Gustavsson L (2012) Time-dependent radiative forcing effects of forest fertilization and biomass substitution. *Biogeochemistry* 109: 203–218.

Schievano A, D'Imporzano G, Corno L, Adani F, Cerino Badone F, Pilu SR (2012) Più biogas a costi inferiori con arundo o doppia coltura. *L'Informatore Agrario: Supplemento Energia rinnovabile* 25: 21–25 [in Italian].

Schievano A, D'Imporzano G, Orzi V, Colombo G, Maggiore T, Adani F (2015) Biogas from dedicated energy crops in Northern Italy: Electric energy generation costs. *Global Change Biol* 7: 899–908.

Schuiling RD, Krijgsman P (2006) Enhanced weathering: An effective and cheap tool to sequester CO2. *Clim Change* 74: 349–354. doi:10. 1007/s10584-005-3485-y.

Searchinger T, Heimlich R, Houghton RA, Dong F, Elobeid A, Fabiosa J, Tokgoz S, Hayes D, Yu T-H (2008) Use of U.S. Croplands for biofuels increases greenhouse gases through emissions from land-use change. *Science* 319: 1238–1240.

Seddon AW, Macias-Fauria M, Long PR, Benz D, Willis KJ (2016) Sensitivity of global terrestrial ecosystems to climate variability. *Nature* 531: 229–232.

Shi Z et al. (2016) Dual mechanisms regulate ecosystem stability under decade-long warming and hay harvest. *Nat Commun* 7: 11973.

Singsaas EL, Lerdau M, Winter K, Sharkey TD (1997) Isoprene increases thermotolerance of isoprene-emitting species. *Plant Physiol* 115: 1413–1420.

Taylor LL, Quirk J, Thorley RMS, Kharecha PA, Hansen J, Ridgwell A, Lomas MR, Banwart SA, Beerling DJ (2016) Enhanced weathering strategies for stabilizing climate and averting ocean acidification. *Nat Clim Change* 6: 402–406. doi:10.1038/nclimate2882.

Tilman D, Hill J, Lehman C (2006) Carbon-negative biofuels from low-input high-diversity grassland biomass. *Science* 314: 1598–1600.

Tilman D, Socolow R, Foley JA, Hill J, Larson E, Lynd L, Pacala S, Reilly J, Searchinger T, Somerville C, Williams R (2009) Beneficial biofuels—The food, energy, and environment trilemma. *Science* 325: 270–271.

Torssonen P, Kilpelainen A, Strandman H, Kellomaki S, Jylha K, Asikainen A, Peltola H (2016) Effects of climate change and management on net climate impacts of production and utilization of energy biomass in Norway spruce with stable age-class distribution. *GCB Bioenergy* 8(2): 419–427.

Torvelainen J, Ylitalo E, Nouro P (2014) Puun energiakaytto 2013. Metsatilastotiedote 31/2014.7 p. (In Finnish).

Trabucco A, Achten WMJ, Bowe C, Aerts R, Van Orshoven J (2010) Global mapping of Jatropha curcas yield based on response of fitness to present and future climate. *GCB Bioenergy* 2: 139–151.

Tulbure MG, Wimberly MC, Boe A, Owens VN (2012) Climatic and genetic controls of yields of switchgrass, a model bioenergy species. *Agric Ecosyst Environ* 146(1): 121–129.

Villamil MB, Alexander M, Silvis AH, Gray ME (2012) Producer perceptions and information needs regarding their adoption of bioenergy crops. *Renew Sust Energ Rev* 16: 3604–3612.

Wang S et al. (2012) Effects of warming and grazing on soil N availability, species composition, and ANPP in an alpine meadow. *Ecology* 93: 2365–2376.

Wang W, Tang X, Zhu Q, Pan K, Hu Q, He M, Li J (2014) Predicting the impacts of climate change on the potential distribution of major native non-food bioenergy plants in China. *PLoS One* 9(11): e111587. doi:10.1371/journal.pone.0111587.

Warren R et al. (2013) Quantifying the benefit of early climate change mitigation in avoiding biodiversity loss. *Nat Clim Change* 3: 678–682.

White JW, Hoogenboom G, Kimball WA, Wall GW (2011) Methodologies for simulating impacts of climate change on crop production. *Field Crop Res* 124(3): 357–368.

Williams PRD, Inman D, Aden A, Heath GA (2009) Environmental and sustainability factors associated with next-generation biofuels in the US: What do we really know? *Environ Sci Technol* 43(13): 4763–4775.

Wipf S, Gottfried M, Nagy L (2013) Climate change and extreme events – their impacts on alpine and arctic ecosystem structure and function. *Plant Ecol Divers* 6: 303–306.

Xiong S, Zhang QG, Zhang DY, Olsson R (2008) Influence of harvest time on fuel characteristics of five potential energy crops in northern China. *Bioresour Technol* 99: 479–485.

Yang H et al. (2011) Community structure and composition in response to climate change in a temperate steppe.*Glob Change Biol* 17: 452–465.

Yang XQ, Kushwaha SPS, Saran S, Xu JC, Roy PS (2013) Maxent modelling for predicting the potential distribution of medicinal plant, *Justicia adhatoda* L. in Lesser Himalayan foothills. *Ecol Eng* 51: 83–87.

Yang Z et al. (2016) Daytime warming lowers community temporal stability by reducing the abundance of dominant, stable species. *Glob Change Biol* 23: 154–163.

Zhuang DF, Jiang D, Liu L, Huang YH (2011) Assessment of bioenergy potential on marginal land in China. *Renew Sust Energ Rev* 15: 1050–1056.

6 Renewable Energy Production from Energy Crops

Effect of Agronomic Practices, Policy, and Environmental and Economic Sustainability

Hossain M. Anawar and Vladimir Strezov

CONTENTS

6.1 INTRODUCTION

Extensive research has been conducted to generate renewable energy as a solution to solving environmental problems, the greenhouse gas (GHG) effect, and the economic and energy security of vast use of fossil fuels (Surendra et al., 2018). Biomass currently contributes a big share of the total renewable energy use in the world (IEA and FAO, 2017). Approximately 4%, or 134 billion L, of renewable transportation fuel came from biofuels in 2015; that number will increase to almost 4.5% by 2020 (IEA and FAO, 2017). Food crops currently produce most of the biofuel, creating a food versus fuel conflict. To avoid this crisis, the bioenergy/biofuel is produced from non-food biomaterials, such as energy crops, crop residues, and waste biomass. The perennial energy crops are one of most important lignocellulosic materials for bioenergy production (Surendra et al., 2018).

Unlike first-generation feedstocks, bioenergy production from lignocellulosic biomass is difficult due to its complex composition that needs mechanical and chemical pre-treatments of biomass materials. These treatments add to the higher cost for bioenergy production. At present, the biofuel production from biomass materials costs several times more than petroleum fuels (Carriquiry et al., 2011). Therefore, the economic and sustainable generation of bioenergy requires important development in both biomass production and conversion technologies. The use of bioenergy has some advantages, such as income generation, creation of new jobs, socio-economic development in rural and regional areas, and mitigation of the carbon footprint (Zafeiriou et al., 2016).

The biogas has high potential for electricity and heat production as part of a cogeneration energy system based on renewable resources (Esen and Yuksel, 2013; Ellabban et al., 2014; Dandikas et al., 2018). However, accurate feedstock assessment is required in anaerobic digestion technology (Mao et al., 2015) for an efficient biogas plant operation and high rates of biogas production.

6.2 ENERGY CROPS: BIOFUELS VERSUS FOOD CONFLICT

Using energy crops to produce biofuels increases the food price and creates a food crisis (Paschalidou et al., 2016). Energy crops can generate different forms of bioenergy, such as combined heat and power (CHP), syngas, electricity, biodiesel, methanol, and ethanol. There are two types of energy crops: One type is grown in agricultural fields, and the other is grown in forests. There are two types of agricultural energy crops: annual and perennial crops (Boukis et al., 2009). Figure 6.1 shows the

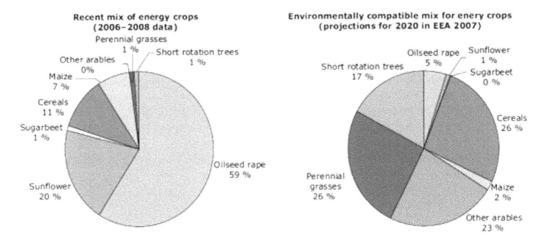

FIGURE 6.1 Mix of energy crops, 2006–2008 (left) and European Economic Area (EEA) scenario for environmentally compatible energy cropping in 2020 (right). (Reprinted from *Renew. Energ.*, 93, Paschalidou, A. et al., Energy crops for biofuel production or for food?—SWOT analysis (case study: Greece), 636–647, Copyright 2016, with permission from Elsevier.)

mix of energy crops, (2006–2008) and the European Economic Area (EEA) scenario for environmentally compatible energy cropping in 2020. Europe's current mix of energy crops is not favorable to the environment. A broader range of crops, particularly perennial crops, such as energy grasses or short-rotation willow plantations, is recommended.

There is a debate on wide-ranging, diverse views regarding biofuel and food price. There are different opinions regarding the biofuel production and food price rise. Various initiatives have been suggested to solve these critical issues. The different socio-economic and environmental impacts and returns that control the price system make this issue complex and uncertain. The use of various types of models and analysis makes the academic side of the debate unclear (HLPE, 2013). The food crops of less-developed billions of peoples are affected due to the increasing rate of bioenergy production without proper policy and management. The biofuels can be promoted if biofuel production does not depend on food crops and the energy crops do not occupy the lands available for food crops (Prusty et al., 2015).

6.2.1 CULTIVATION OF ENERGY CROPS AND DIFFERENT POLICY ISSUES

The adoption of energy crops for bioenergy has some important effects on the land-use change, food security, and encumbrance in carbon footprint emitted by some processes. Based on the different agricultural and renewable energy frameworks, the energy crops should be grown without hampering food security and bioenergy generation by mitigating the GHG effect. The role of income is important in the decision of cultivating energy crops (Zafeiriou and Karelakis, 2016).

Bioenergy production has more socio-economic and environmental benefits when it is compared with the use of fossil fuels, such as reduction of carbon footprint. Despite few environmental adverse effects, European Union (EU) policy supports production and use of energy crops. Therefore, using the linear programming model, Zafeiriou et al. (2016) explored the maximum income generation by cultivating the energy crops in the abandoned and infertile lands in the remote areas. The value of the subsidies and cost of agronomic inputs are the imposed restrictions by different policy scenarios. The results of this study demonstrated that the area and the imports per energy crop did not change and were independent of the policy scenario. In addition, maximum income was obtained by corn cultivation, whereas the implemented Common Agricultural Policy (CAP) had a significant role in deciding on selection of the energy crops. Besides the financial benefits, other benefits should be introduced to the farmers that can encourage their adoption of extensive use of energy crops.

6.2.2 EFFECTS OF PLANT COMPONENTS ON BIOENERGY PRODUCTION BY HIGH-YIELDING TROPICAL ENERGY CROPS

Crop type, crop management, locations, and plant parts determine the composition of lignocellulosic feedstock that significantly controls the type of conversion technology and affects the conversion efficiency of biomass into biofuels and biomaterials (Surendra et al., 2018). The leaves and stems of different energy crops, including Napier grass and energy cane, have a different percentage of fiber content depending on the individual properties of the energy crops. A single conversion technology cannot effectively convert the different plant parts into biofuels and biomaterials for variable composition. However, anaerobic digestion, in combination with thermochemical conversion technologies, was efficient in converting biomass to bioenergy.

Most conversion processes are negatively affected by the high ash content in feedstocks (Surendra et al., 2018), resulting in reduced efficiency of acid pre-treatment (Weiss et al., 2010). High ash content in feedstocks produces fusible slag in thermochemical conversion, resulting in fouling of reactor components, reduced conversion efficiency, and higher production cost (Weiss et al., 2010). The high lignin content in biomass feedstocks decreases the enzymatic hydrolysis and chemical processing during biochemical conversion (Sun and Cheng, 2002).

However, thermochemical conversion can provide a better processing option due to the higher heating value (26.7 MJ kg^{-1} lignin) of lignin.

6.2.3 INCOME VOLATILITY OF ENERGY CROPS

The energy policies for cereal-based bioenergy production are likely to be diminished. Therefore, European farmers can give up their strategy of bioenergy land-use change (Zafeiriou and Karelakis, 2016). The other crops strongly compete with energy crops for uses of agricultural land. The widespread adoption of energy crops is hampered due to primarily low financial returns, while the cultivation of high-income-generating food crops, such as wheat, is more popular (Sherrington et al., 2008).

Furthermore, reliable information regarding the advanced agricultural practices and contract farming of energy crops can determine the farmer's decision (Villamil et al., 2008), where the CAP plays a significant role in farm income, confining income variability. Thus, the current policy schemes present some of the objectives designed for the production of farm-based renewable energy sources, such as ecosystem goods and services, since a sustainable production policy can abolish and replace the system of Single Payment Scheme (SPS) (Convery et al., 2012). The measurement of income volatility can indicate the preliminary situation, provided that the target is already set, and the role of the risk management tool is pre-determined and clear (Velandia et al., 2009).

6.2.4 STAKEHOLDER IDENTIFICATION IN ENERGY CROP CULTIVATION

The serious energy security issues are often found in the developing countries, due to a low diversity in the national energy profile, import-dependent energy supply, and a dominating dependence on single energy sources. Therefore, these countries are paying greater attention to development of non-conventional, renewable energy sources. A range of stakeholders, as well as economic and legislative settings, are key factors to develop energy options. Governments of most countries have a general interest to develop bioenergy options, but other stakeholders, including the private ones, typically have low interest. The limited knowledge and experience on this technology leads to a weak interaction between stakeholders. However, the current economic and legislative settings include some tools to stimulate the renewable energy development, inclusive of energy crops. Such items include reduced environmental tax and indexed tariffs for bioenergy that will provide financial incentives for increase in renewable energy production. The government incentives are key factors to promote growing energy crops (Geletukha et al., 2014).

6.2.5 SMALLHOLDER FARMERS IN BIOENERGY PRODUCTION AND A SUSTAINABILITY ISSUE

Participation of smallholder farmers contributes substantially to the economic sustainability of the energy crop cultivation and biofuel production. In both developed and developing countries, the large-scale companies contribute to the mass production and distribution of biofuels in the world. Production of biomass can be arranged either through plantation estates, through contract farming with smallholders, or by using cooperative institutions, thus empowering the smallholder farmers in the economies of biofuel energy. However, in case of small-scale bioenergy production/development systems for decentralized energy systems, there are a number of technological and socio-economic challenges that urgently require further investigation. If successfully developed, the small-scale bioenergy systems can meet the energy demands of the rural and regional areas, mostly in the developing countries.

Overall, the present tendency of developments in biomass conversion processes and organizations will facilitate the mass production of biofuels in the future. However, we require the congenial policy and regulations that could encourage the smallholders' participation in the cultivation of energy crops and bioenergy production, resulting in creation of new jobs, income, and profit in the rural and regional areas. The use of bioenergy crops for rural progress may differ in various regions, depending on farmers' perception, acceptance, and participation, promoted by the national and local programs for biorenewable energy systems.

6.3 ADVANTAGES AND DISADVANTAGES OF BIOFUELS

Interests in biofuels have grown over the years, due to the concerns about global climate change, environmental and socioeconomic sustainability, and optimum use of natural resources. Although there are advantages to using biofuels, it has some challenges.

6.3.1 POTENTIAL ENERGY CROPS AS SOURCES OF BIOFUEL

A few energy crops are well-known and cultivated on surplus land, including prairie grasses, miscanthus, *Phalaris arundinacea*, willow, poplar, multipurpose trees such as mulberry, *Jatropha*, *Casuarina equisetifolia*, *Eucalyptus globules*, *Leucaena leucephala*, *Melia azadirachta*, *Tamarix dioica*, crassulacean acid metabolism (CAM) plants, perennial herbaceous plants such as giant reed (*Arundo donax*), cardoon (*Cynara curdunculus*), and *Panicum virgatum* (McLaughlin et al., 2002; Hastings et al., 2009; Jurekova et al., 2015). These plants have the potential not only to contribute to the energy security enhancement, but also to yield co-benefits in economic, environmental, and social profits. The economic profitability, environmental sustainability, and social acceptability classify the energy plant species. The *Miscanthus giganteus* is one of the most important energy crops for a commercial and economically sustainable renewable energy business (Daraban et al., 2015). Energy plants provide long-term and four-times-higher profit than normal plants. Therefore, one hectare of energy plant replaces 4,000–7,000 L of oil per year. Furthermore, the *M. giganteus* is suitable for field restoration of metal-contaminated land. This plant may provide many socio-economic benefits of bioenergy production in the contaminated and low fertile marginal lands in the rural and regional areas.

6.3.2 BIOFUEL PRODUCTION FROM ENERGY CROPS

Biogas, biodiesel, and other biofuels are produced from different plant sources. These plants also differ in their performance in biofuel production from place to place.

6.3.2.1 Biogas from Energy Crop Digestion

The biomass from different plant species demonstrated the variable percentage of methane generation capacity. Out of them, the different grass, clover, cereals, rapeseed, and sunflower have high potential for methane generation. Biogas can be produced from plants and crop material, including microalgae to macrophytes. Stewart (1980) has shown that the oats, grass, and straw generated methane of 170–280 $m^3.t^{-1}$ TS. The water hyacinths and freshwater algae produced methane between 150 and 240 $m^3.t^{-1}$ TS (Braun et al., 2016). The maize produced mean methane yields of 348 m^3t^{-1} VS, and barley did 380 m^3t^{-1} VS (KTBL, 2009). The higher cellulose amount and matured plant biomass exhibited the lower methane production capacity.

6.3.2.2 Bioethanol Production

Bioethanol is a biofuel used for vehicles that can reduce the carbon footprint and global warming. The simple sugars starch and lignocellulose can be used to produce bioethanol. Crops that can be used to produce bioethanol include sugarcane, wheat, rice, corn, barley, oat, and sorghum. Bioethanol from sugarcane can provide different environmental benefits when compared with fossil fuels. Different technologies are under development to produce bioethanol from these biomass materials (Balat and Balat, 2009). There are approximately 73.9 Tg of dry wasted crops in the world that could potentially produce 49.1 GL year^{-1} of bioethanol. The total potential bioethanol production from crop residues and wasted crops is 491 GL year^{-1}, which is approximately 16 times higher than the current world ethanol production (Kim and Dale, 2004).

6.3.2.3 Biodiesel Production and Its Use with Biofuel Additive

Although the pure form of bioenergy-derived biodiesel can be used in transportation, it has some limitations, such as fuel properties and its relatively poor cold flow characteristics. Therefore, it is generally mixed with petroleum diesel that can minimize the concentration of particulates, carbon monoxide, hydrocarbons, and air toxins from vehicles. Furthermore, the use of fuel additive can increase the properties of biodiesel. The addition of ethanol, butanol, and diethyl ether augmented the properties of palm oil methyl esters, such as improved acid value, density, viscosity, pour point, and cloud point. However, the energy content slightly decreased with an increasing additive ratio. The biodiesel-additive blend fuels meet the requirements of ASTM D6751 biodiesel fuel standards for the measured properties (Ali et al., 2014).

A number of crops produce more than 15%–50% oil by crushing the seed and squeezing the oil out and then the oil is transesterified to make biodiesel. Biodiesel-producing plants include soya bean, *Elaeis* sp., sorghum, brassicas, sunflower, castor, jatropha, *Euphorbia lathyris*, *Asclepia speciosa*, *Copaifera multijuga*, canola oil, soybeans, algae, palm oil, and sunflower oil.

6.3.3 FACTORS AFFECTING THE DENSIFICATION OF CORN STOVER BRIQUETTES

The heavy weight and bulk density of raw corn stover increases the cost of biomass supply chain and bioenergy production cost (Thoreson et al., 2014). The production of densified corn stover during harvest could decrease harvest and transportation costs, facilitating its use as a biomass feedstock. Figure 6.2 shows the free body diagram of the die region of the densification system (Thoreson et al., 2014). The compression pressure, moisture content, particle size, and material composition have variable roles on the densification and creation of briquettes from raw corn stover. Higher particle sizes and low compression pressures produced the quality briquette with an optimal dry bulk density of 190 kg/m^3. The high moisture content reduced the potential of densification, while the greater cob content showed the beneficial effect on product quality.

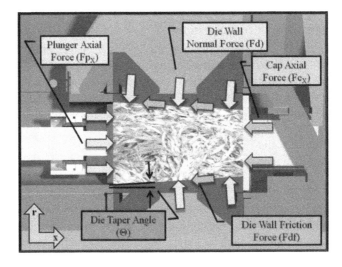

FIGURE 6.2 Free body diagram of the die region of the densification system. (Reprinted from Thoreson, C.P. et al. 2014. *Energies* 7: 4019–4032, under the Creative Commons license.)

6.3.4 EFFECT OF CHEMICAL COMPOSITION ON HYDROLYSIS RATE CONSTANT OF ENERGY CROPS

The hydrolysis controls the rate of anaerobic degradation of complex biomass feedstocks (Weinrich and Nelles, 2015). The hydrolysis rate constant k_h can describe the rate of the overall process if microbial inhibition is diminished during the anaerobic digestion process (Koch and Drewes, 2014).

Biomethane potential (BMP) tests can determine the biogas and methane production potential. The kinetics of biogas production can also indicate the biomass feedstock characteristics. Dandikas et al. (2018) determined how the chemical composition of different energy crops affected the hydrolysis rate constant (k_h). The non-fiber carbohydrates and crude protein of energy crops are good indicators for multiple linear regression models for the prediction of k_h. The first-order kinetic model and the regression models can demonstrate the potential and rate of biogas production for the possible biomass feedstocks.

6.4 TYPES OF SURPLUS LANDS FOR PRODUCTION OF ENERGY CROPS

The lands not used at present for growing food crops, feed, or fiber are called *surplus lands*. These lands are not used due to either poor soil quality or no requirement for production (Faaij, 2007; Rounsevell et al., 2006). The poor-quality agricultural soil and other types of lands can be used for production of energy crops. However, the following factors should be included before energy crops are grown on these lands: production capacity, suitability determination, and environmental and socio-economic issues (Lovett et al., 2009; Jingura et al., 2011). There are different types of surplus lands that depend on their use, soil quality, and capacity of biomass/crop production. These lands are not used for production of food crops, feed, and other essential materials due to political, physical, environmental, technical, social, and economic reasons, but they can be used for production of energy crops (Dale et al., 2010; Alexopoulou et al., 2010; Plieninger and Gaertner, 2011; Dauber et al., 2012). These lands are classified into (1) abandoned land (previously used for crop or grass production, but currently not used), (2) marginal land (crop production is not cost-effective), (3) degraded land (affected by mining or other activities), (4) reclaimed land (industrial- or mining-affected lands that can be remediated and used for cultivation), (5) waste land (unsuitable for human land use; Wiegmann et al., 2008), (6) fallow land (temporarily not used for cropping for some particular time, but it can be used in future, Krasuska et al., 2010), and (7) set-aside land (politically abandoned for crop production for some particular time or permanently).

6.4.1 EXTENSION OF ENERGY CROPS ON SURPLUS LANDS

Land resources are limited in the world. However, there are multiple requirements for their uses. The lands are used not only for the production of food, feed, and fiber, but also for production of biomass and energy crops. These dual uses of land create conflicting demands and land-use changes. The land used for agricultural production per capita has decreased from 0.41 to 0.21 hectare since 1960 (FAO, 2009) due to the increasing demands of land use for livestock grazing and production of bioenergy crops (OECD-FAO, 2009; Tirado et al., 2010). The additional lands should be conserved for production of food crops; and the crop yields can be improved on the existing land by development in agronomic practices and technological development. The appropriate policy framework and regulations should be in place for proper land-use distribution that can overcome the current conflicts of land use in rural areas (EEA, 2010; Kamimura et al., 2011). Bioenergy supplies about 60% of renewable energy in Organisation for economic co-operation and development (OECD) countries and more than 80% in non-OECD countries (IEA, 2010). Bioenergy production increases the share of renewable energy resources, energy security, valuable biobased materials, and socio-economic developments in rural areas and developing regions and thus mitigates the impacts of climate change shown in Figure 6.3 (Rahman et al., 2014).

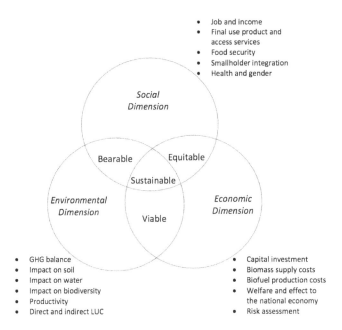

FIGURE 6.3 Scheme for sustainable development of biofuels in developing countries. GHG, greenhouse gases; LUC, land use change. (Reprinted from *Renew. Sust. Energ. Rev.*, 29, Rahman, M.M. et al., Extension of energy crops on surplus agricultural lands: A potentially viable option in developing countries while fossil fuel reserves are diminishing, 108–119, Copyright 2014, with permission from Elsevier.)

A specific bioenergy policy should be developed for the region-specific sustainable bioenergy production systems that can provide the optimal socio-economic and environmental benefits and sustainable regional development (Dauber et al., 2012). Although there are different types of bioenergy crops and biomass plants that can be used to produce bioenergy, they demonstrated variable productivity depending on the land characteristics. Therefore, the agronomic practices should develop the land-specific bioenergy cultivation systems for the different types of surplus land (Berndes et al., 2003; UNEP, 2009). Furthermore, the bioenergy cultivation systems should consider the crop productivity, agronomic inputs, cost of bioenergy production, environmental effects, and socio-economic benefits.

6.4.2 Energy Crop Production through Intercropping

The intercropping of energy crops with other crops in the row crop landscapes can provide a substantial source of biofuel and other bioenergy resources. The proper land management and cropping system will improve the soil and water quality and increase the biodiversity (Bonner et al., 2014). For example, intercropping switchgrass with corn (*Zea mays* L.) could increase the large amount of biomass production from 48% to 99%, while increasing the profit of corn grain. Sometimes the grain production is not cost-effective and incurs a net loss due to the cost of land and inputs. In this case, the intercropping of switchgrass can recover the return.

The appropriate soil management and cropping system, including intercropping, can reduce soil erosion but increase soil fertility, soil organic carbon (SOC) content, and soil health, benefiting the crop productivity. This management system can increase the land productivity by land cover and diversify the crops. Continuous monocropping on degraded land can reduce the SOC due to intensive soil tillage (Lal, 2006). Intercropping of short-rotation woody energy plants into the cropping system can improve the SOC content (Blanco-Canqui, 2010).

Jaya et al. (2014) tested to assess the potential of some intercropping models. The intercropping of food crops with energy crops was investigated to improve the degraded land productivity.

The intercropping models were: (1) castor-hybrid maize, (2) castor–short season maize, (3) castor-mungbean, and (4) castor–short season maize-mungbean. Each crop demonstrated similar productivity in the monoculture and intercropping system. However, the intercropping showed a high land equivalent ratio (LER). The intercropping of castor plants with short-season maize crops showed the highest LER (3.07). These studies indicated that intercropping has great potential to improve the degraded land productivity (Wang et al., 2010).

6.4.3 Impact of Growing Energy Crops on Agricultural Lands

Using the life cycle assessment (LCA) method, Brandao et al. (2010) quantified the effects of different land-use systems on environmental issues for cultivation of energy crops and forestation. The performance of three energy crops, oilseed rape (OSR), miscanthus, and short-rotation coppice (SRC) willow, was compared with that of forest residues. Miscanthus showed the best performance for reducing GHG emissions and increasing the carbon sequestration of these energy crops. But this crop exhibited some adverse environmental impacts, such as acidification and eutrophication. The forest residues are useful by-products for bioenergy production with the least environmental impacts in all categories.

Blanco-Canqui (2010) described the effects of cultivating some energy crops, such as perennial warm-season grasses (WSGs) and short-rotation woody crops (SRWCs), on soil and environment. They increased the soil fertility, SOC, and soil productivity for crop production. They have special positive effects to improve the soil quality of marginal lands compared to croplands or natural forests.

6.5 POTENTIAL OF CONTAMINATED LAND TO PRODUCE ENERGY CROPS

Energy crops cultivated in the heavy metal–contaminated lands show potential in biomass production and heavy-metal accumulation, as well as production of quality biochar.

6.5.1 Energy Crops Production on Trace Element–Contaminated Land

The production of biomass and energy crops on the arable land faces conflicts with food crops and increase of food price. Therefore, the use of industrial- and mining-affected, contaminated lands is highly recommended for the production of biomass and energy crops. These contaminated lands are not suitable for cultivation of food crops, but they can be used to produce biofuels, biochar, non-consumable agricultural products, and wood. Cultivating biomass and energy crops on contaminated land may overcome the bioenergy-food conflicts and provide substantial socio-economic and environmental benefits in the rural and regional communities. The appropriate soil management, addition of amendment to the contaminated soil, and selection of excluder plant species can reduce the uptake of toxic elements into plant biomass and food chain (Pogrzeba et al., 2010). The short-rotation coppice willow, miscanthus, reed canary grass, and switchgrass were grown on brownfield contaminated land amended with green-waste compost (Lord, 2015). They produced annual yields of 4–7 t per hectare, with a gross annual energy yield of 97 GJ per hectare. The contamination levels were acceptable for domestic pellets. Miscanthus has high potential for growth in some marginal and contaminated soils, where they accumulate very low concentrations of metals in the harvested plant part. Therefore, it can contribute significantly to the biofuel production (Pidlisnyuk et al., 2014).

6.5.2 Phytoremediation of Wastewater by Energy Crops

Energy crops have remarkable potential to phytoremediate the contaminated soil and wastewater, and they can simultaneously produce renewable bioenergy (Bonanno et al., 2013). The energy crops can decontaminate the contaminated soil and treat the wastewater in constructed wetlands. Thus, they reduce the land requirement for wastewater/solid waste disposal and bioenergy production, providing significant economic and environmental benefits. The concentration of metals (Cd, Cr, Cu,

Mn, Pb and Zn) in ash is 1.5–3.0 times as high as the values in plant tissues of grasses *Phragmites australis* and *Arundo donax*, grown in wastewater. The short-rotation woody energy crops demonstrated higher environmental benefits than the traditional annual agricultural crops (Simpson et al., 2009). These crops reduce soil erosion, increase the soil's organic carbon, and provide carbon sequestration, thus helping to reduce GHG emissions and climate change. Verma et al. (2007) tested the water hyacinth (*Eichhornia crassipes*) and water chestnut (*Trapa bispinnosa*) to treat and clean up the toxic metal–rich brass and electroplating industrial effluent. Furthermore, they tested the biogas generation potential of these plant species. Biogas production was quicker (maximum from 8 to 12 days) in water hyacinth than in water chestnut (maximum from 12 to 16 days). The biogas production exhibited the correlation with COD, C, N, C/N ratio, and toxic metal contents of the slurry used.

6.5.3 ENERGY CROPS FOR MEDITERRANEAN CONTAMINATED LANDS: VALORIZATION BY PYROLYSIS

Although most of the common energy crops need high water content for growth and are not suitable to establish in the degraded Mediterranean areas, Domínguez et al. (2017) reported two shrub species, *Dittrichia viscosa* (L.) Greuter and milk thistle *Silybum marianum* (L.) Gaertn., that are recognized as potential energy crops grown in these metal-contaminated areas (Robledo and Correal, 2013). These plant species have potential energy value and conversion efficiency.

The thermochemical conversion of biomass is performed using several technologies, combustion, gasification, and pyrolysis (McKendry, 2002). Compared to combustion and gasification, pyrolysis demonstrates more advantages, such as relatively lower temperature and lack of oxygen requirement, and possibility of liquid bio-oil recovery. Pyrolysis produces biogas that, after cleaning, can be used for gas turbines, fuel cells, and production of liquid fuels or chemicals, and generate biochar, which can be used in the agriculture and forestry sectors (Anawar et al., 2015). Therefore, pyrolysis has high potential to produce renewable energy from biomass resources with a lower carbon footprint and higher environmental benefits (Ferreira et al., 2016).

6.6 CONCLUSIONS

Cultivation of energy crops in surplus or contaminated lands is likely to help not only achieve energy security but also attain both environmental and socio-economic sustainability. By increasing cultivation of energy crops and following appropriate measures, global climate change may also be mitigated. Smallholder farmers, mainly of the developing countries, should be considered for inclusion in formulating the policy framework for energy crop cultivation and bioenergy production for socioeconomic and environmental sustainability. However, there is a gap of knowledge regarding cultivation of energy crops and socio-economic, ethical, and environmental issues, particularly in the developing regions of the world.

Government and private stakeholders should invest large-scale funds in (1) the agronomic research for cultivation of suitable energy crop species relating to the land characteristics and regional climate, and (2) technological development for efficient recovery of bioenergy from biomass resources. Furthermore, they should conduct research to develop the appropriate policy framework, organization and institutional capacity for the sustainable biofuel production, economic development, poverty alleviation, and environmental benefits.

REFERENCES

Alexopoulou, E., M. Christou, and I. Eleftheriadis. 2010. Role of 4F cropping in determining future biomass potentials, including sustainability and policy related issues. *Biomass Futures* 2: 1–40.

Ali, O.M., T. Yusaf, R. Mamat, N.R. Abdullah, and A.A. Abdullah. 2014. Influence of chemical blends on palm oil methyl esters' cold flow properties and fuel characteristics. *Energies* 7: 4364–4380.

Anawar, H.M., F. Akter, Z.M. Solaiman, and V. Strezov. 2015. Biochar: An emerging panacea for remediation of soil contaminants from mining, industry and sewage waste. *Pedosphere* 25(5): 654–665.

Balat, M. and H. Balat. 2009. Recent trends in global production and utilization of bio-ethanol fuel. *Applied Energy* 86(11): 2273–2282.

Berndes, G., M. Hoogwijk, and R. van den Broeck. 2003. The contribution of biomass in the future global energy supply: A review of 17 studies. *Biomass and Bioenergy* 25: 1–28.

Blanco-Canqui, H. 2010. Energy Crops and their implications on soil and environment. *Agronomy Journal* 102(2): 403–419.

Bonanno, G., G.L. Cirelli, A. Toscano, R.L. Giudice, and P. Pavone. 2013. Heavy metal content in ash of energy crops growing in sewage-contaminated natural wetlands: Potential applications in agriculture and forestry? *Science of the Total Environment* 52(453): 349–354.

Bonner, I.J., D.J. Jr. Muth, J.B. Koch, and D.L. Karlen. 2014. Modeled impacts of cover crops and vegetative barriers on corn stover availability and soil quality. *BioEnergy Research* 7(2): 576–589.

Boukis, I., N. Vassilakos, G. Kontopoulos, and S. Karellas. 2009. Policy plan for the use of biomass and biofuels in Greece, part I: Available biomass and methodology. *Renewable and Sustainable Energy Reviews* 13(5): 977.

Brandão, M.L., M. Canals, and R. Clift. 2010. Soil organic carbon changes in the cultivation of energy crops: Implications for GHG balances and soil quality for use in LCA. *Biomass & Bioenergy* 35(6): 2323–2336. doi:10.1016/j.biombioe.2009.10.019.

Braun, R., P. Weiland, and A. Wellinger. 2016. Biogas from energy crop digestion. Task 37-energy from biogas and landfill gas. *IEA Bioenergy*. Paris, France.

Carriquiry, M.A., X. Du, and G.R. Timilsina. 2014. Production costs of biofuels. In: Timilsina G., Zilberman D. (Eds.) *The Impacts of Biofuels on the Economy, Environment, and Poverty: Natural Resource Management and Policy* Vol. 41, pp. 33–46, Springer Science+Business Media, New York.

Convery, I., G. Corsane, and P. Davis. 2012. *Making Sense of Place: Multidisciplinary Perspectives*. Boydell & Brewer Ltd, Woodbridge, ON.

Dale, V.H., K.L. Kline, J. Wiens, and J. Fargione. 2010. Biofuels: Implications for land use and biodiversity. *Biofuels and Sustainability Reports* pp. 1–13. http://www.esa.org/biofuelsreports/.

Dandikas, V., H. Heuwinkel, F. Lichti, T. Eckl, J.E. Drewes, and K. Koch. 2018. Correlation between hydrolysis rate constant and chemical composition of energy crops. *Renewable Energy* 118: 34–42.

Daraban, A.E., Ş. Jurcoane, and I. Voicea. 2015. *Miscanthus giganteus*—An overview about sustainable energy resource for household and small farms heating systems. *Romanian Biotechnological Letters* 20(3): 10369–10380.

Dauber, J., C. Brown, A.L. Fernando, J. Finnan, E. Krasuska, J. Ponitka, D. Styles et al. 2012. Bioenergy from "surplus" land: Environmental and socio-economic implications. *BioRisk* 7: 5–50.

Domínguez, M.T., P. Madejon, E. Madejon, and M.J. Diaz. 2017. Novel energy crops for Mediterranean contaminated lands: Valorization of Dittrichia viscosa and Silybum marianum biomass by pyrolysis. *Chemosphere* 186: 968–976.

EEA. 2010. *Mapping the Impacts of Natural Hazards and Technological Accidents in Europe-An Overview of the Last Decade*. European Environment Agency, Copenhagen, Denmark.

Ellabban, O., H. Abu-Rub, and F. Blaabjerg. 2014. Renewable energy resources: Current status, future prospects and their enabling technology. *Renewable and Sustainable Energy Reviews* 39: 748–764.

Esen, M. and T. Yuksel. 2013. Experimental evaluation of using various renewable energy sources for heating a greenhouse. *Energy Build* 65: 340–351.

Faaij, A. 2007. Global outlook on the development of sustainable biomass resource potentials. *Paper Presented at the 1st Conference of the European Biomass Co-firing Network*, July 2–4, 2007, Budapest, Hungary.

FAO. 2009. *Global Agriculture Towards 2050*. High-level expert forum, Rome, Italy. October 12–13, 2009, pp. 1–4. Available at:http://www.fao.org/fileadmin/templates/wsfs/docs/Issues_papers/HLEF2050_Global_Agriculture.pdf.

Ferreira, C.I., V. Calisto, E.M. Cuerda-Correa, M. Otero, H. Nadais, and V.I. Esteves. 2016. Comparative valorisation of agricultural and industrial biowastes by combustion and pyrolysis. *Bioresource Technology* 218: 918–925.

Geletukha, G., T. Zheliezna, and O. Tryboi. 2014. Prospects for the growing and use of energy crops in Ukraine, UABio Position Paper N10, Bioenergy Association of Ukraine. Ukraine. www.uabio.org/activity/uabio-analytics.

Hastings, A., J. Clifton-Brown, M. Wattenbach, C.P. Mitchell, and P. Smith. 2009. The development of MISCANFOR, a new *Miscanthus* crop growth model: Towards more robust yield predictions under different climatic and soil conditions. *Global Change Biology-Bioenergy* 1: 154–170.

HLPE. 2013. *Biofuels and Food Security, a Report by the High-Level Panel of Experts on Food Security and Nutrition of the Committee on World Food Security*, pp. 13–14, Rome, Italy. Retrieved on February 02, 2015, http://www.fao.org/fileadmin/user_upload/hlpe/hlpe_documents/HLPE_Reports/HLPEReport-5_Biofuels_and_food_security.pdf.

IEA and FAO. 2017. *How 2 Guide for Bioenergy: Roadmap Development and Implementation*. Paris, France.

Jay, B.M., D. Howard, N. Hughes, J. Whitaker, and G. Anandarajah. 2014. Modelling socio-environmental sensitivities: How public responses to low carbon energy technologies could shape the UK energy system. *Scientific World Journal* 13. doi: 10.1155/2014/605196.

Jingura, R.M., R. Matengaifa, D. Musademba, and K. Musiyiwa. 2011. Characterisation of land types and agro-ecological conditions for production of Jatropha as a feedstock for biofuels in Zimbabwe. *Biomass and Bioenergy* 35: 2080–2086.

Jurekova, Z., M. Kotrla, M. Prčik, M. Hauptvogl, and Z. Paukova. 2015. Fast-growing energy crops grown in conditions of Slovakia in the context of the EU energy policy. *Acta Regionalia et Environmentalica 1, Nitra, Slovaca Universitas Agriculturae Nitriae* 12: 1–5.

Kamimura, K., H. Kuboyama, and K. Yamamoto. 2009. Estimation of spatial distribution on wood biomass supply potential for three prefectures in the Northern Tohoku region. *Journal of the Japan Institute of Energy* 88: 877–883.

Kim, S. and B.E. Dale. 2004. Global potential bioethanol production from wasted crops and crop residues. *Biomass and Bioenergy* 26(4): 361–375.

Koch, K. and J.E. Drewes. 2014. Alternative approach to estimate the hydrolysis rate constant of particulate material from batch data. *Applied Energy* 120: 11–15.

Krasuska, E., C. Cadórniga, J.L. Tenorio, G. Testa, and D. Scordia. 2010. Potential land availability for energy crops production in Europe. *Biofuels, Bioproducts & Biorefining* 4: 658–673.

KTBL. 2009. *Brochure on Biogas from Energy Crops*. KTBL, D-64289 Darmstadt, Germany.

Lal, R. 2006. Soil and environmental implications of using crop residues as biofuel feedstock. *International Sugar Journal* 108: 161–167.

Lord, RA. 2015. Reed canary grass (*Phalaris arundinacea*) outperforms *Miscanthus* or willow on marginal soils, brownfield and non-agricultural sites for local, sustainable energy crop production. *Biomass and Bioenergy* 78: 110–125.

Lovett, A.A., G.M. Sünnenberg, G.M. Richter, A.G. Dailey, A.B. Riche, and A. Karp. 2009. Land use implications of increased biomass production identified by GIS-based suitability and yield mapping for Miscanthus in England. *Bioenergy Research* 2: 17–28.

Mao, C., Y. Feng, X. Wang, and G. Ren. 2015. Review on research achievements of biogas from anaerobic digestion. *Renewable and Sustainable Energy Reviews* 45: 540–555.

McKendry, P. 2002. Energy production from biomass (part 2): Conversion technologies. *Bioresource Technology* 83: 47–54.

McLaughlin, S.B., D.G. de la Torre Ugarte, C.T. Garten Jr, L.R. Lynd, M.A. Sanderson, V.R. Tolbert, and D.D. Wolf. 2002. High-value renewable energy from Prairie grasses. *Environmental Science and Technology* 36(10): 2122–2129.

Paschalidou, A., M. Tsatiris, and K. Kitikidou. 2016. Energy crops for biofuel production or for food?—SWOT analysis (case study: Greece). *Renewable Energy* 93: 636–647.

Pidlisnyuk, V., L. Erickson, S. Kharchenko, and T. Stefanovska. 2014. Sustainable land management: Growing miscanthus in soils contaminated with heavy metals. *Journal of Environmental Protection* 5(8): 723. doi:10.4236/jep.2014.58073.

Plieninger, T. and M. Gaertner. 2011. Harnessing degraded lands for biodiversity conservation. *Journal for Nature Conservation* 19: 18–23.

Pogrzeba, M., J. Krzyżak, and A. Sas-Nowosielska. 2010. How to grow energy crop on heavy metal contaminated soil. *Proceedings of 15th International Conference on Heavy Metals in the Environment.* pp. 676–679, Hong Kong, China.

Prusty, B.A.K., R. Chandra, and P.A. Azeez. 2015. *Biodiesel: Freedom from Dependence on Fossil Fuels?* Gujarat Institute of Desert Ecology, Bhuj e 370001, 2: 25. Gujarat, India. Retrieved on March 31, 2015 https://www.researchgate.net/publication/36789784_Biodiesel_Freedom_from_Dependence_on_Fossil_Fuels.

Rahman, M.M., S.B. Mostafiz, J.V. Paatero, and R. Lahdelma. 2014. Extension of energy crops on surplus agricultural lands: A potentially viable option in developing countries while fossil fuel reserves are diminishing. *Renewable and Sustainable Energy Reviews* 29: 108–119.

Robledo, A. and E. Correal. 2013. *Cultivos energ_eticos de segunda generaci_on para la producci_on de biomasa lignocelul_osica en tierras de cultivo marginales.* Instituto Murciano de Investigacion y Desarrollo Agrario y Alimentario, Murcia, Spain.

Rounsevell, M.D.A., I. Reginster, M.B. Araujo, T.R. Carter, N. Dendoncker, F. Ewert, J.I. House et al. 2006. A coherent set of future land use scenarios for Europe. *Agriculture, Ecosystems and Environment* 114: 57–68.

Sherrington, C., J. Bartley, and D. Moran. 2008. Farm-level constraints on the domestic supply of perennial energy crops in the UK. *Energy Policy* 36: 2504–2512.

Simpson, T.W., L.A. Martinelli, A.N. Sharpley, and R.W. Howarth. 2009. Impact of ethanol production on nutrient cycles and water quality: The United States and Brazil as case studies. In: *Biofuels: Environmental Consequences and Interactions with Changing Land Use*. R. Howarth and S. Bringezu (Eds.) Cornell University, New York.

Stewart, B.A. 1980. Utilization of animal manures on land: State-of-the-art. *Proceeding 4th International Symposium on Livestock Wastes*. ASAE, St. Joseph, MI, p. 147.

Sun, Y. and J. Cheng. 2002. Hydrolysis of lignocellulosic materials for ethanol production: A review. *Bioresource Technology* 83: 1–11.

Surendra, K.C., R. Ogoshi, H.M. Zaleski, A.G. Hashimoto, and S.K. Khanal. 2018. High yielding tropical energy crops for bioenergy production: Effects of plant components, harvest years and locations on biomass composition. *Bioresource Technology* 251: 218–229.

Thoreson, C.P., K.E. Webster, M.J. Darr, and E.J. Kapler. 2014. Investigation of process variables in the densification of corn stover briquettes. *Energies* 7(6): 4019–4032.

Tirado, M.C., M.J. Cohen, N. Aberman, J. Meerman, and B. Thompson. 2010. Addressing the challenges of climate change and biofuel production for food and nutrition security. *Food Research International* 43(7): 1729–1744.

UNEP. 2009. *Towards Sustainable Production and Use of Resources: Assessing Biofuels*. United Nations Environment programme. Nairobi, Kenya.

Velandia, M., R.M. Rejesus, T.O. Knight, and B.J. Sherrick. 2009. Factors affecting farmers' utilization of agricultural risk management tools: The case of crop insurance, forward contracting, and spreading sales. *Journal of Agricultural and Applied Economics* 41(1): 107–123.

Verma, V.K., Y.P. Singh, and J.P.N. Rai. 2007. Biogas production from plant biomass used for phytoremediation of industrial wastes. *Bioresource Technology* 98: 1664–1669.

Villamil, M.B., A.H. Silvis, and G.A. Bollero. 2008. Potential miscanthus' adoption in Illinois: Information needs and preferred information channels. *Biomass Bioenergy* 32: 1338–1348.

Wang, M.L., J.B. Morris, D.L. Pinnow, J. Davis, and G.A. Pederson. 2010. A survey of the oil content on the entire USDA castor germplasm collection by NMR. *Plant Genetic Resources-Characterization and Utilization* 8: 229–231.

Weinrich, S. and M. Nelles. 2015. Critical comparison of different model structures for the applied simulation of the anaerobic digestion of agricultural energy crops. *Bioresource Technology* 178: 306–312.

Weiss, E., A. Kruse, C. Ceccarelli, and R. Barna. 2010. Influence of phenol on glucose degradation during supercritical water gasification. *The Journal of Supercritical Fluids* 53(1–3): 42–47.

Wiegmann, K., K.J. Hennenberg, and U.R. Fritsche. 2008. Degraded land and sustainable bioenergy feedstock production. *Issue Paper of the Joint International Workshop on High Nature Value Criteria and Potential for Sustainable Use of Degraded Lands*. Oeko-Institute (Darmstadt), Paris, France, pp. 1–10.

Zafeiriou, E. and C. Karelakis. 2016. Income volatility of energy crops: The case of rapeseed. *Journal of Cleaner Production* 122: 113–120.

Zafeiriou, E., K. Petridis, C. Karelakis, and G. Arabatzis. 2016. Optimal combination of energy crops under different policy scenarios; The case of Northern Greece. *Energy Policy* 96: 607–616.

7 Environmental and Energy Potential Assessment of Integrated First and Second Generation Bioenergy Feedstocks

Hannah Hyunah Cho and Vladimir Strezov

CONTENTS

7.1 INTRODUCTION

It is estimated that the global energy demand will rise by 30% between today and 2040 (IEA, 2017), and since the Paris Agreement on climate change entered in force in November 2016, the global energy system has transformed into a cleaner and low-carbon energy production system, particularly with increasing production from renewable energy sources (IEA, 2016). Moreover, concerns about oil depletion, high dependence on fossil fuels, and greenhouse gas (GHG) emissions have led to growing interests in producing energy from biomass (Atabani et al., 2012; Dias De Oliveira et al., 2005; Luo et al., 2011; Tan and Amthor, 2013; Yan et al., 2011). Biomass is a renewable energy source without net GHG emissions, because CO_2 released during the combustion of biofuel can be offset by CO_2 fixed by photosynthesis during the growth of biomass, from which the biofuel is derived (Johnson et al., 2007; Naik et al., 2010b; Strezov, 2015). Biofuel can be used in its pure forms or can be blended with gasoline or diesel, with the minimal modification of currently existing energy use technology and infrastructure (Kaparaju et al., 2009; Koçar and Civaş, 2013; Rajagopal et al., 2007).

First-generation biofuel can be produced from food crops such as corn, sugarcane, or palm oil, and is most widely used at the current level of fuel production systems (WBA, 2017). First-generation feedstocks contain either starch or sugar content that can easily be converted into bioethanol by fermentation, or oil content that can readily be extracted from the crop and converted into biodiesel (de Vries et al., 2010). However, the production of first-generation biofuel has been criticized due to the consumption of water and fossil energy, and the competition with food crops (Davis et al., 2011; Evans et al., 2014; Sharma et al., 2017; Strezov et al., 2008). To avoid these issues, as well as to achieve the sustainability of first-generation biofuel production, it is important to improve the land and water use efficiency by increasing energy yield produced per unit of land and water, with minimal impacts on the environment (Evans et al., 2014).

Globally, agriculture, including biofuel crop production, consumes about 70%–86% of freshwater (Hoekstra and Chapagain, 2007; Pimentel and Patzek, 2005), and it accounts for a large portion of total production cost of biofuel (Holtum et al., 2011). Use of agrochemicals (fertilizer and pesticide) is inevitable to obtain high biomass yield, but the excessive use of such chemicals can adversely affect water and soil quality (Abbasi and Abbasi, 2000; Blottnitz and Curran, 2007). Furthermore, production of agrochemicals consumes fossil fuels, which can contribute to GHG emissions (de Vries et al., 2010; Yan et al., 2011). Therefore, it is important to assess the water and agrochemical use of biofuel production as an indicator of potential environmental impact and the resource use efficiency. For this reason, this study investigates the amount of water required per hectare (ha) of biomass production and per megajoule (MJ) of biofuel production for eight first-generation feedstocks. Agrochemical use per hectare of biomass production, and the associated energy input for using those chemicals, will also be investigated.

Sugarcane, sugar beet, corn, and wheat have been selected as bioethanol feedstocks, and soybean, palm oil, rapeseed, and sunflower have been selected as biodiesel feedstocks based on their current level of production scale and their potential of being commercially utilized in the near term. Bioethanol can be produced by the fermentation process of sugar or starch content in the crops, and it is then distilled to produce fuel-grade ethanol (Halleux et al., 2008). Biodiesel, on the other hand, can be produced through the esterification process using the oil extracted from the crops, and the oil is then purified to produce biodiesel (Leung and Strezov, 2015). The eight selected feedstocks and the major producing countries with their annual productions are summarized in Table 7.1. Sugarcane was the crop with the largest annual production rates, followed by corn, wheat, soybean, and sugar beet. The remaining crops had much smaller annual production rates.

In order to improve the resource use efficiency of biofuel production, increasing total energy yield without increasing input resources can be one of the most desirable options (Börjesson and Tufvesson, 2011). Agricultural crop residues, such as wheat straw and corn stover, have been considered as potential sources that can produce additional energy along with the energy produced from the main crops (Koga, 2008). The crop residues are classified as lignocellulosic biomass that can produce second-generation biofuel (Sindhu et al., 2016; Strezov, 2015). Lignocellulosic biomass is the most abundant renewable resource in the world (Naik et al., 2010a; Saritha and Arora, 2012), and it cannot be used as a food source for human consumption; thus, competition with food crops can be avoided (Stephen et al., 2012). However, there are some technical challenges that delay the scale-up of commercial production of second-generation biofuel due to the highly recalcitrant structure of lignocellulosic biomass (Moreno et al., 2017). Pre-treatment is essential to disrupt this structure so that higher-fermentable sugar, and ultimately more biofuel, can be produced (Sharma et al., 2017; Sindhu et al., 2016). Lignocellulosic biomass is well known for the production of bioethanol (Sindhu et al., 2016), but other types of energy, such as biogas or electricity, can also be produced (Sharma et al., 2017). The pre-treatment process employed for second-generation biofuel production can affect the final energy yield; thus, it is difficult to estimate the energy yield produced from different types of crop residues. Calorific value (CV, presented as megajoule of energy produced per kilogram of residues) or heating value of crop residues has been widely used to estimate the potential energy yield produced from the different types of crop residues (Demirbaş and Demirbaş, 2004). Therefore, in this study, the energy yield (presented as GJ/ha) that can potentially be produced from the residues will be estimated based on their CV, together with the energy yield (GJ/ha) that can be produced from the main crops. The aim of this study is to investigate the agrochemical and water-use efficiency of eight biofuel feedstocks as indicators of potential environmental impacts, and to estimate the total energy yield produced from both crops and residues in order to identify the most efficient first-generation feedstock, and to investigate the potential for producing additional energy from the residues.

TABLE 7.1

Eight Selected Feedstocks, and Major Producing Countries and Their Productions

	Major Producing Country	2014 Production (million tonne)	% of Global Total Production	2014 Total Global Production (million tonne)
Sugarcane	Brazil	736.1	37	2,010
	India	352.1	18	
	China, mainland	125.6	6	
	Thailand	103.7	5	
	Pakistan	62.8	3	
Corn	USA	361.1	29	1,254
	China, mainland	215.6	17	
	Brazil	79.9	6	
	Argentina	33.1	3	
	Ukraine	28.5	2	
Wheat	China, mainland	126.2	15	855
	India	95.9	11	
	Russian Federation	59.7	7	
	USA	55.1	6	
	France	39.0	5	
Sugar beet	France	37.8	14	278
	Russian Federation	33.5	12	
	Germany	29.7	11	
	USA	28.4	10	
	Turkey	16.7	6	
Soybean	USA	106.9	34	319
	Brazil	86.8	27	
	Argentina	53.4	17	
	China, mainland	12.2	4	
	India	10.5	3	
Rapeseed	Canada	15.6	18	89
	China, mainland	14.8	17	
	India	7.9	9	
	Germany	6.2	7	
	France	5.5	6	
Palm oil	Indonesia	29.3	51	58
	Malaysia	19.7	34	
	Thailand	1.9	3	
	Colombia	1.1	2	
	Nigeria	0.9	2	
Sunflower	Ukraine	10.1	23	44
	Russian Federation	8.5	19	
	China, mainland	2.4	5	
	Romania	2.2	5	
	Argentina	2.1	5	

Source: FAOSTAT, World crop production statistics from FAOSTAT Database. Food and Agriculture Organization of the United Nations, 2014. Available at: http://www.fao.org/faostat/en/#data/QC. Accessed on October 4, 2017.

7.2 RESOURCE USE EFFICIENCY OF BIOMASS PRODUCTION

7.2.1 AGROCHEMICAL USE EFFICIENCY

Intensive use of agrochemicals can help to obtain favorable biomass yield, but it can result in a number of negative impacts on the environment, and those impacts cannot be completely eliminated. Therefore, it is important to investigate optimum amount of agrochemicals required to produce the unit of biomass in order to reduce any adverse impacts on the environment, as well as to achieve more efficient chemical use while maintaining the favorable biomass yield for energy production. Some of the nutrients or pollutants resulting from the agrochemical application can enter the soil and water systems (Blottnitz and Curran, 2007), and can ultimately alter the natural ecosystem, as well as affect drinking-water quality (Turner et al., 2007). Continuous use of chemical fertilizer can cause deficiency of other micronutrients in the soil and can decrease soil organic matter, which can lead to a decline of biomass yield (Singh et al., 2007). Although fertilizer requirements for each crop can vary based on the types of soil and crop, growing season and region, and agricultural practices (Iriarte et al., 2010), it could be useful to investigate the optimal quantity of fertilizer requirement to avoid any excess use of fertilizers and the associated energy input.

Each crop has its own optimal ratio of nitrogen (N), phosphorus (P), and potassium (K) fertilizer application, but N fertilizer typically has the highest application rate among N, P, and K fertilizers (Chen and Chen, 2011; Iriarte et al., 2010). N fertilizer has received increasing attention due to its environmental impacts and sustainability issues (de Vries et al., 2010). Production of N fertilizer is energy intensive (Mortimer et al., 2003), during which considerable amounts of GHG can be emitted (Macedo et al., 2008), and this makes N fertilizer have high value of energy content (EC). EC embodied in the fertilizers represents the fossil energy consumption and GHG emissions occurred during the production and distribution of the fertilizers, and it is presented as megajoules per kilogram of fertilizer. Therefore, EC can be used as an indicator for assessing the environmental impact of the fertilizer use (de Vries et al., 2010; Koga, 2008). For this reason, the embodied EC of each N, P, and K fertilizer was used to estimate the total energy input for the fertilizer use, which was calculated as described in Equation 7.1. The values of EC employed in this paper were 58.2 MJ/kg, 14.1 MJ/kg, and 10.2 MJ/kg for N, P and K fertilizers, respectively (Angarita et al., 2009; Erdal et al., 2007; Hülsbergen et al., 2001; Mrini et al., 2001).

Energy input for fertilisers (GJ/ha)

$$= \frac{\begin{aligned}&\left(\text{Amount of N applied}\left(kg/ha\right)*\text{EC of N}(MJ/kg)\right)+\left(\text{Amount of P applied}*\text{EC of P}\right)+ \\ &\qquad\qquad\left(\text{Amount of K applied}*\text{EC of K}\right)\end{aligned}}{1000} \quad (7.1)$$

Since the amount of fertilizer applied for crop production (kilograms per hectare) can differ every year based on the crop development stages, the total amount of fertilizer applied during the entire growing period should be annualized to a year (Dominguez-Faus et al., 2009; Pleanjai and Gheewala, 2009). For this reason, annualized fertilizer application data for each N, P, and K fertilizer were collected from the previous studies and used for this analysis. For the types of fertilizers used, it was assumed that the most widely used fertilizers had been applied for the eight feedstock productions, those are; urea as N fertilizer, super phosphate as P fertilizer, and muriate of potassium as K fertilizer (Iriarte et al., 2010; Mrini et al., 2001; Pryor et al., 2017; Seabra et al., 2011).

Pesticide use for the eight selected biofuel feedstocks was also derived from the previous studies. However, the data for the amount of each herbicide (H), insecticide (I), and fungicide (F) application were not readily available; thus the sum of H and I application was used to estimate the total energy input for the pesticide use because few crops required F, and only occasionally if so the amount of F application was negligible. Glyphosate as H and aldicarb as I were assumed to be the most commonly used pesticides (Iriarte et al., 2010; Pleanjai and Gheewala, 2009; Tzilivakis et al., 2005). The energy input values for the pesticide applications were calculated following Equation 7.2, with the average EC of 212.7 MJ/kg for H and I.

$$\text{Energy input for pesticides } (\text{GJ/ha})$$

$$= \frac{\text{Sum of H and I applied } (\text{kg/ha}) * \text{Average EC of H and I } (\text{MJ/kg})}{1000} \quad (7.2)$$

Table 7.2 presents the average amount and deviation of the fertilizer and pesticide application, including the energy input associated with production of the fertilizer and pesticides, for the selected first-generation biofuel feedstocks. Among the eight selected feedstocks, sugar beet required the largest total amount of fertilizers for the crop production (Table 7.2), with the amounts of 127.5, 112.5, and 206.3 kg/ha for N, P, and K fertilizers, respectively, which resulted in a total 11.1 GJ/ha of energy input for fertilizer use. However, rapeseed ranked at the top in energy input for fertilizer use (11.3 GJ/ha) among eight feedstocks, despite the much smaller amount of fertilizer applied for its production than sugar beet. This could be due to the higher rate of N fertilizer application in rapeseed (N:P:K = 53.7:27.6:18.7) than in sugar beet (N:P:K = 28.6:25.2:46.2), since N fertilizer has the highest EC compared to the P and K fertilizers. Despite the relatively small difference in the amount of fertilizer application in corn and palm oil (309.2 and 282.7 kg/ha, respectively), the energy input for these two crops showed a much greater difference (10.9 GJ/ha for corn and 6.4 GJ/ha for palm oil) (Figure 7.1), which also could be due to a higher N fertilizer use in the corn production. The smallest total amount of fertilizer application and the associated energy input were observed in the case of soybean production, with the smallest ratio of N fertilizer application of 8.6. For the pesticide use, sugar beet demonstrated the significantly high pesticide requirement among all eight crops, with the highest energy input of 4.7 GJ/ha for the pesticide use (Table 7.2). The second-highest pesticide requirement was observed in sugarcane, but the difference in the energy inputs for pesticide use between the two crops was substantial (4.7 GJ/ha for sugar beet, versus 0.9 GJ/ha for sugarcane). Overall, the agrochemical energy inputs, both for fertilizers and pesticides, were slightly higher in bioethanol feedstocks than in biodiesel feedstocks.

7.2.2 WATER USE EFFICIENCY

High water consumption for biofuel feedstock cultivation can be a major obstacle; thus, it is important to evaluate the water requirement of biofuel production in order to achieve high water use efficiency (WUE). The water requirement for biofuel crop production can be defined as the total annual volume of water consumption during the entire growing period of the crop (Gerbens-Leenes et al., 2009; Hoekstra et al., 2009). Providing data for the amount of water required per hectare of crop production (liter of water per hectare of crop production) may be useful for more efficient water use during feedstock cultivation, but it may be more appropriate to assess the water requirement per unit of energy produced from the crop (e.g., liter of water per megajoule of energy produced) for comparing the WUE of biofuel production derived from each crop. Therefore, in this work, the WUE of the selected first-generation biofuel feedstocks was estimated using Equation 6.3 and presented as the amount of water required per unit of energy produced (L/MJ). Water consumption during the biofuel conversion process is almost negligible (around 0.2% of total water consumption, from Singh and Kumar, 2011); thus, it was excluded in this study.

TABLE 7.2

Average Quantity of Fertilizer and Pesticide Application for Crop Production and the Associated Energy Input

	Sugarcane			Sugar Beet			Corn			Wheat		
	N	P	K	N	P	K	N	P	K	N	P	K
Fertilizer application (kg/ha)	92.1 (±34.3)	41.7 (±17.3)	79.8 (±29.9)	127.5 (±10.6)	112.5 (±17.7)	206.3 (±8.8)	155.6 (±13.1)	74.9 (±23.4)	78.7 (±8.5)	104.1 (±48.4)	28.3 (±8.6)	34.5 (±29.0)
N:P:K application ratio	43.1:19.5:37.4			28.6:25.2:46.2			50.3:24.2:25.5			62.4:17.0:20.6		

	Soybean			Palm Oil			Rapeseed			Sunflower		
	N	P	K	N	P	K	N	P	K	N	P	K
Fertilizer application (kg/ha)	4.7 (±1.4)	27.5 (±14.6)	22.5 (±10.8)	68.6 (±38.7)	54.2 (±37.0)	159.9 (±17.1)	164.3 (±31.9)	84.5 (±44.6)	57.0 (±31.0)	130.0 (±28.3)	85.6 (±20.6)	86.0 (±19.8)
N:P:K application ratio	8.6:50.3:41.1			24.3:19.2:56.5			53.7:27.6:18.7			43.1:28.4:28.5		

	Sugarcane	Sugar Beet	Corn	Wheat	Soybean	Palm Oil	Rapeseed	Sunflower
Pesticide application (kg/ha)[a]	3.8 (±1.2)	22.0 (±4.5)	3.2 (±1.2)	2.7 (±1.4)	1.3 (±0.1)	3.1 (±0.5)	2.4 (±0.1)	2.0 (±1.4)
Energy input for fertilizer use (GJ/ha)	6.8 (±2.0)	11.1 (±1.0)	10.9 (±1.1)	6.8 (±3.1)	0.9 (±0.0)[b]	6.4 (±1.9)	11.3 (±2.2)	9.7 (±1.7)
Energy input for pesticide use (GJ/ha)	0.8 (±0.3)	4.7 (±1.0)	0.7 (±0.3)	0.6 (±0.3)	0.3 (±0.0)[b]	0.7 (±0.1)	0.5 (±0.0)[b]	0.4 (±0.3)
Total energy input for agrochemical use (GJ/ha)	**7.6 (±2.0)**	**15.8 (±1.0)**	**11.6 (±1.1)**	**7.5 (±3.1)**	**1.2 (±0.0)[b]**	**7.1 (±1.9)**	**11.8 (±2.2)**	**10.1 (±1.7)**

Note: All values are presented as average value ± standard deviation.

For fertilizer application: Chen and Chen, 2011; De Souza et al., 2010; Dias De Oliveira et al., 2005; Elsgaard et al., 2013; Foteinis et al., 2011; Halleux et al., 2008; Hill et al., 2006; Iriarte et al., 2010; Macedo et al., 2008; Pimentel and Patzek. 2008; Seabra et al., 2011; Singh et al., 2011; Singh and Kumar, 2011; USDA, 2016; Wayagari et al., 2001; Yusoff and Hansen, 2007.

For pesticide application: De Souza et al., 2010; de Vries et al., 2010; Dias De Oliveira et al., 2005; Dominguez-Faus et al., 2009; Foteinis et al., 2011; Halleux et al., 2008; Hill et al., 2006; Iriarte et al., 2010; Macedo et al., 2008; Pimentel and Patzek, 2008; Pleanjai and Gheewala, 2009; Pryor et al., 2017; Seabra et al., 2011; Tzilivakis et al., 2005; USDA, 2016.

For energy content of agrochemicals: Angarita et al., 2009; Erdal et al., 2007; Hülsbergen et al., 2001; Mrini et al., 2001.

[a]   The amount of pesticide application (kg/ha) is the sum of H and I.

[b]   All the values presented in this study have been rounded up to one decimal place; thus, the presented value "0" can be actually higher than zero.

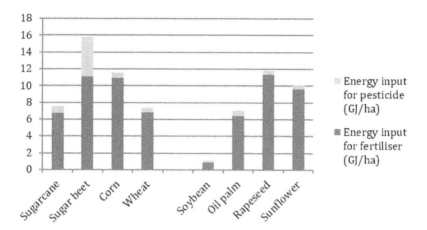

FIGURE 7.1 Total energy input for agrochemical (fertilizer + pesticide) use for eight feedstocks.

Total volume of water required per unit of energy produced (L/MJ)

$$= \frac{\text{Total volume of water required per hectare of crop production} \, (L/ha)}{\text{Total energy produced per hectare} \, (MJ/ha)} \quad (7.3)$$

Among the eight feedstocks, palm oil required the largest volume of water per hectare of crop production (13.5 mL/ha), followed by sugarcane (11 mL/ha) (Figure 7.2a). However, if the water requirement was compared based on the unit of energy produced, palm oil was the sixth and sugarcane was the fourth in WUE, with WUEs of 90.8 and 80.7 L/MJ, respectively (Figure 7.2b). Since WUE is expressed as the volume of water required per unit of energy produced, lower values in WUE indicate that the crops are more water-use efficient. As seen in the case of palm oil, the crop that requires a large volume of water per hectare of crop production does not always mean that the crop has a low WUE, because the high water requirement during crop production can be offset by the favorable energy yield. The second-largest water consumption in the case of sugarcane could be due to the large volume of water required to remove soils attached to the stalks during the harvesting (Dias De Oliveira et al., 2005), but the high crop and energy yield could allow sugarcane to become a water-use efficient feedstock (WUE of 77.8 L/MJ; see Figure 7.2b). Similarly, corn ethanol production could also be water-use efficient (WUE of 64.3 L/MJ) when the whole life cycle, from crop cultivation to energy production, is considered. However, substantial amounts of irrigation and groundwater mining for corn production have been criticized, since this could adversely affect the environment (Pimentel, 2003). As mentioned earlier, the high-water requirement for biofuel production can be one of the biggest shortfalls that hinder expansion of the production scale and puts additional pressures on water resources (Gerbens-Leenes et al., 2009). Recently, it has been reported that some feedstocks, such as agave (e.g., *Agave salmiana*) can be grown in naturally water-limited or marginal land areas (Davis et al., 2011), so criticism on the intensive water use may be alleviated.

Overall, bioethanol feedstocks were relatively more water-use efficient than biodiesel feedstocks, with the highest WUE found in the case of sugar beet, partly due to the higher energy yield than other feedstocks. Similar results can be found in the studies by Mekonnen and Hoekstra (2011) and Singh and Kumar (2011). Rockstrom et al. (2003) also suggested that the WUE can be improved by high energy yield.

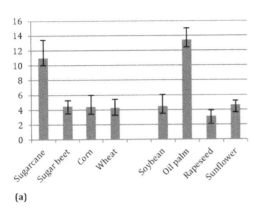 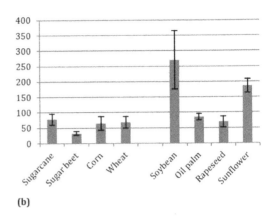

(a)   (b)

FIGURE 7.2 Water use assessment for crop cultivation; (a) water requirement of crop production (ML/ha); (b) water use efficiency (WUE) (L/MJ of energy produced). WUE was calculated by using Equation 3, and total energy yields (GJ/ha) used to calculate WUE were 141.5 for sugarcane, 136.7 for sugar beet, 68.9 for corn, 62.3 for wheat, 16.6 for soybean, 157.5 for palm oil, 44.9 for rapeseed, and 25.1 for sunflower. Error bars indicate the standard deviation of data. (From de Vries, S.C. et al., *Biomass Bioenergy, 34*, 588–601, 2010; Dominguez-Faus, R. et al., *Environ. Sci. Technol.*, 43, 3005–3010, 2009; Foteinis, S. et al., *Energy Policy, 39*, 4834–4841, 2011; Gerbens-Leenes, P.W. et al., *Ecolog. Econ.*, 68, 1052–1060, 2009; Iriarte, A. et al., *J. Clean. Prod.*, 18, 336–345, 2010; Kongboon, R and Sampattagul, S., *Procedia-Soc. Behav. Sci.*, 40, 451–460, 2012; Mekonnen, M. and Hoekstra, A.Y., *Hydrol. Earth Sys. Sci. Discuss.*, 8, 763–809, 2011; Pryor, S.W. et al., *J. Clean. Prod.*, 141, 137–145, 2017; Singh, S. and Kumar, A., *Bioresour. Technol.*, 102, 1316–1328, 2011.)

7.3 TOTAL BIOMASS AND ENERGY YIELD

7.3.1 BIOMASS (CROP AND RESIDUE) YIELD

Increasing the biomass yield is one of the key aims to achieve high total energy yield. Higher biomass yield may lead to an intensive use of input materials, such as agrochemicals and water, but it still could be beneficial to attempt to increase biomass yield, because the consumption of the input materials is not directly proportional to the biomass yield (Börjesson and Tufvesson, 2011).

Total biomass yield is defined as the sum of the crop yield and the crop residue yield. Unlike food production, in which crop yield is the only yield parameter of interest, total biomass yield can be applied for the purpose of biofuel production because the residues could be utilized for bioenergy production (Gerbens-Leenes et al., 2009). A large quantity of crop residues is produced annually (Lal, 2005); thus, it could be beneficial to consider the crop residues as the potential energy sources together with the main crops. In this study, the data for the total crop yield (t/ha) and the residue yield (t/ha) were collected from literature to estimate total potential energy yield that can be produced from both crops and residues. The residue of each feedstock in this study was limited to the agricultural crop residues, which included leaves, stems, shells, and straws, except sugarcane due to data availability. In the case of sugarcane, bagasse, the by-product produced after the extraction of sugarcane juice, was considered as the residue. Any other by-products produced during the biofuel production processes (e.g., stillage from sugarcane processing or glycerol from rapeseed oil production) were excluded in this study. Conventionally, a large portion of agricultural residue has been retained on the field for the purpose of soil conservation and fertilization or has been used as animal bedding (Börjesson and Tufvesson, 2011; Herr et al., 2010; Ji, 2015). Also, the harvestable amount of residues can be limited by technical constraints. For instance, the crop residues cannot be cut lower than 12.5 cm, depending on the harvest machinery used (Herr et al., 2010). Thus, the residue yield in this paper only considered the amount of harvestable residue that was available for energy production after considering the environmental and technical constraints.

TABLE 7.3

Total Biomass (Crop + Residue) Yield and the Proportion of Residues to the Total Biomass Yield of Eight Feedstocks

	Sugarcane	Sugar Beet	Corn	Wheat	Soybean	Palm Oil	Rapeseed	Sunflower
Crop yield (t/ha)	73.4 (± 8.6)[a]	56.3 (± 9.6)	7.3 (± 2.2)	3.8 (± 2.1)	2.6 (± 0.3)	19.5 (± 2.1)	2.7 (± 1.2)	1.6 (± 0.4)
Residue yield (t/ha)	14.6 (± 2.4)	4.6 (± 1.6)	3.2 (± 1.1)	4.1 (± 0.7)	2.4 (± 0.1)	6.7 (± 2.2)	3.3 (± 1.0)	1.5 (± 0.3)
Total biomass yield (t/ha)[b]	88.0 (± 8.6)	60.9 (± 9.6)	10.5 (± 2.2)	7.9 (± 2.1)	5.0 (± 0.3)	26.2 (± 2.2)	6.0 (1.2)	3.1 (± 0.4)
Proportion of residue to total biomass yield (%)[c]	16.6	7.6	30.5	51.9	48.0	25.6	55.0	48.4

For crop yield: Angarita et al., 2009; De Souza et al., 2010; de Vries et al., 2010; Dias De Oliveira et al., 2005; Dominguez-Faus et al., 2009; FAO, 2008; Foteinis et al., 2011; García et al., 2011; Iriarte et al., 2010; Kim and Day, 2011; Koga, 2008; Kongboon and Sampattagul, 2012; Lal, 2005; Luo et al., 2011; Macedo et al., 2008; Maung and Gustafson, 2011; Mekonnen and Hoekstra, 2011; Naylor et al., 2007; OECD/FAO, 2017; Pimentel and Patzek, 2008; 2005; Pleanjai and Gheewala, 2009; Pryor et al., 2017; Rajagopal et al., 2007; Seabra et al., 2011; Singh and Kumar, 2011; Thamsiriroj and Murphy, 2009; Tzilivakis et al., 2005; Unkovich et al., 2010; Visser et al., 2011; Yusoff and Hansen, 2007.

For residue yield: Börjesson and Tufvesson, 2011; De Souza et al., 2010; Farine et al., 2012; Gerbens-Leenes et al., 2009; Halleux et al., 2008; Koga, 2008; Lal, 2005; Luo et al., 2011; Macedo et al., 2008; Singh and Kumar, 2011; Thomsen and Haugaard-Nielsen, 2008; Unal and Alibas, 2007; Visser et al., 2011; Yusoff and Hansen, 2007.

[a] '±' values in the bracket denote the standard deviation of each data.
[b] Total biomass yield is the sum of crop and residue yield of each feedstock.
[c] Proportion of residue to the total biomass yield was calculated by (residue yield/total biomass yield) × 100.

Table 7.3 shows the crop and residue yield of the selected feedstocks. Among them, sugarcane demonstrated considerably high average yield in both crop and residue, with 73.4 and 14.6 t/ha, respectively. Sugar beet crop yield was the second highest (56.3 t/ha), but the residue yield was only approximately 4.6 t/ha. On the contrary, soybean, rapeseed, and sunflower showed significantly low crop yields, because only small fractions of the total plant can be used for bioenergy. The proportion of residue yield to the total biomass yield varied greatly depending on the types of crops, ranging from 7.6% to 55.0% of total biomass yield (Table 7.3). In the case of soybean and sunflower, the residue yield was similar to their crop yield, but in wheat and rapeseed, the residue yield was slightly higher than their crop yield, indicating great potential to increase total energy yield by producing additional energy from the residues. Among the eight feedstocks, sunflower demonstrated the lowest yield in both crop and residue, suggesting that the productivity should be improved. In general, the total average biomass yield showed high variation according to the types of crops, with the significantly high biomass yield of 88 t/ha for sugarcane and 60.9 t/ha for sugar beet (Figure 7.3).

7.3.2 POTENTIAL ENERGY YIELD FROM CROPS AND RESIDUES

Due to the different fuel conversion process used to produce bioethanol or biodiesel, as well as the different sugar or oil content in each crop, total biofuel yield (liters of bioethanol or biodiesel produced per hectare) is not exactly proportional to the total crop yield. For instance, bioethanol can be produced through the fermentation process of sugar content in the usable part of the sugarcane plant; thus the bioethanol yield can be affected by the amount of sugar content in the sugarcane rather than the total crop yield (Bessou et al., 2011). Generally, sugar crops (e.g., sugarcane,

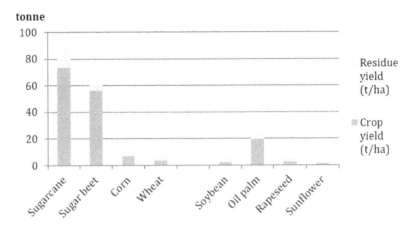

FIGURE 7.3   Total biomass yield (crop + residue yield) of eight feedstocks.

sugar beet) produce more ethanol than starch crops (e.g., corn, wheat), because the sugar yield per hectare is usually higher than the starch yield, and the sugar content in the crops can be directly fermented to produce ethanol (Cardona and Sánchez, 2007; Leiper et al., 2006), whereas starch content requires additional pre-treatment processes before it is fermented (Barros-Rios et al., 2015). Bioethanol can also be produced from various second-generation lignocellulosic feedstocks, such as agricultural crop residues (Koçar and Civaş, 2013). Similarly, biodiesel can be produced from the oil extracted from the crops; thus the potential energy yield can be determined by the oil content in the crops and the employed fuel conversion processes (Díaz et al., 2010). Therefore, in this paper, the total potential energy yield (GJ/ha) of the selected feedstocks was calculated based on the bioethanol and biodiesel yield (L/ha). The biofuel yield data was firstly collected from the relevant previous studies, and the data multiplied by the energy content of 23.4 MJ/L for bioethanol and 34.6 MJ/L for biodiesel (Department of Environment, 2015) to estimate the total potential energy yield produced per hectare (GJ/ha) (Table 7.4).

Although the pre-treatment processes for producing biofuel from the lignocellulosic feedstocks are still challenging, it is important to estimate the energy content and the quantity of crop residues for expanding the production scale of biofuel (Unal and Alibas, 2007). The crop residue is one of the most abundant biomass resources, and its utilization can be advantageous to increase overall efficiency of biofuel production by producing additional energy from the residues. However, estimating the potential energy yield can be complex, because the types of energy produced from the residues can vary (e.g., biogas, electricity) according to the types of residues, and the fuel conversion processes employed (Sharma et al., 2017). Most of the energy produced from the residues is currently used as an operating energy during the biofuel production processes (de Vries et al., 2010), and these factors make it difficult to compare the energy yield produced from different types of residues. For this reason, CV (presented as megajoules of energy content per kilogram of dry residues) of each type of residue was employed to estimate the total potential energy yield produced from the residues. The CV can be obtained by the complete combustion of dried crop residues using a bomb calorimeter (Demirbaş and Demirbaş, 2004), and because the CV can be reduced by the moisture content present in the residues (McKendry, 2002), the dry weight of residue was used to estimate the energy yield.

The results of this study showed great variation in the bioethanol and biodiesel yields, with the highest bioethanol yield in sugarcane among the four bioethanol feedstocks, followed by sugar beet, with the ethanol yield of 6,045 and 5,840 L/ha, respectively (Table 7.4). Among the other four biodiesel feedstocks, palm oil demonstrated incomparably high biodiesel yield (4,552 L/ha), whereas soybean and sunflower only had biodiesel yields of 479 and 725 L/ha, respectively. Sugarcane also

TABLE 7.4

Total Biomass Yield (Crop + Residue Yield) and Total Energy Yield (Produced from Crops and Residues) for Eight Feedstocks

	Sugarcane	Sugar Beet	Corn	Wheat	Soybean	Palm Oil	Rapeseed	Sunflower
Crop yield (t/ha)	73.4 (± 8.6)	56.3 (± 9.6)	7.3 (±2.2)	3.8 (± 2.1)	2.6 (± 0.3)	19.5 (± 2.1)	2.7 (± 1.2)	1.6 (± 0.4)
Biofuel yield (L/ha)	6044.7 (± 747.9)	5840.3 (± 543.0)	2942.9 (± 617.1)	2660.1 (± 339.3)	478.8 (± 49.9)	4551.6 (± 631.4)	1298.2 (± 147.0)	725.0 (± 37.8)
Energy yield from crops (GJ/ha)	**141.5** (± 17.5)	**136.7** (± 12.7)	**68.9** (± 14.4)	**62.3** (± 7.9)	**16.6** (± 1.7)	**157.5** (± 21.9)	**44.9** (± 5.1)	**25.1** (± 1.3)
	Sugarcane Bagasse	**Sugar Beet Residue**	**Corn Stover**	**Wheat Straw**	**Soybean Straw**	**Palm Residue**	**Rapeseed Straw**	**Sunflower Residue**
Residue yield (t/ha)	14.6 (± 2.4)	4.6 (± 1.6)	3.2 (± 1.1)	4.1 (± 0.7)	2.4 (± 0.1)	6.7 (± 2.2)	3.3 (± 1.0)	1.5 (± 0.3)
Ratio of residue to total biomass yield (%)	16.6	7.6	30.5	51.9	48.0	25.6	55.0	48.4
Calorific value of residues (MJ/dry kg)	18.3 (± 1.0)	16.5 (± 0.3)	18.0 (± 0.3)	17.9 (± 1.4)	18.6 (± 0.6)	19.8 (± 1.3)	18.0 (± 0.5)	16.8 (± 2.3)
Energy surplus from residues (GJ/ha)	**266.0** (± 44.0)	**74.9** (± 26.1)	**57.2** (± 20.4)	**72.8** (± 11.9)	**44.5** (± 1.6)	**133.2** (± 43.5)	**58.5** (± 17.6)	**25.7** (± 5.6)

For crop yield: Angarita et al., 2009; De Souza et al., 2010; de Vries et al., 2010; Dias De Oliveira et al. 2005; Dominguez-Faus et al., 2009; FAO, 2008; Foteinis et al., 2011; García et al., 2011; Iriarte et al. 2010; Kim and Day, 2011; Koga, 2008; Kongboon and Sampattagul, 2012; Lal, 2005; Luo et al., 2011; Macedo et al., 2008; Maung and Gustafson, 2011; Mekonnen and Hoekstra, 2011; Naylor et al., 2007; OECD/FAO, 2017; Pimentel and Patzek, 2008; 2005; Pleanjai and Gheewala, 2009; Pryor et al., 2017; Rajagopal et al., 2007; Seabra et al., 2011; Singh and Kumar, 2011; Thamsiriroj and Murphy, 2009; Tzilivakis et al., 2010; Visser et al., 2011; Yusoff and Hansen, 2007.

For residue yield: Börjesson and Tufvesson, 2011; De Souza et al., 2010; Farine et al., 2012; Gerbens-Leenes et al., 2009; Halleux et al., 2008; Koga, 2008; Lal, 2005; Luo et al., 2011; Macedo et al., 2008; Singh and Kumar, 2011; Thomsen and Haugaard-Nielsen, 2008; Unal and Alibas, 2007; Visser et al., 2011; Yusoff and Hansen, 2007.

For biofuel yield: de Vries et al., 2010; Dias De Oliveira et al., 2005; Dominguez-Faus et al., 2005; Farine et al., 2012; Groom et al., 2008; Hill et al., 2006; Koga, 2008; Macedo et al., 2008; Mekonnen and Hoekstra, 2011; Naylor et al., 2007; Rajagopal et al., 2007; Rosenberger et al., 2001.

For calorific value: Ahiduzzaman et al., 2014; Begum et al., 2013; Channiwala and Parikh, 2002; Demirbaş and Demirbaş, 2004; Dodić et al., 2010; Hülsbergen et al., 2001; Karaosmanoğlu et al., 1999; Koçar and Civaş, 2013; Koga, 2008; Masiá et al., 2007; McKendry et al., 2002; Motghare et al., 2016; Mu et al., 2010; Naik et al., 2010a; Şensöz and Kaynar, 2006; Tsai et al., 2006; Yin, 2011.

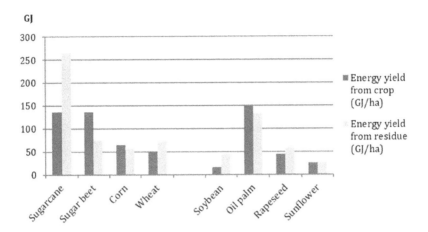

FIGURE 7.4   Total energy yield (GJ/ha) produced from crops and potential energy surplus produced from residues.

had the highest total potential energy yield (Figure 7.4) produced from both crop and residue among the eight feedstocks. The favorable energy yield of sugarcane is usually obtained in Brazil, where the sugarcane industry is in its mature stage (Macedo et al., 2008). Moreover, the recent reduction in sugarcane burning at the field has increased its biomass and the energy yields, and has consequentially decreased GHG emissions resulting from the burning. Increased levels of bagasse utilization can also contribute to the higher net energy yield, since most of the energy required during the fuel production can be supplied by the bagasse-derived energy, thereby reducing the energy input during the fuel production process (Dias De Oliveira et al., 2005; Macedo et al., 2008). Palm oil showed the second-highest total energy yield produced from the crop and residues (empty fruit bunches, shells, and fibers) among the eight feedstocks, due to the higher energy content in biodiesel than in bioethanol, and the high cellulose and hemicellulose contents in the residues, which represent a great potential to produce substantial energy surplus from the residues (Kelly-Yong et al., 2007). Rapeseed cake and straw have been suggested as potential energy sources that can produce various types of bioenergy, such as bioethanol, biohydrogen, and biomethane, although this could be only viable when the efficiency of residue-derived energy production is improved (Luo et al., 2011). Wheat straw also showed its potential to produce usable heat, due to its relatively low moisture content in the straw (Koga, 2008).

It is important to note that the additional energy produced from the residues indicates the potential and experimental energy yield surplus; therefore, the potential energy yield estimated here may not always be the same with the volume of energy that can actually be produced from the residues. This could be one of the reasons that can explain the higher energy yield produced from the residues than the energy produced from the crops in some cases (e.g., in the case of sugarcane, wheat, soybean, rapeseed, and sunflower; see Figure 7.4). Also, any energy losses during the fuel conversion processes have not been considered in this energy yield analysis. In addition, a considerable proportion of crop residues is being used as organic fertilizers at the current farming practices in order to reduce the use of chemical fertilizers, to stabilize the soil structure, and to conserve the soil carbon content (Farine et al., 2012; Lal, 2005). Therefore, it is important to make a balance between retaining crop residues on the field for environmental purposes and utilizing them for producing additional energy to maximize the total energy yield.

7.4   CONCLUSIONS

This study aimed to compare eight selected first-generation biofuel feedstocks in terms of their agrochemical and WUE, and the energy yield from both crops and residues in order to investigate the efficiency and the sustainability of an integrated biofuel production system, and to estimate the potential energy yield of each feedstock. Soybean, palm oil, and wheat cultivation seemed to be efficient in agrochemical use, whereas sugar beet–, corn–, and rapeseed-derived biofuel production appeared to be relatively water-use efficient. The bioethanol yield of sugarcane and sugar beet, and the biodiesel yield of palm oil were favorable, and the energy surplus that can be produced from their residues provided a potential for the use of residues as energy sources.

Use of the crop residues can also increase land use efficiency by producing additional energy per unit of area. In addition, expanding the area for energy crop cultivation from the agricultural land to less-optimal areas for food crop production could be another option to avoid competition with food crops and to add value to the areas with a relatively low fertility. Since the crop residues are non-edible and left-over materials after the harvesting of food crops, using the residues for energy production could be a feasible option to avoid food-versus-fuel criticism (Strezov, 2015). However, the pre-treatment technology for producing energy from residues has not reached its optimal level; therefore, it should be improved to achieve high overall efficiency of the biofuel production system. The combined production system for producing biofuel from the main feedstock, and for generating heat and electricity from the residues (e.g., combined heat and power [CHP] generation system) has been introduced as a great option to increase the overall efficiency (Börjesson and Tufvesson, 2011; Punter et al., 2004).

In addition, it could be important to consider the spatial distribution and seasonal availability of crop residues, since the stable supply of feedstock can be one of the crucial factors to determine the completeness and success of an integrated biofuel production system (Herr and Dunlop, 2011). Besides, long transport distance from the residues to the fuel-producing facilities may not be economically and environmentally sustainable, since additional fossil fuel consumption during the transport can occur and consequently increase total energy input and GHG emissions (Farine et al., 2012).

It is expected that the results of this study could aid in deciding the optimal level of agrochemicals and water use at the farm-gate so that excessive use of those resources can be avoided. Although the energy yield analysis in this study has shown great potential of producing energy using the crops and residues, future studies that seek opportunities to improve technical challenges in utilizing the crop residues and to sustainably increase total energy yield could be necessary to scale up the current level of biofuel production.

REFERENCES

Abbasi, S.A. and Abbasi, N., 2000. The likely adverse environmental impacts of renewable energy sources. *Applied Energy*, 65(1), 121–144.

Ahiduzzaman, M., Islam, A.A., Yaakob, Z., Ghani, J.A. and Anuar, N., 2014. Agricultural residues from crop harvesting and processing: A renewable source of Bio-Energy. In *Biomass and Bioenergy* (pp. 323–337). Springer International Publishing, Cham, Switzerland.

Angarita, E.E.Y., Lora, E.E.S., da Costa, R.E. and Torres, E.A., 2009. The energy balance in the palm oil-derived methyl ester (PME) life cycle for the cases in Brazil and Colombia. *Renewable Energy*, 34(12), 2905–2913.

Atabani, A.E., Silitonga, A.S., Badruddin, I.A., Mahlia, T.M.I., Masjuki, H.H. and Mekhilef, S., 2012. A comprehensive review on biodiesel as an alternative energy resource and its characteristics. *Renewable and Sustainable Energy Reviews*, 16(4), 2070–2093.

Barros-Rios, J., Romaní, A., Garrote, G. and Ordas, B., 2015. Biomass, sugar, and bioethanol potential of sweet corn. *Gcb Bioenergy*, 7(1), 153–160.

Begum, S., Kumaran, P. and Jayakumar, M., 2013. Use of oil palm waste as a renewable energy source and its impact on reduction of air pollution in context of Malaysia. In *IOP Conference Series: Earth and Environmental Science* (Vol. 16, No. 1, p. 012026). IOP Publishing, Bristol, UK.

Bessou, C., Ferchaud, F., Gabrielle, B. and Mary, B., 2011. Biofuels, greenhouse gases and climate change. A review. *Agronomy for Sustainable Development*, 31(1), 1–79.

Blottnitz, H.V. and Curran, M.A., 2007. A review of assessments conducted on bio-ethanol as a transportation fuel from a net energy, greenhouse gas, and environmental life cycle perspective. *Journal of Cleaner Production*, 15(7), 607–619.

Börjesson, P. and Tufvesson, L.M., 2011. Agricultural crop-based biofuels–resource efficiency and environmental performance including direct land use changes. *Journal of Cleaner Production, 19*(2), 108–120.

BP, 2017. Statistical Review of World Energy 2017. Available at: https://www.bp.com/content/dam/bp/en/corporate/pdf/energy-economics/statistical-review-2017/bp-statistical-review-of-world-energy-2017-full-report.pdf. Accessed on October 3, 2017.

Cardona, C.A. and Sánchez, O.J., 2007. Fuel ethanol production: Process design trends and integration opportunities. *Bioresource Technology, 98*, 2415–2457.

Channiwala, S.A. and Parikh, P.P., 2002. A unified correlation for estimating HHV of solid, liquid and gaseous fuels. *Fuel, 81*(8), 1051–1063.

Chen, H. and Chen, G.Q., 2011. Energy cost of rapeseed-based biodiesel as alternative energy in China. *Renewable Energy, 36*(5), 1374–1378.

Davis, S.C., Dohleman, F.G. and Long, S.P., 2011. The global potential for Agave as a biofuel feedstock. *Gcb Bioenergy, 3*(1), 68–78.

Dawson, L. and Boopathy, R., 2007. Use of post-harvest sugarcane residue for ethanol production. *Bioresource Technology, 98*(9), 1695–1699.

De Souza, S.P., Pacca, S., De Avila, M.T. and Borges, J.L.B., 2010. Greenhouse gas emissions and energy balance of palm oil biofuel. *Renewable Energy, 35*(11), 2552–2561.

de Vries, S.C., van de Ven, G.W.J., van Ittersum, M.K. and Giller, K.E., 2010. Resource use efficiency and environmental performance of nine major biofuel crops, processed by first-generation conversion techniques. *Biomass and Bioenergy, 34*(5), 588–601.

Demirbaş, A. and Demirbaş, A.H., 2004. Estimating the calorific values of lignocellulosic fuels. *Energy Exploration & Exploitation, 22*(2), 135–143.

Department of the Environment and Energy, 2014. Australian government response to the climate change authority's 2014. *Renewable Energy Target Review*, Canberra, Australia. Available at: http://www.environment.gov.au/system/files/resources/79393683-bef7-4e70-86f5-88b853d3b9a6/files/government-response-cca-2014-ret-review.pdf. Accessed on September 22, 2017.

Department of the Environment, 2015. National greenhouse accounts factors. Department of the Environment, Australian Government August, 2015. Available at: https://www.environment.gov.au/system/files/resources/3ef30d52-d447-4911-b85c-1ad53e55dc39/files/national-greenhouse-accounts-factors-august-2015.pdf. Accessed on September 22, 2017.

Dias De Oliveira, M.E., Vaughan, B.E. and Rykiel, E.J., 2005. Ethanol as fuel: Energy, carbon dioxide balances, and ecological footprint. *AIBS Bulletin, 55*(7), 593–602.

Díaz, M.J., Cara, C., Ruiz, E., Romero, I., Moya, M. and Castro, E., 2010. Hydrothermal pretreatment of rapeseed straw. *Bioresource Technology, 101*, 2428–2435.

Dodić, S.N., Zekić, V.N., Rodić, V.O., Tica, N.L., Dodić, J.M. and Popov, S.D., 2010. Situation and perspectives of waste biomass application as energy source in Serbia. *Renewable and Sustainable Energy Reviews, 14*(9), 3171–3177.

Dominguez-Faus, R., Powers, S.E., Burken, J.G. and Alvarez, P.J., 2009. The water footprint of biofuels: A drink or drive issue? *Environmental Science and Technology, 43*(9), 3005–3010.

Egües, I., Alriols, M.G., Herseczki, Z., Marton, G. and Labidi, J., 2010. Hemicelluloses obtaining from rapeseed cake residue generated in the biodiesel production process. *Journal of Industrial and Engineering Chemistry, 16*, 293–298.

Elsgaard, L., Olesen, J.E., Hermansen, J.E., Kristensen, I.T. and Børgesen, C.D., 2013. Regional greenhouse gas emissions from cultivation of winter wheat and winter rapeseed for biofuels in Denmark. *Acta Agriculturae Scandinavica, Section B–Soil & Plant Science, 63*(3), 219–230.

EPA, 2017. Renewable fuel standard program: Standards for 2018 and biomass-based diesel volume for 2019. Proposed rules. *Environmental Protection Agency, Federal Register, 82*(139), 34206–34245.

Erdal, G., Esengün, K., Erdal, H. and Gündüz, O., 2007. Energy use and economical analysis of sugar beet production in Tokat province of Turkey. *Energy, 32*(1), 35–41.

European Commission, 2016. *EU Energy in Figures. Statistical pocketbook 2016*. Office of the European Union, Luxembourg, Germany.

Evans, A., Strezov, V. and Evans, T.J., 2015. Sustainability considerations for electricity generation from biomass. In: Strezov V. and Evans T.J., (Eds.) *Biomass Processing Technologies*. CRC Press, Boca Raton, FL, pp. 1–31.

FAO, 2008. *The State of Food and Agriculture: Biofuels: Prospects, Risks and Opportunities*. Food and Agriculture Organization of the United Nations, Rome, Italy.

FAOSTAT, 2014. World crop production statistics from FAOSTAT Database. Food and Agriculture Organization of the United Nations. Available at: http://www.fao.org/faostat/en/#data/QC. Accessed on October 4, 2017.

Farine, D.R., O'Connell, D.A., John Raison, R., May, B.M., O'Connor, M.H., Crawford, D.F., Herr, A. et al. 2012. An assessment of biomass for bioelectricity and biofuel, and for greenhouse gas emission reduction in Australia. *Gcb Bioenergy, 4*(2), 148–175.

Foteinis, S., Kouloumpis, V. and Tsoutsos, T., 2011. Life cycle analysis for bioethanol production from sugar beet crops in Greece. *Energy Policy, 39*(9), 4834–4841.

García, C.A., Fuentes, A., Hennecke, A., Riegelhaupt, E., Manzini, F. and Masera, O., 2011. Life-cycle greenhouse gas emissions and energy balances of sugarcane ethanol production in Mexico. *Applied Energy, 88*(6), 2088–2097.

Gerbens-Leenes, P.W., Hoekstra, A.Y. and Van der Meer, T.H., 2009. The water footprint of energy from biomass: A quantitative assessment and consequences of an increasing share of bio-energy in energy supply. *Ecological Economics, 68*(4), 1052–1060.

Groom, M.J., Gray, E.M. and Townsend, P.A., 2008. Biofuels and biodiversity: Principles for creating better policies for biofuel production. *Conservation Biology, 22*(3), 602–609.

Halleux, H., Lassaux, S., Renzoni, R. and Germain, A., 2008. Comparative life cycle assessment of two biofuels ethanol from sugar beet and rapeseed methyl ester. *The International Journal of Life Cycle Assessment, 13*(3), 184.

Herr, A. and Dunlop, M., 2011. Bioenergy in Australia: An improved approach for estimating spatial availability of biomass resources in the agricultural production zones. *Biomass and Bioenergy, 35*(5), 2298–2305.

Herr, A., Farine, D., Poulton, P., Baldock, J., Bruce, J., Braid, A., Dunlop, M., O'Connell, D. and Poole, M., 2010. *Harvesting Stubble for Energy in Australia—Take it or Leave it. Phase II of Opportunities for Energy Efficiency, Regional Self-sufficiency and Energy Production for Australian Grain Growers.* CSIRO, Canberra, Australia.

Hill, J., Nelson, E., Tilman, D., Polasky, S. and Tiffany, D., 2006. Environmental, economic, and energetic costs and benefits of biodiesel and ethanol biofuels. *Proceedings of the National Academy of Sciences, 103*(30), 11206–11210.

Hoekstra, A.Y. Chapagain, A.K., 2007. Water footprints of nations: Water use by people as a function of their consumption pattern. *Water Resources Management, 21*, 35–48.

Hoekstra, A.Y., Chapagain, A.K., Aldaya, M.M. and Mekonnen, M.M., 2009. *Water Footprint Manual: State of the Art 2009.* Water Footprint Network, Enschede, the Netherlands.

Holtum, J.A.M., Chambers, D., Morgan, T. and Tan, D.K.Y., 2011. Agave as a biofuel feedstock in Australia. *Gcb Bioenergy, 3*(1), 58–67.

Hülsbergen, K.J., Feil, B. and Diepenbrock, W., 2002. Rates of nitrogen application required to achieve maximum energy efficiency for various crops: Results of a long-term experiment. *Field Crops Research, 77*(1), 61–76.

Hülsbergen, K.J., Feil, B., Biermann, S., Rathke, G.W., Kalk, W.D. and Diepenbrock, W., 2001. A method of energy balancing in crop production and its application in a long-term fertilizer trial. *Agriculture, Ecosystems & Environment, 86*(3), 303–321.

IEA, 2016. World Energy Outlook 2016—Executive summary. International Energy Agency. Available at: https://www.iea.org/publications/freepublications/publication/WorldEnergyOutlook2016ExecutiveSummary English.pdf. Accessed on September 14, 2017.

IEA, 2017. World energy outlook 2017. International Energy Agency, November 2017. Available at: https://www.iea.org/publications/freepublications/publication/WEO_2017_Executive_Summary_English_version.pdf. Accessed on September 14, 2017.

Iriarte, A., Rieradevall, J. and Gabarrell, X., 2010. Life cycle assessment of sunflower and rapeseed as energy crops under Chilean conditions. *Journal of Cleaner Production, 18*(4), 336–345.

Ji, L.Q., 2015. An assessment of agricultural residue resources for liquid biofuel production in China. *Renewable and Sustainable Energy Reviews, 44*, 561–575.

Johnson, J.M.F., Barbour, N.W. and Weyers, S.L., 2007. Chemical composition of crop biomass impacts its decomposition. *Soil Science Society of America Journal, 71*(1), 155.

Kaparaju, P., Serrano, M., Thomsen, A.B., Kongjan, P. and Angelidaki, I., 2009. Bioethanol, biohydrogen and biogas production from wheat straw in a biorefinery concept. *Bioresource Technology, 100*, 2562–2568.

Karaosmanoğlu, F., Tetik, E. and Göllü, E., 1999. Biofuel production using slow pyrolysis of the straw and stalk of the rapeseed plant. *Fuel Processing Technology, 59*(1), 1–12.

Kelly-Yong, T.L., Lee, K.T., Mohamed, A.R. and Bhatia, S., 2007. Potential of hydrogen from oil palm biomass as a source of renewable energy worldwide. *Energy Policy, 35*(11), 5692–5701.

Kim, M. and Day, D.F., 2011. Composition of sugar cane, energy cane, and sweet sorghum suitable for ethanol production at Louisiana sugar mills. *Journal of Industrial Microbiology & Biotechnology, 38*(7), 803–807.

Koçar, G. and Civaş, N., 2013. An overview of biofuels from energy crops: Current status and future prospects. *Renewable and Sustainable Energy Reviews, 28*, 900–916.

Koga, N., 2008. An energy balance under a conventional crop rotation system in northern Japan: Perspectives on fuel ethanol production from sugar beet. *Agriculture, Ecosystems & Environment, 125*(1), 101–110.

Kongboon, R. and Sampattagul, S., 2012. The water footprint of sugarcane and cassava in northern Thailand. *Procedia-Social and Behavioral Sciences, 40*, 451–460.

Lal, R., 2005. World crop residues production and implications of its use as a biofuel. *Environment International, 31*(4), 575–584.

Leiper, K.A., Schlee, C., Tebble, I. and Stewart, G.G., 2006. The fermentation of beet sugar syrup to produce bioethanol. *Journal of the Institute Brewing 112*, 122–133.

Leung, G. and Strezov, V., 2015. Esterification. In: Strezov V. and Evans T.J. (Eds.) *Biomass Processing Technologies.* CRC Press, Boca Raton, FL, pp. 213–255.

Luo, G., Talebnia, F., Karakashev, D., Xie, L., Zhou, Q. and Angelidaki, I., 2011. Enhanced bioenergy recovery from rapeseed plant in a biorefinery concept. *Bioresource Technology, 102*(2), 1433–1439.

Macedo, I.C., Seabra, J.E. and Silva, J.E., 2008. Green house gases emissions in the production and use of ethanol from sugarcane in Brazil: The 2005/2006 averages and a prediction for 2020. *Biomass and Bioenergy, 32*(7), 582–595.

Masiá, A.T., Buhre, B.J.P., Gupta, R.P. and Wall, T.F., 2007. Characterising ash of biomass and waste. *Fuel Processing Technology, 88*(11), 1071–1081.

Maung, T.A. and Gustafson, C.R., 2011. The economic feasibility of sugar beet biofuel production in central North Dakota. *Biomass and Bioenergy, 35*(9), 3737–3747.

McKendry, P., 2002. Energy production from biomass (part 1): Overview of biomass. *Bioresource Technology, 83*(1), 37–46.

Mekonnen, M. and Hoekstra, A.Y., 2011. The green, blue and grey water footprint of crops and derived crops products. *Hydrology and Earth System Sciences Discussions, 8*(47), 763–809.

Moreno, A.D., Alvira, P., Ibarra, D. and Tomás-Pejó, E., 2017. Production of ethanol from lignocellulosic biomass. In: Fang Z., Smith, Jr. R., Qi X (Eds.) *Production of Platform Chemicals from Sustainable Resources* (pp. 375–410). Biofuels and Biorefineries. Springer, Singapore.

Mortimer, N.D., Cormack, P., Elsayed, M.A. and Horne, R.E., 2003. Evaluation of the comparative energy, global warming and socio-economic costs and benefits of biodiesel. Report to the Department for Environment, Food and Rural Affairs Contract Reference No. CSA, 5982.

Motghare, K.A., Rathod, A.P., Wasewar, K.L. and Labhsetwar, N.K., 2016. Comparative study of different waste biomass for energy application. *Waste Management, 47*, 40–45.

Mrini, M., Senhaji, F. and Pimentel, D., 2001. Energy analysis of sugarcane production in Morocco. *Environment, Development and Sustainability, 3*(2), 109–126.

Mu, D., Seager, T., Rao, P.S. and Zhao, F., 2010. Comparative life cycle assessment of lignocellulosic ethanol production: Biochemical versus thermochemical conversion. *Environmental Management, 46*(4), 565–578.

Naik, S.N., Goud, V.V., Rout, P.K., Jacobson, K. and Dalai, A.K., 2010a. Characterization of Canadian biomass for alternative renewable biofuel. *Renewable Energy, 35*(8), 1624–1631.

Naik, S.N., Goud, V.V., Rout, P.K. and Dalai, A.K., 2010b. Production of first and second generation biofuels: A comprehensive review. *Renewable and Sustainable Energy Reviews, 14*(2), 578–597.

Naylor, R.L., Liska, A.J., Burke, M.B., Falcon, W.P., Gaskell, J.C., Rozelle, S.D. and Cassman, K.G., 2007. The ripple effect: Biofuels, food security, and the environment. *Environment: Science and Policy for Sustainable Development, 49*(9), 30–43.

OECD/FAO, 2017. OECD-FAO Agricultural Outlook 2017-2026, OECD Publishing, Paris. Available at: http://dx.doi.org/10.178 7/agr_outlook-2017-en. Accessed on September 5, 2017.

Owen, N.A. and Griffiths, H., 2014. Marginal land bioethanol yield potential of four crassulacean acid metabolism candidates (Agave fourcroydes, Agave salmiana, Agave tequilana and Opuntia ficus-indica) in Australia. *Gcb Bioenergy, 6*(6), 687–703.

Peterson, J.B.D., 2006. Ethanol production from agricultural residues. *International Sugar Journal, 108*(1287), 177–180.

Pimentel, D. and Patzek, T., 2008. Ethanol production using corn, switchgrass and wood; biodiesel production using soybean. In *Biofuels, Solar and Wind as Renewable Energy Systems* (pp. 373–394). Springer Netherlands, Dordrecht, the Netherlands.

Pimentel, D. and Patzek, T.W., 2005. Ethanol production using corn, switchgrass, and wood; biodiesel production using soybean and sunflower. *Natural Resources Research, 14*(1), 65–76.

Pleanjai, S. and Gheewala, S.H., 2009. Full chain energy analysis of biodiesel production from palm oil in Thailand. *Applied Energy, 86*, S209–S214.

Pryor, S.W., Smithers, J., Lyne, P. and van Antwerpen, R., 2017. Impact of agricultural practices on energy use and greenhouse gas emissions for South African sugarcane production. *Journal of Cleaner Production, 141*, 137–145.

Punter, G., Rickeard, D., Larivé, J.F., Edwards, R., Mortimer, N., Horne, R., Bauen, A. and Woods, J., 2004. Well-to-wheel evaluation for production of ethanol from wheat. *Fuels Working Group of the Low Carbon Vehicle Partnership*, 1–40. Available at: http://www.rms.lv/bionett/Files/BioE-2004-001%20 Ethanol_WTW_final_report.pdf. Accessed on September 13, 2017.

Rajagopal, D., Sexton, S.E., Roland-Holst, D. and Zilberman, D., 2007. Challenge of biofuel: Filling the tank without emptying the stomach? *Environmental Research Letters, 2*(4), 044004.

Rathke, G.W. and Diepenbrock, W., 2006. Energy balance of winter oilseed rape (Brassica napus L.) cropping as related to nitrogen supply and preceding crop. *European Journal of Agronomy, 24*(1), 35–44.

Rettenmaier, N., Reinhardt, G., Gärtner, S. and Münch, J., 2008. Bioenergy from grain and sugar beet: Energy and greenhouse gas balances. IFEU–Institute for Energy and Environmental Research, Heidelberg, Germany.

Rockström, J., Barron, J. and Fox, P., 2003. Water productivity in rain-fed agriculture: Challenges and opportunities for smallholder farmers in drought-prone tropical agroecosystems. *Water Productivity in Agriculture: Limits and Opportunities for Improvement, 85199*(669), 8.

Rosenberger, A., Kaul, H.P., Senn, T. and Aufhammer, W., 2001. Improving the energy balance of bioethanol production from winter cereals: The effect of crop production intensity. *Applied Energy, 68*(1), 51–67.

Saritha, M. and Arora, A., 2012. Biological pretreatment of lignocellulosic substrates for enhanced delignification and enzymatic digestibility. *Indian Journal of Microbiology, 52*(2), 122–130.

Seabra, J.E., Macedo, I.C., Chum, H.L., Faroni, C.E. and Sarto, C.A., 2011. Life cycle assessment of Brazilian sugarcane products: GHG emissions and energy use. *Biofuels, Bioproducts and Biorefining, 5*(5), 519–532.

Şensöz, S. and Kaynar, İ., 2006. Bio-oil production from soybean (Glycine max L.); fuel properties of Bio-oil. *Industrial Crops and Products, 23*(1), 99–105.

Sharma, H.K., Xu, C. and Qin, W., 2017. Biological pretreatment of lignocellulosic biomass for biofuels and bioproducts: An overview. *Waste and Biomass Valorization, 9*(51), 1–17.

Sindhu, R., Binod, P. and Pandey, A., 2016. Biological pretreatment of lignocellulosic biomass–An overview. *Bioresource Technology, 199*, 76–82.

Singh, K.P., Suman, A., Singh, P.N. and Lal, M., 2007. Yield and soil nutrient balance of a sugarcane plant–ratoon system with conventional and organic nutrient management in sub-tropical India. *Nutrient Cycling in Agroecosystems, 79*(3), 209–219.

Singh, S. and Kumar, A., 2011. Development of water requirement factors for biomass conversion pathway. *Bioresource Technology, 102*(2), 1316–1328.

Smeets, E.M.W. and Junginger, H.M., 2005. *Supportive Study for the OECD on Alternative Developments in Biofuel Production Across the World*. Department of Science, Technology and Society, Copernicus Institute, Utrecht, the Netherlands.

Strezov, V., 2015. Properties of biomass fuels. In: Strezov V. and Evans T.J. (Eds.). *Biomass Processing Technologies*. CRC Press, Boca Raton, FL, pp. 1–31.

Suhartini, S., Heaven, S. and Banks, C.J., 2014. Comparison of mesophilic and thermophilic anaerobic digestion of sugar beet pulp: Performance, dewaterability and foam control. *Bioresource Technology, 152*, 202–211.

Tan, D.K. and Amthor, J.S., 2013. Chapter 12. Bioenergy. In *Photosynthesis*. InTech. pp. 299–330. Available at: doi:10.5772/55317. Accessed on December 1, 2017.

Thamsiriroj, T. and Murphy, J.D., 2009. Is it better to import palm oil from Thailand to produce biodiesel in Ireland than to produce biodiesel from indigenous Irish rape seed? *Applied Energy, 86*(5), 595–604.

Thomsen, M.H. and Haugaard-Nielsen, H., 2008. Sustainable bioethanol production combining biorefinery principles using combined raw materials from wheat undersown with clover-grass. *Journal of Industrial Microbiology & Biotechnology, 35*(5), 303–311.

Tomás-Pejó, E., Alvira, P., Ballesteros, M. and Negro, M.J., 2011. Pretreatment technologies for lignocellulose-to-bioethanol conversion. *In Biofuels*, A. Pandey, C. Larroche, S. C. Ricke, C-G. Dussap, and E. Gnansounou (Eds.), Amsterdam, the Netherlands: Elsevier, Academic Press, pp. 149–176.

Tomei, J. and Upham, P., 2009. Argentinean soy-based biodiesel: An introduction to production and impacts. *Energy Policy, 37*(10), 3890–3898.

Tsai, W.T., Lee, M.K. and Chang, Y.M., 2006. Fast pyrolysis of rice straw, sugarcane bagasse and coconut shell in an induction-heating reactor. *Journal of Analytical and Applied Pyrolysis, 76*(1), 230–237.

Turner, B.T., Plevin, R.J., O'Hare, M. and Farrell, A.E., 2007. Creating markets for green biofuels. *Transportation Sustainability Research Center.* University of California, Berkeley, CA.

Tzilivakis, J., Warner, D.J., May, M., Lewis, K.A. and Jaggard, K., 2005. An assessment of the energy inputs and greenhouse gas emissions in sugar beet (Beta vulgaris) production in the UK. *Agricultural Systems, 85*(2), 101–119.

Unal, H. and Alibas, K., 2007. Agricultural residues as biomass energy. *Energy Sources*, Part B, *2*(2), 123–140.

Unkovich, M., Baldock, J. and Forbes, M., 2010. Variability in harvest index of grain crops and potential significance for carbon accounting: Examples from Australian agriculture. *Advances in Agronomy, 105*, 173–219.

USDA, 2017. EU Biofuels Annual 2017. Global Agricultural Information Network (GAIN) Report, No: NL7015. Foreign Agriculturle Service, U.S. Department of Agriculture. June, 2017. Available at: https://gain.fas.usda.gov/Recent%20GAIN%20Publications/Biofuels%20Annual_The%20Hague_EU-28_6-19-2017.pdf. Accessed on September 13, 2017.

USDA, 2016. Agricultural Chemical Use Survey 2015—Wheat. United States Department of Agriculture. National Agricultural Statistics Service. No. 2016-5. Available at: https://www.nass.usda.gov/Surveys/Guide_to_NASS_Surveys/Chemical_Use/2015_Cotton_Oats_Soybeans_Wheat_Highlights/ChemUseHighlights_Wheat_2015.pdf. Accessed on September 13, 2017.

Visser, E.M., Oliveira Filho, D., Martins, M.A. and Steward, B.L., 2011. Bioethanol production potential from Brazilian biodiesel co-products. *Biomass and Bioenergy, 35*(1), 489–494.

Wayagari, J.W., Amosun, A. and Misari, S.M., 2001. Economic optimum NPK fertilizer ratios and time of application for high yield and good quality sugarcane production. *Sugar Technology, 3*(1), 34–39.

WBA, 2017. WBA Global bioenergy statistics 2017. World Bioenergy Association. Available at: http://www.worldbioenergy.org/uploads/WBA%20GBS%202017_hq.pdf. Accessed on September 16, 2017.

World Energy Council, 2016. World Energy Resources; Bioenergy 2016. World Energy Council. Available at: https://www.worldenergy.org/wp-content/uploads/2017/03/WEResources_Bioenergy_2016.pdf. Accessed on September 16, 2017.

Yan, X., Tan, D.K., Inderwildi, O.R., Smith, J.A.C. and King, D.A., 2011. Life cycle energy and greenhouse gas analysis for agave-derived bioethanol. *Energy & Environmental Science, 4*(9), 3110–3121.

Yılgın, M., Duranay, N.D. and Pehlivan, D., 2010. Co-pyrolysis of lignite and sugar beet pulp. *Energy Conversion and Management, 51*(5), 1060–1064.

Yin, C.Y., 2011. Prediction of higher heating values of biomass from proximate and ultimate analyses. *Fuel, 90*(3), 1128–1132.

Yusoff, S. and Hansen, S.B., 2007. Feasibility study of performing a life cycle assessment on crude palm oil production in Malaysia. LCA Case Studies. *The International Journal of Life Cycle Assessment, 12*(1), 50–58.

8 System Approach to Bio-Oil Production from Microalgae

Margarita Rosa Albis Salas, Vladimir Strezov, and Hossain M. Anawar

CONTENTS

8.1 INTRODUCTION

The world's growing population has resulted in an increasing demand for biomass for food and animal feed. Additionally, due to climate change and the general impact of fossil fuels on the environment, initiatives are being undertaken to move toward an economy where biomass replaces petroleum. As it is unlikely that agricultural biomass production will meet these demands, increasing attention has been given to microalgae as a possible solution (Xu et al. 2011b). Some of microalgae's attributes that make it attractive are its lower energy consumption, lower water usage and lower nutrient demand, lower land demand per unit area, higher biomass production, and potential to grow in wastewater, brackish water, or seawater (González-González et al. 2018).

As microalgae grow in a liquid environment, specific cultivation, harvesting, and processing techniques are required for biodiesel to be produced effectively (Choi et al. 2017). The production of biodiesel follows a series of stages that include selection of a strain with potential for production of biodiesel (i.e., high biomass and lipid production), selection of a strategic site with features that mitigate cost and enhance productivity, identification of cultivation conditions for algae to grow efficiently (with consideration to variables such as algae metabolism, nutrient, salinity, temperature, and mixing), harvesting of biomass, drying of biomass, extraction of oil (by chemical, biological, or mechanical methods), and finally, biodiesel production (Mata et al. 2010). Algae cell walls and membranes have a high mechanical strength and chemical resistance that limit the extraction of lipids. Several mechanical, biological, and chemical techniques are in practice to solve this problem. Thermochemical conversion of biomass, such as pyrolysis and hydrothermal liquefaction (HTL), has been presented as the most suitable options. However, oxidative stress, using free nitrous acid, has been introduced as a new, viable alternative due to the use of greener and more renewable chemicals for lipid recovery (Bai et al. 2014).

Recent efforts have focused on improving the methods in each production stage to reduce the cost and increase the yields as biodiesel production is rapidly moving from laboratories to pilot and commercial scales (Vandamme et al. 2013). Additionally, integration of cultivation with power plants and wastewater treatments is now being considered in order to feed cultures with nutrients and reduce cost, while also performing bioremediation (Mata et al. 2010).

The following chapter focuses on the main stages for production of biodiesel from microalgae and emphasizes the advances in technologies that aim to improve overall performance. Additionally, the chapter discusses integration of biodiesel production with flue gas from power plants by highlighting recommendations in terms of algae strains and technology in the biodiesel production stages that adjust to this scenario.

8.2 ALGAE STRAIN SELECTION

The number of available algae species range between 70,000 and 1 million species. Only 44,000 have been described and few have been studied carefully to verify their performance for the amount of lipids and other important commercial products they can produce (Neofotis et al. 2016). Novel algae strains represent a significant potential for discovery that can find use in production of energy. There has been a recent effort to discover species with high lipid content and also algae with adaptations to specific climates and cultivation conditions (e.g., tolerance to salinity, flue gas, and wastewater conditions) (Tale et al. 2014). Additionally, algae strains are also being genetically modified in order to improve cultivation or harvesting performance, lipid yield, quality of oil, and other algae products (e.g., carotenoids) (Neofotis et al. 2016).

Tale et al. (2014) studied the response of several strains of *Chlorella* sp., *Scenedesmus* sp., and *Monoraphidium* sp. to flue gas power plant conditions. They found that *Chlorella* sp. KMN3 and *Monoraphidium* sp. KMN5 are excellent candidates for biodiesel production under these conditions, as they had high biomass accumulation (1.59 ± 0.05 g L^{-1}) and high lipid content (35%), respectively. In addition, the oil had high quality due to the fatty acid methyl esters composition (C-16:0, C-18:0, C-18:1, and C-18:2). Likewise, Aslam et al. (2017) subjected an algae community to 100% flue gas supplementation and phosphate buffering at higher concentrations and found that, after a considerable biodiversity loss, *Desmodesmus* spp. was the dominant alga. This identified *Desmodesmus* spp. as a potential strain to be cultured under these conditions.

Currently, there are more than 1,000 microalgae strains in specialized collections around the world. They have been studied and genetically analyzed. They work as "starter cultures" to industries, research organizations, and universities (Li et al. 2012). The collection has been constantly updated by the research carried out by universities, institutes, and industry (e.g., Neofotis et al. 2016, Aslam et al. 2017). For example, the National Alliance for Advanced Biofuels and Bioproducts (NAABB) developed a project to identify new potential platform strains with high growth rates and lipid productivities. The selection of potential strains followed the steps shown in Figure 8.1. Algae collected from different habitats were passed through several filters or screens to select the most suitable algae for commercial purposes. During the screens, algae that survived standard laboratory medium conditions were selected and characterized biochemically and genetically to later determine those potential algae for biofuel production. The strains were then added to the collections to expand existing germplasm. Coccoid green algae were some of the best-performing algae. Examples of those are *Acutodesmus (Scenedesmus) obliquus*, *Chlorella sorokiniana, Desmodesmus* sp., and *Ankistrodesmus* sp. (Neofotis et al. 2016).

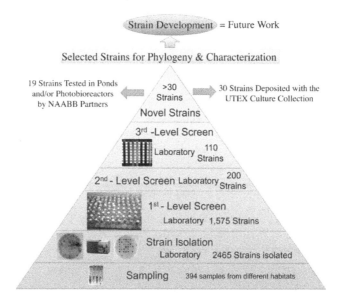

FIGURE 8.1 Summary of the phycoprospecting process performed by the National Alliance for Advanced Biofuels and Bioproducts Consortium to identify the best-performing microalgae strains for biofuel production. More than 394 algae samples from different habitats were collected. In the first screen, the algae that survive the media cultures were classified as potential producers and later characterized. During the second screen, three media culture were analyzed to determine the optimum media in which each algae culture could grow in. In the third screen, potential algae for biodiesel production are determined by characterizing biomass and lipid content. These strains are later added to the collection. UTEX: University of Texas. (Reprinted from Neofotis, P. 2016. *Algal Res* 15: 164–178, with permission from Elsevier.)

8.3 SITE SELECTION

Algae cultivation has low requirements of land per unit area, nutrients, water, and energy consumption when compared to other sources of biodiesel (González-González et al. 2018). Despite these properties, farms for algae culture require strategic locations to make viable industrial production. This is because a considerable amount of land is still required to install raceway ponds (open systems) to maintain biodiesel industrial capacity (Table 8.1). Although there are close system cultivation options (e.g., photobioreactors) that occupy less space, they are expensive. However, Bravo-Fritz et al. (2015)

TABLE 8.1
Land Sizes Required for Different Scales of Biodiesel Production

Biodiesel Scale (m³/yr.)	Scale Size of Plant	Surface Area (ha)
<8,000	Pilot scale	<530
8,000–40,000	Demonstration (medium) scale	530–2670
>42,000	Industrial scale	>2,670

Source: Reprinted from Bravo-Fritz, C. 2015. *Algal Res* 11: 343–349, with permission from Elsevier.

TABLE 8.2

Important Factors to Consider When Locating Microalgae Cultivation Locations

Land	Climate	Water	CO$_2$	Nutrients
Slope	Solar radiation	Ocean	Power plants	Food production facilities
Land use/Land cover	Temperature (maximum and minimum)	Saline aquifers	Industrial plants	Confined animal facilities
Ownership	Growing season length Precipitation	Lakes/rivers	Chemical production facilities cement plants	Agricultural runoff
Economic, cultural and environmental value	Wind	Wastewater treatment plant	Wastewater treatment plants	Wastewater treatment plants
Soils	Evaporation Storms		Petroleum and natural gas processing	Fertilizers

Source: Reprinted from Boruff, B. et al. 2015. *Appl Energy* 149: 379–391, with permission from Elsevier.

recommend a mix of the two cultivation methods, as close systems were suggested to provide uncontaminated and constant algae inoculum for the open systems.

Various studies have used geographic information systems (GIS) as a tool to integrate different variables considered strategic for optimal results in terms of alga productivity and cultivation cost. GIS integrates and weighs the variables to create maps that show the most suitable areas to implement microalgae cultivation. The most common variables considered are land, impact on natural and agricultural resources, transport cost, energy consumption, nutrient availability, public perception, and climate (Boruff et al. 2015, Bravo-Fritz et al. 2015) (Table 8.2).

8.4 CULTIVATION

Microalgae are fast-growing organisms that depend on carbon and light for photosynthesis (Liang et al. 2009). Although they are mainly autotrophic, they can adjust their metabolism according to environmental conditions. Some organisms can grow photoautotrophically by using light as a sole energy source; heterotrophically by using organic compounds; mixotrophically by using light as an energy source, although organic compounds and CO$_2$ are essential; and photoheterotrophically by using light to metabolize organic compounds (Wan et al. 2012)

In photoautotrophic cultures, the significant fluctuations in biomass and lipid content, due to climate conditions, make biodiesel production unviable (Liang et al. 2009). This is the reason alternative cultures are considered. For instance, as heterotrophic cultures are easier to manipulate, better results can be obtained for algae growth, biomass and lipid accumulation, and scalability (Rodolfi et al. 2009, Zhou et al. 2017) (Table 8.3). For instance, Wan et al. (2012) showed that heterotrophic cultures of *Chlorella sorokiniana* achieved a lipid accumulation of 56% (w/w) dry weight after seven days in high glucose concentrations. In contrast, they registered 19% lipid accumulation in 30 days of photoautotrophic culture. However, Wang et al. (2017) showed that the effect of the type of culture also depends on the algae strain and nitrogen source, when comparing photoautotrophic, heterotrophic, and mixotrophic cultures of *Tribonema* sp. (strain EA903 and EA904). They found the highest lipid productivity and biomass concentrations in mixotrophic cultivations with NaNO$_3$ additions. They found the lowest lipid productivity and biomass concentrations in heterotrophic cultivations with peptone additions (Table 8.4).

The source of energy for cultivation plays an important role in creating optimal conditions for specific alga strains (Zhan et al. 2014). As important as the carbon source is, other parameters, such as pH, nutrients, oxygen, light intensity, temperature, and by-product removal, need to be quantified and managed properly to control culture conditions, promote optimal microalgae growth, and avoid contamination (Chen et al. 2011).

TABLE 8.3

Biomass Productivity, Lipid Content, and Lipid Productivity under Different Types of Cultures

Species	Cultivation Type	Biomass Productivity (g L⁻¹ d⁻¹)	Lipid Content (%DCW)	Lipid Productivity (mg L⁻¹ d⁻¹)	Author
Tribonema minus	Heterotrophic	0.31	—	730	Zhou et al. (2017)
Chlorella vulgaris #259	Heterotrophic	0.08–0.15	23.0–36.0	27.0–35.0	Liang et al. (2009)
Chlorella protothecoides	Heterotrophic	4.0–4.4	43.0–46.0	1881.3–1840.0	Cheng et al. (2009)
Scenedesmus obliquus	Mixotrophic	0.10–0.51	6.6–11.8	11.6–58.6	Mandal and Mallick (2009)
Chlorella vulgaris #259	Mixotrophic	0.09–0.25	21–34	22.54	Liang et al. (2009)
Chlorella sorokiniana	—	—	56	—	Wan et al. (2012)
Botryococcus braunii UTEX 572	Phototrophic	0.03	20.8	5.5	Yoo et al. (2010)
Chlorella vulgaris CCAP 211/11B	Phototrophic	0.17	19.2	32.6	Rodolfi et al. (2009)
Chlorella vulgaris F&M-M49	Phototrophic	0.20	18.4	36.9	Rodolfi et al. (2009)
Chlorococcum sp. UMACC 112	Phototrophic	0.28	19.3	53.7	Rodolfi et al. (2009)

TABLE 8.4

Lipid Contents of EA903 and EA904 with Autotrophic, Heterotrophic, and Mixotrophic Cultivation

Strains		Culture Mode			
	Autotrophic	Heterotrophic (NaNO₃)	Heterotrophic (peptone)	Mixotrophic (NaNO₃)	Mixotrophic (peptone)
EA903	49.24 ± 1.2	40.17 ± 0.92	27.07 ± 0.52	42.35 ± 1.4	30.07 ± 1.62
EA904	45/9 ± 2.71	35.45 ± 1.14	27.22 ± 1.03	38.86 ± 2.04	38.11 ± 0.5

Source: Reprinted from *Algal Res.*, 24, Wang, H. et al., A comparative analysis of biomass and lipid content in five *Tribonema* sp. strains at autotrophic, heterotrophic and mixotrophic cultivation, 284–289, Copyright 2017, with permission from Elsevier.

Particularly, when flue gas from power plants is integrated to microalgae culture, a large amount of CO_2 and NOx need to be metabolized. Chen et al. (2018) and Zhang et al. (2014) recommend the use of photoautotrophic cultivation, as CO_2 and NOx fixation are light-dependent processes.

8.5 HARVESTING

Harvesting is a challenging process, as microalgae are too small for conventional filtration systems (unicellular eukaryotic algae 3–30 μm and cyanobacteria 0.2–2 μm). Additionally, cultures are relatively diluted; thus large amounts of water need to be removed. This makes the process time-consuming and expensive. Harvesting accounts for 20%–30% of the total cost of bio-oil production (Uduman et al. 2010). Thus, it is important to develop techniques that result in lower-cost systems that can lead to commercialization (Schlesinger et al. 2012).

Harvesting is comprised of several liquid-solid separation stages. These include the thickening of the solution to a slurry of about 2%–7% of total suspended solids (TSS) and a further dewatering to a cake of about 15%–25% TSS (Brenna and Owende 2010). They are comprised of chemical (coagulation/flocculation), biological (bioflocculation), and physical (gravity sedimentation, flotation, and electrical procedures) methods. Dewatering incorporates physical techniques, such as filtration and centrifugation (Barros et al. 2015).

The pertinence of the use of a specific method depends on several variables. The first is the desired product quality. Centrifuges, for example, are suitable to avoid contamination in food or aquaculture products, while gravity, enhanced by flocculation, is more suitable for biodiesel production (Vandamme et al. 2018). Another criterion is the adjustment of the necessary moisture level, as some oil extraction can use wet algae biomass with water contents of up to 95% (Ghasemi Naghdi et al. 2016). The type of algae is also critical for selection of the harvesting method. For example, filtration better suits microalgae of larger dimensions. Examples include *Coelastrum proboscideum*, *Scenedesmus platensi*, and *Arthros piraplatensis* (Rossi et al. 2008).

8.5.1 THICKENING

Coagulation/flocculation is widely used for thickening of the solution, as it allows for rapid treatment of large cultures. During the process, algae are aggregated into larger particles that form conglomerates, followed by agglomeration into large flocs that settle to the bottom (Gerde et al. 2014).

There are several approaches to induce flocculation/coagulation. Chemical approaches involve the use of salts, synthetic polyacrylamide polymers, and natural biopolymers. Of these approaches, using salts is commonly effective. For instance, Papazi et al. (2010) found that $Al_2(SO_4)_3$, $AlCl_3$, $Fe_2(SO_4)_3$, $FeCl_3$, $ZnSO_4$, and $ZnCl_2$ act as coagulants of *Chlorella minutissima* cultures. Specifically, sulphate and zinc salts presented better coagulation efficiencies, with values of 0.75 and 0.5 g L^{-1}. However, contamination with heavy metals can limit additional use of algae biomass. This is important for mitigation of the overall cost of bio-oil production (Wyatt et al. 2011).

Biopolymers are a safer alternative to synthetic polymers. Chitosan is an effective compound; however, its efficiency at high pH levels is countered by the low pH conditions normally present in algae cultures (Şirin et al. 2012). Alternatively, cationic starch allows these compounds to work within a broader pH range, due to the addition of quaternary ammonium groups to the starch (Vandamme et al. 2009). Cationic starch efficiency is limited to freshwater algae. Vandamme et al. (2010) found that a cationic starch–to-algal biomass ratio of 0.1 and 0.03 to flocculate 80% of the biomass of the freshwater species *Parachlorella kessleri* and *Scenedesmus obliquus*. On the contrary, the ratio for the marine species *Nannochloropsis salina* and *Phaeodactylum tricornotum* was 1.0.

Autoflocculation is another approach of flocculation/coagulation. This is the spontaneous precipitation of algae at pH values above 9. Carbonate phosphate and calcium carbonate have been successfully used. An excess of positive ions neutralizes the negative charge of microalgae cells (Vandamme et al. 2010). Wastewater, brackish water, or seawater can be used as a source of phosphates, carbonates, calcium, and magnesium ions, as these compounds can normally be expensive. However, the concentrations are variable, which does not always guarantee the concentration required for optimum results (Smith et al. 2012).

Recent studies have explored more viable physical coagulation/flocculation methods using electromagnetic nanoparticles (Vandamme et al. 2018). These methods avoid common issues of other physical separation techniques, such as scalability, biomass contamination, and cost. During magnetic separation, magnetite (Fe_2O_3) nanoparticles are attached to the algae cells, and later, the formed conglomerates are separated from the medium when a magnetic field is applied. Cerf et al. (2012) reported high separation efficiencies in freshwater (*Chlamydomonas reinhardtii* and

Chlorella vulgaris) and marine algae (*Phaeodactylum tricornutum* and *Nannochloropsis salina*) using magnetite (>95%) in only five minutes in a high-gradient magnetic filter system.

Bioflocculation is an inexpensive and cleaner way for algae harvesting. It occurs spontaneously, possibly because of the production of chemicals present in some algae and fungi (Zhou et al. 2012). Algae with poor or null autoflocculate capacities are frequently mixed with other known algae and fungi species with autoflocculate capacity (Vandamme et al. 2013). The high concentrations of carbon in wastewater treatment confer a great deal of potential for this harvesting method (Zhou et al. 2012).

Flocculation/coagulation normally precedes other thickening methods in order to improve harvesting performance. Chemical flocculation is recommended before applying the flotation method. During flotation, colloidal particles are lifted by the generation of microbubbles that attach to hydrophobic portions of algae (Barros et al. 2015). The formation of these hydrophobic surfaces is promoted by chemical flocculation. Hanotu et al. (2012) found that the chemical flocculation (aluminum sulfate, ferric III chloride, and ferric sulphate) and algal recovery improves at low pH at 5, and with an increasing coagulant dose (95% recovery at 150 mg L^{-1} with aluminum sulphate and 92.7% recovery at 50, 75, and 100 mg L^{-1} for ferric sulphate).

Flotation presents various advantages, such as scalability, low space requirements, short operating times, flexibility and low initial equipment costs (Hanotu et al. 2012, Barros et al. 2015). However, when the final product has low value, as in biofuel, gravity sedimentation is recommended. Different coagulation/flocculation methods have been successfully applied to speed settling times and increase recovery efficiencies (80%–99%) (Papazi et al. 2010, Şirin et al. 2012, Rashid et al. 2013, Xu et al. 2013). In particular, Chatsungnoen and Chisti (2016) evaluated the recovery by flocculation-sedimentation using aluminum sulphate and ferric chloride as flocculants in five algae species of different sizes, morphology, and ionic strength (*Chlorella vulgaris, Choricystis minor, Cylindrotheca fusiformis, Neochloris* sp., *Nannochloropsis salina*). Although the recovery was effective in all species, the minimum flocculant dose required to reach a 95% removal depended mainly on the cell size of the algae, followed by the initial concentration of biomass and the ionic strength of the culture medium. Specifically, the flocculant dose increased linearly with the biomass concentration. Likewise, aluminum sulphate had an increased effectiveness and reduced cost due to its high surface charge density of the Al^{3+} ion compared to Fe^{3+}.

8.5.2 Dewatering

Dewatering facilitates posterior drying and bio-oil processing. As mentioned above, filtration and centrifugation are the current mechanical methods that are employed. During filtration, microalgae deposits in the filtration membranes. The resistance increases throughout the process, raising operational maintenance and costs. Filtration is sustainable for filamentous algae or for those forming colonies. The high coating rate, membrane replacement, and general maintenance make filtration economically viable only for volumes lower than 2 $m^3 d^{-1}$. On the contrary, centrifugation can be more cost-effective at larger scales (more than 20 $m^3 d^{-1}$) (Rossignol et al. 2009).

The harvesting of biomass by centrifugation utilizes centrifugal forces to enhance the rate of sedimentation. The main advantages of centrifugation are high recovery rates, predictability and ease of concentration, adaptability to all alga strains' absence of chemicals, and speed to achieve results (Gerardo et al. 2015). However, it is expensive and energy consuming, and has low flow rates. To minimize cost, one common practice is the pre-concentration of algal slurry using thickening methods, such as coagulation/flocculation. This reduces the algae volume by up to 65% during centrifugation (Schlesinger et al. 2012). For instance, Collotta et al. (2017) performed a lifecycle analysis (LCA) of *Chlorella vulgaris* and found that, when only centrifugation is used for harvesting, the impact on different environmental aspects is considerably higher than the scenario in which flocculant/coagulant precedes centrifugation. However, Igou et al. (2012) found that centrifugation is the best alternative to water recycling, as it reduces or removes inhibitors, such as fungus and bacteria. This is especially important with integration of wastewater treatment and algae cultures where it was possible to recycle the used water up to 10 times. This resulted in an average rate of 82%–84% of water recovery per growth cycle. Yang et al. (2011) also showed that the water footprint could be reduced from 3,279 L/L to 520 L/L of biodiesel with water recycling.

Likewise, Dassey and Theegala (2013) found that capture efficiencies lower than 90% can be counteracted by increasing flow rates (>1 L/min) as larger volumes of culture water are managed. For instance, they showed a reduction in energy consumption by 82% when the algal biomass was harvested at 18 L/min. When applying these changes to algae with high lipid content and high culture density, harvesting cost can drop from $4.52/L oil when using a low flow rate to $0.864/L oil when using a high flow rate.

8.6   OIL EXTRACTION

The extraction of lipids from algae for biodiesel production involves applications of organic solvents (e.g., chloroform:methanol and hexane), or by the pressing of dry algae, followed by transesterification reactions (Balasubramanian et al. 2011). Although high lipid yields can be obtained, this method has drawbacks in terms of the costs and impacts on the environment. For instance, 60% of minerals from algae solid residue turn into waste, with only 10% of this being recyclable (Umdu et al. 2009, Babich et al. 2011). There are currently several pre-treatment methods, which aim to break the cell wall and membrane barriers of microalgae cells to enable lipid release. Solvent-based techniques are still used in conjunction with these pre-treatments; however, less amounts of chemicals are used when compared with untreated biomass (Lee et al. 2010).

The lipid extraction from algae can follow a wet or a dry route. The wet route has the advantage of suppressing the dewatering and drying process steps, which could reduce the final product cost (Xu et al. 2011a). It comprises mechanical, biological, and chemical approaches, such as microwave-assisted extraction, ultrasound-assisted extraction, hydrothermal liquefaction, osmotic shock, enzymatic disruption, oxidative stress, electroporation, and supercritical carbon dioxide extraction (Ghasemi Naghdi 2016) (Table 8.5).

Pyrolysis is a method that follows the dry route of thermal extraction of bio-oils from algae, and pyrolysis presents several attributes that make it attractive. The first is the conversion of not only the lipid from cells, but also the entire cell composition to produce energy. This is because pyrolysis involves high temperatures to decompose biomass. Pyrolysis also produces two co-products, biochar and biogas, that can be used as a fertilizer and as a carbon sink, and for further energy production. Additionally, the biochar has high heating values and lipid yields (Brennan & Owende 2010, Chiaramonti et al. 2017) (Table 8.4).

HTL is also an attractive technology, despite its high cost. Similar to pyrolysis, HTL also decomposes all components of cells to produce energy. The decomposition of all components results in contamination with nitrogen and oxygenates of the final product, as well as emission of considerable amounts of NOx. The quality of the oil can be improved by using a catalyst, as they can extend conversion time periods (Babich et al. 2011, Du et al. 2013). For example, Thangalazhy-Gopakumar (2012) found that zeolite HZSM-5 increases the carbon yield of aromatic hydrocarbons by a factor of 25 (from 0.9% to 25.8%) in *Chlorella vulgaris* when comparing catalytic to non-catalytic pyrolysis.

Microwave-assisted extraction is a promising conversion technology for its safety, short reaction times, high oil yields, and high-quality lipids (Balasubramanian et al. 2011). High lipid yields using 10% NaCl were reported by Lee et al. (2010) for *Scenedesmus* spp., *Botryococcus* sp., and *Chorella vulgaris* when comparing with other pre-treatment methods, including bead beating, sonication, autoclaving, and osmotic shock. Although it is recommended to identify technologies that can be easily applied to all algae strains, osmotic shock, which is considered species specific, could work successfully in integration with flue gas from power plants, where only a few strains are able to tolerate the specific conditions.

TABLE 8.5

Comparison of the Performance of Oil Extraction Methods

	Lipid Productivity (mg L⁻¹ d⁻¹)	Relative Oil Extraction Efficiency %	Extractable Yield (% per dry mass)	Solvent	Species	Author
Microwave-assisted	—	77.11 ± 5.06	62.04 ± 2.42	Ethanol	*Scenedesmus obliquus*	Balasubramanian et al. (2011)
			40.71 ± 4.46	Hexane		
	10.2	—	30	Chloroform and methanol	*Botryococcus* sp.	Lee et al. (2010)
	7.4	—	10	Chloroform and methanol	*Scenedesmus* sp.	
	7.4	—	10	Chloroform and methanol	*Chlorella vulgaris*	
Osmotic shock	—	—	10	50–100 g/L of NaCl with 30% methanol (v/v)	*Botryococcus* sp.	
	—	—	7	50–100 g/L of NaCl with 30% methanol (v/v)	*Scenedesmus* sp.	
	—	—	8	50–100 g/L of NaCl with 30% methanol (v/v)	*Chlorella vulgaris*	
Enzymatic disruption	—	56%	—	Cellulose	*Chlorella* sp.	Fu et al. (2010)
	—	24%	—	Cellulose	*Chlorella vulgaris*	Zheng et al. (2011)
Oxidative stress	—	—	24	—	*Tetraselmis striata* M8	Bai et al. (2014)
Supercritical carbon dioxide extraction	—	—	7.1	n-hexane	*Chlorococcum* sp.	Halim et al. (2011)
	—	—	25	Hexane	*Nannochloropsis* sp.	Andrich et al. (2005);
Hydrothermal liquefaction	—	—	29	Dichloromethane	*Chlorella vulgaris*	Hu et al. (2017)
	—	—	33.2	Dichloromethane	*Spirulina* sp.	Vardon et al. (2011)
	—	—	38	Cyclohexane	*Nannochloropsis* sp.	Patel and Hellgardt (2015)
Pyrolysis	—	—	55	Acetone	*Scenedesmus* sp.	Harman-Ware et al. (2013).
	—	—	40	Methanol: chloroform	*Chlorella* sp.	Babich et al. (2011)
	—	—	16.8	—	*Chlorella vulgaris*	Du et al. (2013)

8.7 INTEGRATION OF ALGAE CULTIVATION WITH WASTEWATER TREATMENT AND POWER STATION FLUE GAS

Commercialization of bio-oil from algae is limited due to the high costs for its production. Several studies aim to improve efficiency, scalability, processing times, natural resources, and related materials (Chen et al. 2018). Reduction of the cost was proposed by integration of microalgae cultivation with wastewater treatment and flue gas from power plants to nourish the microalgae with nutrients. Besides improving productivity of cultures, this approach provides an alternative to reduce the cost of wastewater and air pollutant treatments (Giostri et al. 2016). In particular, the elevated amount of NOx and CO_2 produced by power plants results in environmental impacts to climates, biodiversity, and human health. Current technologies to remove CO_2, such as cryogenic distillation, membrane separation, and chemical and physical absorption, are expensive and consume large amounts of energy, as they require CO_2 disposal. Likewise, NOx removal treatments considerably increase costs and release additional wastes that require further treatment (Chen et al. 2018).

NOx and CO_2 in power plants are produced when nitrogen (N) and carbon (C) are oxidized during biomass and coal combustion. Other traces of elements (e.g., Mg, K, Ca, P, and Fe) are available in the ash and could also be used as a fertilizer during microalgae cultivation (Chen et al. 2018). Constant mixing and extensive land reserves are required for open pond cultivation systems to ensure contact of algae with the CO_2 in the air. With the influx of CO_2 from power plants, the cost of mixing and land acquisition declines, while there is an increase in biomass growth. For instance, Zhang et al. (2014) found that cultures fed with CO_2 from flue gas reach a biomass plateau growth in half of the time (seven days) when compared to those without CO_2 injection (14 days).

Microalgae cultivation could be used as a carbon sink for CO_2 produced by power plants due to their greater capacity to fix carbon. The rate is approximately 10–50 times greater than that of plants. For instance, 180 tonnes of CO_2 is fixed to produce 100 tonnes of microalgal biomass. Chen et al. (2018) demonstrated successfully that *Chlorella* sp. C2 is an excellent alga strain for bioremediation, with an ash denutrition rate of 13.33 g L^{-1} d^{-1}, a NOx reduction (DeNOx) efficiency of ~ 100%, and a CO_2 sequestration rate of 0.46 g L^{-1} d^{-1}. *Chlorella* sp. C2 also benefits from injection of nutrients from the flue gas and ash of a power plant. There was an increase of 39% of lipid productivity at 99.11 mg L^{-1} d^{-1} and 35% of biomass productivity at 0.31 g L^{-1} d^{-1} when biomass power plant ash and flue gas are used as a nutrient source for microalgae cultivation.

Several studies demonstrated that culture conditions, such as nitrogen starvation, promote lipid content, accompanied by a decrease in biomass as protein production for cell growth is inhibited (Rodolfi et al. 2009; Zhu et al. 2016). However, Chen et al. (2018) found that, under supplementary CO_2, an increase in both biomass and lipid is possible, as carbon seems to be fixed during photosynthesis and channeled into the biolipid synthesis pathway in *Chlorella* sp. C2. Under another study, Zhang et al. (2014) reported neutral accumulation of lipids without compromising biomass increase. Meanwhile, Zhu et al. (2016) found that adding trace amounts of urea (18 mg dl ll) in microalgae in *Chlorella* sp. A2 culture under laboratory conditions can produce 416% (w/w) more neutral lipid than when cells are cultured with nitrogen starvation media. In outdoor environments, they found that the lipid productivity of cells is 88% higher than those with nitrogen starvation media.

One limitation of supplementary CO_2 into cultures is acidification. This may result in inhibited cell growth. However, a bicarbonate and phosphate buffer system is a viable option to maintain an ideal pH without compromising cost. For instance, under autotrophic outdoor conditions, Choi et al. (2017) found that the buffer enhances biomass (105%) and astaxanthin (103%) in *Haematococcus pluvialis*. From a mixed culture of oleaginous microalgae, Aslam et al. (2018) found that a phosphate buffer reduces fatty acid methyl esters yield, but instead, it increases biodiesel quality at high CO_2 concentration (5.5%) with saturated fatty acids at 36.28% and unsaturated fatty acids at 63.72%.

8.8 CONCLUSIONS

There is currently a large amount of effort being invested in alga biodiesel in order to meet future demands while reducing the impact to the environment. Although the cost, scalability, and purity of the derived biodiesel still has limitations, a huge effort is taking place on behalf of institutes, industries, and universities to improve the methods and technologies in order to move from laboratory-scale to commercial-scale biodiesel production (Mata et al. 2010, Li et al. 2012, Gonzales 2018).

Expenses can be reduced when cultured microalgae is combined with flue gas from power plants. The mitigation of CO_2 emissions is added to the final value of algae biodiesel. New findings of algae adapted to high CO_2 concentrations and with high biomass and lipid productivity are promising (Neofotis et al. 2016). Additionally, alternative harvesting methods may reduce the overall cost (e.g., combination of flocculation/sedimentation before centrifugation) (Vandamme et al. 2013). Harvesting issues may be solved if the selected oil extraction methods follow a wet route. For example, HTL can be used on wet algal biomass with water content as high as 80%–95% (Ghasemi Naghdi et al. 2016). Mechanical oil extraction methods also provide high lipid yields, together with the reuse or recycling of nutrients, water, or co-products (Babich et al. 2011).

REFERENCES

Andrich, G, Nesti, U, Venturi, F, Zinnai, A & Fiorentini. R 2005, Supercritical fluid extraction of bioactive lipids from the microalgae nannochloropsis sp., *European Journal of Lipid and Science Technology*, vol. 107, pp. 381–386.

Aslam, A, Thomas-Hall, S, Manzoor, M, Jabeen, F, Iqbal, M, uz Zaman, Q, Schenk, P & Asif Tahir, M 2018, Mixed microalgae consortia growth under higher concentration of CO2 from unfiltered coal fired flue gas: Fatty acid profiling and biodiesel production, *Journal of Photochemistry and Photobiology B: Biology*, vol. 179, pp. 126–133.

Aslam, A, Thomas-Hall, S, Mughal, T & Schenk, P 2017, Selection and adaptation of microalgae to growth in 100% unfiltered coal-fired flue gas, *Bioresource Technology*, vol. 233, pp. 271–283.

Babich, I, van der Hulst, M, Lefferts, L, Moulijn, J, O'Connor, P & Seshan, K 2011, Catalytic pyrolysis of microalgae to high-quality liquid bio-fuels, *Biomass and Bioenergy*, vol. 35, no. 7, pp. 3199–3207.

Bai, X, Ghasemi-Naghdi, F, Ye, L, Lant, P & Pratt, S 2014, Enhanced lipid extraction from algae using free nitrous acid pretreatment, *Bioresource Technology*, vol. 159, pp. 36–40.

Balasubramanian, S, Allen, J, Kanitkar, A, & Boldor, D 2011, Oil extraction from Scenedesmus obliquus using a continuous microwave system–design, optimization, and quality characterization, *Bioresource Technology*, vol. 102, pp. 3396–3403.

Barros, A, Gonçalves, A, Simões, M & Pires, J 2015, Harvesting techniques applied to microalgae: A review *Renewable and Sustainable Energy Reviews*, vol. 41, pp. 1489–1500.

Boruff, B, Moheimani, N & Borowitzka, M 2015, Identifying locations for large-scale microalgae cultivation in Western Australia: A GIS approach, *Applied Energy*, vol. 149, pp. 379–391.

Bravo-Fritz, C, Sáez-Navarrete, C, Herrera Zeppelin, L & Ginocchio Cea, R 2015, Site selection for microalgae farming on an industrial scale in Chile, *Algal Research*, vol. 11, pp. 343–349.

Brennan, L & Owende, P 2010, Biofuels from microalgae A review of technologies for production, processing, and extractions of biofuels and co-products, *Renewable and Sustainable Energy Reviews*, vol. 14, no. 2, pp. 557–577.

Cerf, M, Morweiser, M, Dillschneider, R, Michel, A, Menzel, K & Posten, C 2012, Harvesting fresh water and marine algae by magnetic separation: Screening of separation parameters and high gradient magnetic filtration, *Bioresource Technology*, vol. 118, pp. 289–295.

Chatsungnoen, T & Chisti, Y 2016, Harvesting microalgae by flocculation–sedimentation *Algal Research*, vol. 13, pp. 271–283.

Chen, M, Tang, H, Ma, H, Holland TC, Ng, KY, Salley & SO 2011, Effect of nutrients on growth and lipid accumulation in the green algae Dunaliella tertiolecta, *Bioresource Technology*, vol. 102, pp. 1649–1655.

Chen, H, Wang, J, Zheng, Y, Zhan, J, He, C & Wang, Q 2018, Algal biofuel production coupled bioremediation of biomass power plant wastes based on chlorella sp. C2 cultivation, *Applied Energy*, vol. 211, pp. 296–305.

Cheng, Y. Zhou, W, Gao, C, Lan, K, Gao, Y & Wu, Q 2009, Biodiesel production from Jerusalem artichoke (Helianthus tuberosus L.) tuber by heterotrophic microalgae Chlorella protothecoides, *Journal of Chemical Technology & Biotechnology*, vol. 84, pp. 777–781.

Chiaramonti, D, Prussi, M, Buffi, M, Rizzo, A & Pari, L 2017, Review and experimental study on pyrolysis and hydrothermal liquefaction of microalgae for biofuel production, *Applied Energy*, vol. 185, pp. 963–972.

Choi, Y, Joun, J, Lee, J, Hong, Pham, H, Chang, W & Sim, S 2017, Development of large-scale and economic pH control system for outdoor cultivation of microalgae *Haematococcus pluvialis* using industrial flue gas, *Bioresource Technology*, vol. 244, pp. 1235–1244.

Collotta, M, Champagne, P, Mabee, W, Tomasoni, G, Leite, G, Busi, L & Alberti, M 2017, Comparative LCA of flocculation for the harvesting of microalgae for biofuels production *Procedia CIRP*, vol. 61, pp. 756–760.

Dassey, AJ & Theegala CS 2013, Harvesting economics and strategies using centrifugation for cost effective separation of microalgae cells for biodiesel applications, *Bioresource Technology*, vol. 128, pp. 241–245.

Du, Z, Hu, B, Ma, X, Cheng, Y, Liu, Y, Lin, X, Wan, Y, Lei, H, Chen, P & Ruan, R 2013, Catalytic pyrolysis of microalgae and their three major components: Carbohydrates, proteins, and lipids, *Bioresource Technology*, vol. 130, pp. 777–782.

Fu, C, Hung, T, Chen, J, Su, C & Wu, W 2010, Hydrolysis of microalgae cell walls for production of reducing sugar and lipid extraction, *Bioresource Technology*, vol. 101, pp. 8750–8754.

Gerardo, M, Van Den Hende, S, Vervaeren, H, Coward, T & Skill, S 2015, Harvesting of microalgae within a biorefinery approach: A review of the developments and case studies from pilot-plants *Algal Research*, vol. 11, pp.248–262.

Gerde, J, Yao, L, Lio, J, Wen, Z & Wang, T 2014, Microalgae flocculation: Impact of flocculant type, algae species and cell concentration. *Algal Research*, 3, pp. 30–35.

Ghasemi Naghdi, F, González, L, Chan, W and Schenk, P 2016, Progress on lipid extraction from wet algal biomass for biodiesel production, *Microbial Biotechnology*, vol. 9, no. 6, pp. 718–726.

Giostri, A, Binotti, M & Macchi, E 2016, Microalgae cofiring in coal power plants: Innovative system layout and energy analysis, *Renewable Energy*, vol. 95, pp. 449–464.

González-González, L, Correa, D, Ryan, S, Jensen, P, Pratt, S & Schenk, P 2018, Integrated biodiesel and biogas production from microalgae: Towards a sustainable closed loop through nutrient recycling, *Renewable and Sustainable Energy Reviews*, vol. 82, pp. 1137–1148.

Halim, R, Gladman, B, Danquah, MK & Webley, PA 2011, Oil extraction from microalgae for biodiesel production, *Bioresource Technology*, vol. 102, no. 1, pp. 178–85.

Hanotu, J, Bandulasena, H & Zimmerman, W 2012, Microflotation performance for algal separation, *Biotechnology and Bioengineering*, vol. 109, no. 7, pp. 1663–1673.

Harman-Ware, A, Morgan, T, Wilson, M, Crocker, M, Zhang, J, Liu, K. Stork, J & Debolt, S 2013, Microalgae as a renewable fuel source: Fast pyrolysis of scenedesmus sp., *Renewable Energy*, vol. 60, pp. 625–632.

Hu, Y, Gong, M, Xu, C & Bassi, A 2017 Investigation of an alternative cell disruption approach for improving hydrothermal liquefaction of microalgae, *Fuel*, vol. 197, pp. 138–144.

Lee, J, Yoo, C, Jun, S, Ahn, C & Oh, H 2010, Comparison of several methods for effective lipid extraction from microalgae, *Bioresource Technology*, vol. 101, no. 1, pp. S75–S77.

Li, Y, Moheimani, N & Schenk, P 2012, Current research and perspectives of microalgal biofuels in Australia, *Biofuels*, vol. 3, no. 4, pp. 427–439.

Liang, YN, Sarkany, N, Cui, Y 2009, Biomass and lipid productivities of chlorella vulgaris under autotrophic, heterotrophic and mixotrophic growth conditions, *Biotechnology Letters*, vol. 31, pp. 1043–1049.

Mandal, S & Mallick, N 2009, Microalga Scenedesmus obliquus as a potential source for biodiesel production, *Applied Microbiology and Biotechnology*, vol. 84, pp. 281–291.

Mata, T, Martins, A & Caetano, N 2010, Microalgae for biodiesel production and other applications: A review, *Renewable and Sustainable Energy Reviews*, vol. 14, no.1, pp. 217–232.

Neofotis, P, Huang, A, Sury, K, Chang, W, Joseph, F, Gabr, A, Twary, S, Qiu, W, Holguin, O & Polle, J 2016, Characterization and classification of highly productive microalgae strains discovered for biofuel and bioproduct generation, *Algal Research*, vol. 15, pp. 164–178.

Papazi, A, Makridis, P & Divanach, P 2010, Harvesting chlorella minutissima using cell coagulants, *Journal of Applied Phycology*, vol. 22, no. 3, pp. 349–355.

Patel, B & Hellgardt, K 2015, Hydrothermal upgrading of algae paste in a continuous flow reactor *Bioresource Technology*, vol. 191, pp. 460–468.

Rashid, N, Rehman, MSU & Han, J-I, 2013, Use of chitosan acid solutions to improve separation efficiency for harvesting of the microalga Chlorella vulgaris, *Chemical Engineering Journal*, vol. 226, pp. 238–242.

Rodolfi, R, Chini Zittelli, G, Bassi, N, Padovani, G, Biondi, N, Bonini, G & Tredici, M 2009, Microalgae for oil: Strain selection, induction of lipid synthesis and outdoor mass cultivation in a low-cost photobioreactor, *Biotechnology and Bioengineering*, vol. 102, no. 1, pp. 100–112.

Rossi, N, Derouiniotchaplain, M, Jaouen, P, Legentilhomme, P & Petit, I 2008, Arthrospira platensis harvesting with membranes: Fouling phenomenon with limiting and critical flux, *Bioresource Technology*, vol. 99, no. 14, pp. 6162–6167.

Rossignol, N, Vandanjon, L, Jaouen, P, Quéméneur, F 2009, Membrane technology for the continuous separation microalgae/culture medium: Compared performances of cross-flow microfiltration and ultrafiltration, *Aquacultural Engineering,* vol. 20, pp. 191–208.

Schlesinger, A, Eisenstadt, D, Bar-Gil, A, Carmely, H, Einbinder, S & Gressel, J. 2012, Inexpensive non-toxic flocculation of microalgae contradicts theories; overcoming a major hurdle to bulk algal production, *Biotechnology Advances,* vol. 30, no. 5, pp. 1023–1030.

Şirin, S, Trobajo, R, Ibanez, C & Salvadó, J 2012, Harvesting the microalgae phaeodactylum tricornutum with polyaluminum chloride, aluminium sulphate, chitosan and alkalinity-induced flocculation, *Journal of Applied Phycology,* vol. 24, no. 5, pp. 1067–1080.

Smith, BT & Davis, RH 2012, Sedimentation of algae flocculated using naturally- available, magnesium-based flocculants, *Algal Research,* vol. 1, pp. 32–39.

Tale, M, Ghosh, S, Kapadnis, B & Kale, S 2014, Isolation and characterization of microalgae for biodiesel production from Nisargruna biogas plant effluent, *Bioresource Technology,* vol. 169, pp. 328–335.

Thangalazhy-Gopakumar, S, Adhikari, S, Chattanathan, SA & Gupta, RB 2012, Catalytic pyrolysis of green algae for hydrocarbon production using H+ZSM-5 catalyst *Bioresource Technology,* vol. 118, pp. 150–157.

Uduman, N, Qi, Y, Danquah, M, Forde, G & Hoadley, A 2010, Dewatering of microalgal cultures: A major bottleneck to algae-based fuels, *Journal of Renewable and Sustainable Energy,* vol. 2, no. 1, pp. 012701.

Umdu, E, Tuncer, M & Seker, E 2009, Transesterification of Nannochloropsis oculata microalga's lipid to biodiesel on Al_2O_3 supported CaO and MgO catalysts, *Bioresource Technology,* vol. 100, pp. 2828–2831.

Vandamme, D, Foubert, I, Fraeye, I, Meesschaert, B & Muylaert, K 2010, Flocculation of Chlorella vulgaris induced by high pH: Role of magnesium and calcium and practical implications, *Bioresource Technology,* vol. 105, pp. 114–119.

Vandamme, D, Foubert, I, Meesschaert, B & Muylaert, K 2009, Flocculation of microalgae using cationic starch *Journal of Applied Phycology,* vol. 22, no. 4, pp. 525–530.

Vandamme, D, Foubert, I & Muylaert, K 2013, Flocculation as a low-cost method for harvesting microalgae for bulk biomass production, *Trends in Biotechnology,* vol. 31, no. 4, pp. 233–239.

Vandamme, D, Gheysen, L, Muylaert, K & Foubert, I 2018, Impact of harvesting method on total lipid content and extraction efficiency for phaeodactylum tricornutum, *Separation and Purification Technology,* vol. 194, pp. 362–367.

Vardon, D, Sharma, B, Scott, J, Yu, G. Wang, Z, Schideman, L, Zhang, Y & Strathmann, T 2011, Chemical properties of biocrude oil from the hydrothermal liquefaction of spirulina algae, swine manure, and digested anaerobic sludge *Bioresource Technology,* vol. 102, no. 17, pp. 8295–8303.

Wan, M, Wang, R, Xia, J, Rosenberg, J, Nie, Z, Kobayashi, N, Oyler, G & Betenbaugh, M 2012, Physiological evaluation of a new chlorella sorokiniana isolate for its biomass production and lipid accumulation in photoautotrophic and heterotrophic cultures *Biotechnology and Bioengineering,* vol. 10, no. 8, pp. 1958–1964.

Wang, H, Zhou, W, Shao, H & Liu, T 2017, A comparative analysis of biomass and lipid content in five Tribonema sp. strains at autotrophic, heterotrophic and mixotrophic cultivation, *Algal Research,* vol. 24, pp. 284–289.

Wyatt, N, Gloe, L, Brady, P, Hewson, J, Grillet, A, Hankins, M & Pohl, P 2011, Critical conditions for ferric chloride-induced flocculation of freshwater algae, *Biotechnology and Bioengineering,* vol. 109, no. 2, pp. 493–501.

Xu, L, Brilman, D, Withag, J, Brem, G & Kersten, S 2011b, Assessment of a dry and a wet route for the production of biofuels from microalgae: Energy balance analysis, *Bioresource Technology,* vol. 102, no. 8, pp. 5113–5122

Xu, L, Guo, C, Wang, F, Zheng, S & Liu, C 2011a, A simple and rapid harvesting method for microalgae by in situ magnetic separation *Bioresource Technology,* vol. 102, no. 21, pp. 10047–10051.

Xu, Y, Purton, S & Baganz, F 2013, Chitosan flocculation to aid the harvesting of the microalga chlorella sorokiniana *Bioresource Technology,* vol. 129, pp. 296–301.

Yang, J, Xu, M, Zhang, X, Hu, Q, Sommerfeld, M & Chen, Y 2011, Lifecycle analysis on biodiesel production from microalgae: Water footprint and nutrients balance, *Bioresource Technology,* vol. 102, pp. 159–165.

Yoo, C, Jun, S-Y, Lee, J-Y, Ahn, C-H & O, H-M 2010, Selection of microalgae for lipid production under high levels carbon dioxide, *Bioresource Technology,* vol. 101, no. 1, pp. S71–S74.

Zhang, X, Chen, H, Chen, W, Qiao, Y, He, C & Wang, Q 2014, Evaluation of an oil-producing green alga chlorella sp. C2 for biological DeNOx of industrial flue gases, *Environmental Science & Technology,* vol. 48 no.17, pp. 10497–10504.

Zheng, H, Yin, J, Gao, Z, Huang, H, Ji, X, & Dou, C 2011, Disruption of Chlorella vulgaris cells for the release of biodiesel-producing lipids: A comparison of grinding, ultrasonication, bead milling, enzymatic lysis, and micro- waves, *Apply Biochemistry Biotechnology*, vol. 164, pp 1215–1224.

Zhou, W, Cheng, Y, Li, Y, Wan, Y, Liu, Y, Lin, X & Ruan, R 2012, Novel Fungal Pelletization-Assisted Technology for Algae Harvesting and Wastewater Treatment *Applied Biochemistry and Biotechnology*, vol. 167, no. 2, pp. 214–228.

Zhou, W, Wang, H, Chen, L, Cheng, W & Liu, T 2017, Heterotrophy of filamentous oleaginous microalgae Tribonema minus for potential production of lipid and palmitoleic acid *Bioresource Technology*, vol. 239, pp. 250–257.

Zhu, J, Chen, W, Chen, H, Zhang, X, He, C, Rong, J & Wang, Q 2016, Improved productivity of neutral lipids in *chlorella* sp. A2 by minimal nitrogen supply, *Frontiers in Microbiology*, vol. 7.

9 Properties, Applications, and Prospects of Carbonaceous Biomass Post-processing Residues

Suraj Adebayo Opatokun, Vladimir Strezov, and Hossain M. Anawar

CONTENTS

9.1 INTRODUCTION

Substantial quantities of biomass and organic wastes are redirected from landfills and other "end-of-pipe" treatments to usher in integrated solid waste management (ISWM) systems, primarily for environmental protection and recycling. Through ISWM systems, waste avoidance and minimization are given priority, followed by the three Rs (reduce, reuse, and recycle [3Rs]) to ensure efficiency in the appropriation of these outplaced resources, termed *waste*. This widely adopted strategy provides tangible throughput across the sustainability scorecards (ecology, economic, and social). This initiative thus not only attracts standards and regulations, but also makes evident the role of process assessment and optimization in biomass management. Organically rich wastes are now considered resources for the production of renewable energy, biogas, biofertilizers, compost/soil amendments, and, recently, have been considered as liquid and solid fuel sources (Ni et al., 2006). This quantum leap provides insight into smart utilization of scarce resources, while environmental impacts are significantly ameliorated. For instance, the global annual 11.2 billion tonnes of solid wastes require attention to avert approximately 5% of the greenhouse gas (GHG) emissions (UNEP, 2011) that are generated due to unstructured decay of the waste's organic constituents.

Sources of these wastes include agriculture (Holm-Nielsen et al., 2009) (animal dumps, plant and animal remains), municipal solid waste (food waste, yard trimmings), and commercial and industrial sectors (food and pharmaceutical industries, wholesale and retail food distribution, sales outfits, and hospitality and catering industries) (Brown & Li, 2012; Macias-Corral et al., 2008). The nascent biochemical conversion processes (fermentation, aerobic and anaerobic digestion), thermochemical conversion processes (biochar, charcoal, bio-oil), and physicochemical (esterification, extraction, and/or separation) treatment processes (Appels et al., 2011) offer varying advantages and disadvantages, with waste composition and treatment goals as pre-requisite for process adoption. These treatment processes equally produce residues, most of which pose greater or equivalent challenges for management (Alburquerque et al., 2012; Appels et al., 2011). Thus, the environmental issues, constraints, and risks of these residues are associated with their environmental safety; lack of standards and assessment framework for their management, such as treatment (if further treatments are necessary) and disposal procedures or policies; and application directions, including their sustainable use as fertilizer substitutes.

Biomass-related energy makes more than 14% of world's final energy consumption (Parikka, 2004). The choice and effect of biomass treatment processes are critical to unlock about 470 KJ of energy captured through every gram mole carbon fixed during photosynthesis (Klass, 2004; Zheng et al., 2010). Biomass-related energy, unlike solar, wind, and geothermal energy, remains the most used form of renewable energy, partly due to its abundance and provision of material feedstock alongside energy. Thus, harnessing the cyclic and biospheric carbon from these organic wastes requires adequate characterization and careful selection of treatment technique to ensure equilibrium in the socioeconomic and environmental stance of the waste system (Grierson, 2012). Treatment process design is pivoted often on the quantity of biomass, desired form of product (such as energy, chemicals), environmental standards, and economic viability, especially at commercial scale (McKendry, 2002).

This chapter aims to review the properties and uses of the carbonaceous solid products of biomass processing (digestates, compost, fermentation residue, and biochar) from various organic-oriented treatment processes with the view to: (1) expose or reflect the potential resourcefulness of biomass post-processing products beyond their current deployment as fertilizers or soil amendments and (2) present the implication or effects of treatment processes on the resultant product characteristics and applications.

9.2   PROPERTIES AND USES OF THE BIOMASS POST-PROCESSING RESIDUES

Organic waste substrate's degradation path can broadly be classified into three categories, as indicated in Figure 9.1. The physicochemical processes explore extraction techniques to generate target products; thermochemical processes reflect temperature treatment on substrates in the absence or presence of oxygen (Wang et al., 2012), while biochemical processes express microbial processing of substrates (Li et al., 2011). The latter process is often characterized by solubilization of substrates through a hydrolytic stage, wherein a substrate's macromolecules are mineralized through enzymatic activities into monomers and, consequently, gases and other products (Figure 9.2).

Biodegradation of substrates is mostly deployed due to its environmental and cost benefits, as indicated in Clarke (2002), wherein operations and cost effectiveness of implementing various in-vessel anaerobic digestions in the United Kingdom and United States are reviewed.

Waste treatment residues as solids (digestates with total solid (TS) >15%), liquids (digestates with TS range of 0.5%–15%) (Li et al., 2011) and gaseous products, as shown in Figure 9.1, is a function of the inputs and processing technique, coupled with other key factors. Subsequent products of the digestion process are channelled towards either aerobic or anaerobic pathways, which equally determine the microbial constituents, performance, and output features.

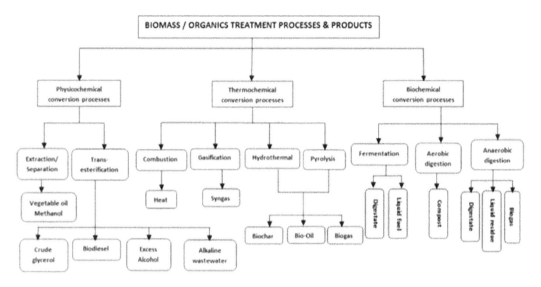

FIGURE 9.1   Substrate degradation processes and products. (Adapted and modified from Appels, L. et al., *Renew. Sust. Energy Rev.*, 15, 4295–4301, 2011; Fukuda, H. et al., *J. Biosci. Bioeng.*, 92, 405–416, 2001; Santibáñez, C. et al., *Chil. J. Agric. Res.*, 71, 3, 2011.)

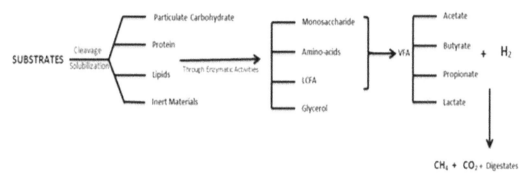

FIGURE 9.2   Substrate digestion sequence. LCFA: long chain fatty acid. (Adapted from Vavilin, V. et al., *Waste Manag.*, 28, 939–951, 2008; Weiland, P., *Appl. Microbiol. Biotechnol.*, 85, 849–860, 2010.)

9.2.1   Residues of the Biochemical Treatments of Food Wastes

9.2.1.1   Fermentation Residues

Fermentation residues are produced during microbial conversion of simple sugars to ethanol, enzymes, and CO_2, especially in pharmaceutical, food, brewery, waste treatment, and other related industries. The residues consist of incomplete fermented fibers, water, microbial cells, organic constituents (glycols), and other components useful as fertilizer, animal feed, and other purposes (Boruff, 1947; Dennis, 1945). Although fermentation has a long history associated with ethanol and its related products (wine, beer, drugs), the demand for ethanol as a substitute for gasoline strengthens the industrial relevance of the process. In this process, sugar or starch-oriented feeds and, in more advanced processes, lignocellulosic biomass are subjected to four serial but complex biochemical pathways (pre-treatment, hydrolysis and saccharification, fermentation of monomers

TABLE 9.1

Elemental and Nutrient Constituents of Fermentation Residues from Selected Biomass

	HHV (MJ/kg)	C (wt%)	H (wt%)	N (wt%)	O (wt%)	P[a] (%)	K[a] (%)	Protein (wt%)
Red maple	24.2	56.8	5.6	1	35	0.14	1.5	6.1
Switchgrass	20.8	50.1	5.5	1.5	36.5	0.14	2	9.6
Miscanthus	19	45.6	4.7	1.4	34.1	0.11	1.1	8.5

Source: Sannigrahi & Ragauskas (2011).

[a] Approximated and converted from histograph values.

and purification) with adequate consideration of substrate component and structure (Thommes & Strezov, 2015). Generally, the major challenges associated with fermentation techniques are the formation of waste streams (acid pre-treatment materials and toxic compounds), high cost of enzymes, and economic subsidies (Datar et al., 2004).

Fermentation process yields are often high (>90%); however, the quantity of residues cannot be underestimated due to disposal conditions, cost, and the associated environmental impacts. For instance, Juang et al. (2011) reported 75%–80% conversion efficiency for the production of ethanol, thus about 20%–25% organic waste residues (pH of 4.0) with 1,481 mg/L volatile solids, 22,600 mg/L carbohydrate, 4,400 mg/L organic nitrogen, organic acids, and other forms of alcohols. Enzyme compatibility, inhibitory consequences of pre-treatment, and different temperature demands are other disadvantages of simultaneous saccharification and fermentation, despite the yield height and low enzyme requirement (Khan et al., 2009). Nutrients and mineral richness of selected biomass residues after fermentation are illustrated in Table 9.1.

Valorization pathways and information related to nutrients, minerals, and other complex constituents of fermentation residues are now facilitated through characterization using analytical tools (Fenández et al., 2007). Brewery malt residue was reported to contain 28.4% hemicellulose, 27.8% lignin, 16.8% cellulose, 15.25% protein, and a small quantity of extractives and ashes (Khan et al., 2009). The fermentation residues are now considered for gas production, fertilizer, and animal feed inputs. Lignin-rich wet residues (77.3 wt% moisture content) were gasified in the presence of catalyst to produce synthesis gas (H_2S and COS) (Koido et al., 2013), which is required for some industrial operations. Compared to unfermented slurry (animal feces, urine, and slaughterhouse liquid), fermented slurries are richer in nutrients and considered more suitable for agricultural use, except for higher NH_4-N and pH, which necessitate hygiene during application (Pötsch et al., 2004).

9.2.1.2 Digestate

Anaerobically digested organic wastes produce renewable energy (methane) and nutrient-rich residues (Chynoweth et al., 2001; Holm-Nielsen et al., 2009) called *digestate*. Anaerobic digestion (AD) is a widely deployed, relatively effective treatment method for processing of organic wastes. Biogas and other by-products are produced using single- or two-stage digestion of agricultural wastes, such as animal manure, sewage sludge and/or industrial effluent, food and vegetable solid wastes, and organic fractions of municipal solid waste (OFMSW). AD produces a high quantity of digestate residues due to varying conditions associated with the process parameters (Chen et al., 2008; Igoni et al., 2008; Sung & Liu, 2003; Ward et al., 2008), which are a function of the process aims and objectives. Moreover, the heterogeneity of the co-digestated wastes attracts facultative and anoxic microbial diversity, which equally limit its efficiency, especially in the absence of proper trade-off process conditions (Juang et al., 2011). Studies indicate a methane yield range of 40%–70% of entire biogas, while CO_2 ranges between 32% and 38% (Li et al., 2011). However, about 15%–40% of the organic inputs are often utilized for biogas production, especially on large scales (Table 9.2), while the remaining liquid and or solid digestates constitute the

TABLE 9.2
Selected Process Performance/Efficiency

Digestate Source(s)	Digestion Process	Energy/Biogas Yield	Substrate Utilization (%)[a]	Process Scale	Ref.
Beet leaf & potato	Two-stage AD	3.9 kWh/kg VS	—	Pilot	(Parawira et al., 2008)
Pharmaceutical industry sludge	Mesophilic AD	0.36 L/g	72.84	Laboratory	(Gómez et al., 2007)
Primary sludge & OFMSW	Thermophilic AD	0.67 L/g	53.40	Laboratory	(Gómez et al., 2007)
Cattle manure	Mesophilic AD	0.14 L/g	53.40	Laboratory	(Gómez et al., 2007)
Pig manure, industrial waste, & biowaste	Two-stage AD & aerobic system	19 GWh/yr.	13.11	Large scale	(Bioenergy, 2012)
Dairy cattle & pig slurry[b]	AD	—	38.96	Large scale	(Smith et al., 2010)
Pig slurry[b]	AD	—	40.88	Large scale	(Smith et al., 2010)
Dairy cattle slurry (in Kent, England, & Scotland)[b]	AD	—	33.33 & 17.18	Pilot scales	(Smith et al., 2007)
Cow slurry[b]	Plug flow AD	93,501 ft³/day	38.50	Large scale	(Martin, 2005)

[a] Values are calculated from the data reported, NB: (Gómez et al., 2007) values determination was based on the VS as a function of the organic constituents.

[b] Studies' utilization strengths are determined using the chemical oxygen demand of input and output, respectively.

AD: anaerobic digestion; OFMSW: organic fractions of municipal solid waste; VS: volatile solid.

effluent (Bioenergy, 2012; Cooney et al., 2007; Xu et al., 2012). Consequently, biogas, slurry, and solid residues (digestate) are produced and invariably need management, mostly due to their toxicity.

Several studies provide significant information on the performance patterns of AD with respect to substrate use and biogas yield, while emphasis is laid on scale, substrate type, and digestion process adopted (Bouallagui et al., 2009; Bougrier et al., 2006; Stroot et al., 2001), as illustrated in Table 9.2. The isolated high-percentage substrate utilization expressed by Gómez et al. (2007) can be credited to the process scale, especially the extremely low working volume of substrates. Furthermore, most laboratory-scale digestion processes reported significantly effective yields, while single-stage AD ensures efficient utilization of substrate and cost-benefit and commercialization improvements relative to the two-stage AD system. However, this may not be unconnected to several other factors previously highlighted. Although degradation process evaluation cannot be isolated from the loading rate and the system hydraulic retention time (Salminen & Rintala, 2002; Tao et al., 2006), adequate optimization of process conditions is basic requirement for bioconversion system setup.

Generally, large-scale organic digestion significantly increases the net farm income and reduces odor, while substantially reducing the GHGs (CO_2 and CH_4) produced through uncontrolled disposal. However, there are no significant reductions in ammonia emission and water-quality potentials in the system outputs. Digestates often are comprised of a partially degraded organic matter coupled with microbial biomass and other inorganic constituents (Alburquerque et al., 2012). Unlike compost, digestate is an immature product of anaerobic processes (Bioenergy, 2012) that may require post-treatment measures and management. The physicochemical constituents of digestates, as summarized in Table 2.2, are greatly influenced by substrate source, microbial composition, pre-treatment measures, and process techniques and parameters (Raposo et al., 2006; Salminen & Rintala, 2002; Singh et al., 2011).

TABLE 9.3
Physical Properties of Digestates

Digestate Source(s)	Digestion Process	pH	Total Solid (%)[a]	Volatile Solid (%)[a]	Ref.
Dairy manure & Biowaste	Two-stage AD	7.4	4.1	3.0	(Paavola & Rintala, 2008)
Pharmaceutical industry sludge	Mesophilic AD	7.8	22.5	15.7	(Gómez et al., 2007)
Primary sludge & OFMSW	Thermophilic AD	7.5	23.6	16.5	(Gómez et al., 2007)
Cattle manure	Two-stage AD	7.6	122.6	105.4	(Gómez et al., 2007)
Food waste	Thermophilic AD	7.87	33.5	45	(Forster-Carneiro et al., 2008)
Wastewater sludge	Mesophilic AD	7.9	3.9	2.66	(Forster-Carneiro et al., 2008)

[a] Original values of these data are converted to percentage for coherence.
AD: anaerobic digestion; OFMSW: organic fractions of municipal solid waste.

The distribution and mineralization dynamics of nutrients in digestates are rarely discussed in the literature (Parawira et al., 2008). Phosphorus in the digestate is reposited in the solid fraction, while the liquid residue hosts most of the mineralized nitrogen, especially in a two-stage AD system. These digestion residues are mainly characterized (Table 9.3) by a slightly neutral pH, except for few substrates, such as sugarcane, which indicates an acid state due to its homogeneity (Demattê et al., 2004).

Similarly, the pH of digestates was reported to be equally influenced by the substrate composition, microbial constituents, and other process parameters adopted for the system (Alburquerque et al., 2012; Parawira et al., 2008; Smith et al., 2010). Although volatile solid (VS) and total solid (TS) constituents of most digestates are proportional, VSs are often used to estimate the organic concentration of the substrate (Parawira et al., 2008). This parameter is considered as a vital degradation measure and thus an indicator for microbial activities within the system. Nevertheless, critical evaluation of the VS percentage on TS basis, as presented in Table 9.3, shows VS content at an average of above 70%, is a characteristic of higher biodegradability, resulting in more digestate yields (Nallathambi Gunaseelan, 1997).

Organic constituents and their degradation pattern are also reflected with parameters such as neutral detergent fiber (NDF), volatile fatty acid (VFA), and both the chemical and biological oxygen demand (Bossen et al., 2008; Dogan et al., 2009; Liew et al., 2012; Perez et al., 2001). Most of these parameters are considered to evaluate not just the microbial metabolic strength and pattern on residues produced, but also to determine the effectiveness of the process and, consequently, the quality of the yield and other products. Similarly, the organic fraction behavior in the system provides information on the relationship between the substrates, microbial performance, and inhibitory features. For instance, VFA accumulation in the system results in low pH and consequently increases the concentration of ammonium, especially at the methanogenesis stage of the system (Chen et al., 2008; Izumi et al., 2010; Nallathambi Gunaseelan, 1997). Although increased retention time (RT), air stripping, and chemical precipitation are possible solutions to these accumulations, further initiatives, such as immobilization of organisms using inert materials, ion exchangers, or adsorbents, may be necessary for an effective output.

The fate of micro- and macro-elements is crucial in the residue formation and equally determines the post-treatment nature and eventual use of these products. The nutrients in Table 9.4 have significance in the use of residues as soil amendment and fertilizer. The main constraint of the AD product is the abundance of readily available NH_4-N, which can easily be converted into NO_3-N and

TABLE 9.4
Nutrient and Macro-Element Constituents of Digestates

Digestate Source(s)	Digestion Process	TN (%)	pH	TP (%)	TK (%)	C (%)	C:N (%)	Ref.
Pig manure & rapeseed residue	AD	0.36	7.82	0.11	0.31	1.47	4.083	(Alburquerque et al., 2012)
Pig manure & sunflower residue	AD	0.35	7.92	0.11	0.31	1.22	3.486	(Alburquerque et al., 2012)
Cattle manure & maize oat silage	Mesophilic AD	0.397	7.50	0.08	0.31	3.38	8.5	(Alburquerque et al., 2011)
Slaughterhouse wastewater & biodiesel wastewater	Mesophilic AD	0.396	8.20	0.02	0.2	0.59	1.5	(Alburquerque et al., 2011)
Pig slurry & slaughterhouse sludge & biodiesel wastewater	Mesophilic industrial AD	0.38	8.3	0.05[a]	0.24[b]	0.47	1.2	(Alburquerque et al., 2012)
Pig slurry	Industrial thermophilic AD	2.67	6.54	3.1	1.14	35.2	13.18	(Bustamante et al., 2012)

Note: Values are calculated from the data reported.
AD: anaerobic digestion.
[a] As P_2O_5.
[b] As K_2O.

N_2O through nitrification and denitrification processes by soil organisms. Although NH_4-N could be readily available for plants absorption, its excess, as reflected in the C:N ratio, may result into emissions of N_2O and NH_3. Moreover, residues with low C:N ratios (<25) provide significant quantities of nitrogen, which mineralize in the soil.

Elemental constituents, such as heavy metals and other metallic elements reported in Table 9.5, are significantly low—perhaps due to substrate sources, which are largely food crops, animal feces,

TABLE 9.5
Heavy Metal and Micro-Element Constituents of Digestates

Digestate Sources	Digestion Process	Ca (%)	Mg (%)	Na (%)	Ni (%)	Zn (%)	Cu (%)	Ref.
Food waste	Fermentation	7.74	0.23	2.36	—	—	—	(Su et al., 2012)
Willow	Two-stage AD	—	—	—	—	0.018	0.002	(Lehtomäki & Björnsson, 2006)
Sugar beet	Two-stage AD	—	—	—	0.004	0.019	0.010	(Lehtomäki & Björnsson, 2006)
Pig manure & rapeseed	AD	0.20	0.06	0.07	—	—	—	(Alburquerque et al., 2012)
Grass	Two-stage AD	—	—	—	0.001	0.011	0.006	(Lehtomäki & Björnsson, 2006)
Maize	Two-stage AD	—	—	—	0.001	0.0034	0.003	(Selling et al., 2008)
Horse manure	Two-stage AD	—	—	—	0.0004	0.004	0.0014	(Selling et al., 2008)

Note: Values are calculated from the data reported.
AD: anaerobic digestion.

and source-separated organic fraction of municipal solid wastes. Co-mingled wastes are reported to contain higher concentration of nutrients when compared to sorted or source-separated wastes.

The study of Parawira et al. (2008) shows the significance of co-digestion as compared to single-substrate digestion, wherein co-digestion yield is 60% higher. Synergetic performance of heterogeneous substrates is equally reported for OFMSW and wastewater (Viotti et al., 2004), industrial products of potato and pig manure (Kaparaju & Rintala, 2005), and energy crops and animal manure (Alburquerque et al., 2012).

9.2.1.3 Compost

The composting process transforms and practically stabilizes organics or biomass into nutrients and mineral-rich materials. This technique not only recycles a substantial segment of wastes, but also safely and beneficially conditions and amends soil structure. Moreover, degradation of large organic molecules by composting also ensures energy management (Sonesson et al., 2000) and influences disinfection of the organic matter through the heat produced while CO_2, leachate, and other products are equally produced, as indicated in the equation below.

$$\text{Organic matter} + O_2 \longrightarrow \text{Compost} + CO_2 + H_2O + \text{Mineral products} + \text{Heat}$$

Recent awareness about health, bioavailability, metal constituents, and organic loads of compost and residues poses a challenge for direct application of the latter and thus demands stringent standards (Paavola & Rintala, 2008).

Compost, unlike organic degradation residues, is considered mature and fit for agronomical use, with no lagging physicochemical characteristics. Its relatively stable biodegradable organic matter is measured as the process evolution index (Brewer & Sullivan, 2003). The GHG effect, free radicals intrusion, and leachate concentration from composting could have environmental impacts significant enough for assessment and review of the process. Hao et al. (2001) posited the effect of composting methods on GHG emissions. The study showed that active treatment (turning for aeration) of cattle feedlot manure accounts for more than 200% and 100% carbon lost in the form of CO_2 and CH_4, respectively, when compared to passive treatment (no turning). Similarly, nitrogen lost in the form of N_2O is equally more than 150% higher for the active method, even though compost produced through a passive system results in unstable manure (Hao et al., 2001). In-vessel composting was also reported to be 68% cheaper operationally compared to sanitary landfill systems, but the latter provides greater energy yield (Cabaraban et al., 2008).

Unstable compost or digestion residue is characterized by a high proportion of biodegradable matter, which further sustains an elevated microbial activity in the soil (Ferreras et al., 2006). The application of immature organic constituents to soil also increases nutrients' immobilization, especially nitrogen, and may spread animal and plant pathogens (Bustamante et al., 2008). Physicochemical and microbiological properties are often considered when setting up standards and regulations for typical processes and products, such as composting across various countries (Lasaridi & Stentiford, 1998; Manungufala et al., 2008), a situation currently rare in the use of digestates for similar purpose. The institutionalization of biomass or organic digestion product standards should be broadened and widened to accommodate the entire life cycle, rather than only focusing on the application points, as currently experienced by the sector. Established indicators, shown in Table 9.6, such as C:N ratio, microbial activity, germination index, cation exchange capacity (CEC), humic substances, water-soluble carbon (WSC), dissolved organic matter, NH_4^+ -N and NO_3^- -N; ratios of NO_4^+ -N to NO_3^- -N, WSC to TN, and WSC to organic-N (Benito et al., 2003; Goyal et al., 2005; Said-Pullicino et al., 2007; Smith & Hughes, 2004; Tang et al., 2006) are possible inventory data sources to be considered. However, due to differences in feedstock, coupled with the wide variety of process conditions

TABLE 9.6
Properties of Compost from Organic Wastes

Source(s) of Wastes	pH	DM (%)	TN (%)	TOC (%)	C:N	Germination Index	Ref.
Rice husk	4.51–8.91	—	1.13	40.62	36	85	(Chang & Chen, 2010)
Sawdust	6.09–8.6	—	0.87	44.98	51.5	45.2	(Chang & Chen, 2010)
Rice bran	Initial 4.18	—	1.59	49.33	31	—	(Chang & Chen, 2010)
Food waste & straw	4.47	—	3.4	34.8	10.24	—	(Zheljazkov & Warman, 2002)
Pig slurry + slaughterhouse sludge + biodiesel wastewater	—	1.9	0.38	0.47	1.2	—	(Alburquerque et al., 2012)
Digestate + wheat straw + almond shell	—	—	3.93	49.06	12.5	98	(Bustamante et al., 2012)

AD: anaerobic digestion.

(e.g., facility scale, aeration, temperature, pH, and moisture content), no single maturity indicator can be applied (Ishii & Takii, 2003).

Another major constraint associated with the recycling system is the loss of minerals during composting, which includes ammonia volatilization, nutrients leached through runoff or rainwater (in a large-scale windrow system), and methane or nitrous oxide emissions. Peigne and Girardin (2004) reported negligible nitrogen loss of 0.5% to leachate water, while 19%–42% of initial manure nitrogen was forfeited by gas emissions during feedlot beef manure composting. However, the environmental and sustainability quotient of these impacts should be considered to establish impact assessment in all spheres. Generally, the equilibria and rates of nutrient dynamics are influenced mostly by interaction of the process conditions and substrate physicochemical properties (Tiquia, 2002). The environmetal and health importance of heavy metals makes the dynamic in composts and behavior in soil significant. Smith (2009) revealed increased heavy metals complexation when organic waste residues are applied directly to soil. This invariable limit metals solubility and bioavailability in soil because of its strong interaction with compost matrix. Although the metal sorption properties of compost from municipal solid waste (MSW) and sewage sludge can be of advantage in the remediation of metals-contaminated soils (Businelli et al., 2009), the application of compost to agricultural soil with relatively stable metals distribution may equally contribute to the degree of metal bioaccumulation in crops.

9.2.2 Thermochemical Treatment: Pyrolysis and Products

9.2.2.1 Charcoal (Biochar)

Charcoal is the primary product of thermochemical conversion and is one of the earliest forms of synthetic fuel. The properties of charcoal depend on the thermochemical process conditions. High temperatures typically reduce the volatile matter and enhance the carbon content of charcoal. Charcoal with high carbon concentrations may be theoretically desired; however, higher heating temperatures would seriously reduce the production levels and mechanical strength of the charcoal. Lower heating temperatures would leave larger amounts of volatile matter in the charcoal. The recommended maximum heating temperature to achieve maximized charcoal yields is 400°C

(Antal & Grønli, 2003). Charcoal transformation is almost complete at this temperature, and it typically contains 20%–25% of volatile matter and 3%–4% ash, with the remaining 75%–80% being fixed carbon. The maximum heating temperature has a strong impact on the charcoal heat of combustion (Fuwape, 1996). The heat of combustion is almost constant at 23 MJ/kg for charcoals produced at temperatures below 250°C, while in charcoals produced at temperatures above 300°C, the heat of charcoal combustion increases by 45% (Strezov et al., 2007).

Due to the loss of large amounts of mass from liberation of volatiles and liquids, the charcoal contains dangling carbon bonds, making it a highly reactive material (Antal & Grønli, 2003). Particle size was found to affect the carbon macromolecular structure, increasing the homogeneity of the porous carbon material and reducing the intra-particle pore size (Treusch et al., 2004). The specific surface area of charcoal increases at temperatures above 450°C and can reach high surface areas of over 250 m^2/g at 700°C, making it very suitable material for filtration as an activated carbon. Char porosity and surface area depend on the heating conditions, and higher heating rates generate chars with more open pore structures and larger macropore surface areas (Demirbas, 2001). Mochidzuki et al. (2003) determined the physical and electrical properties of charcoal materials produced in laboratory conditions. They found predominantly alkyl aromatic structure with oxygen enriched C-O-H, C=O and C-O-C functional groups. Upon thermal treatment at 650°C, these groups decompose, forming condensed aromatic C-H structures, which further break down at 750°C and, consequently, evolve hydrogen at elevated temperature range.

Heating rate has a profound effect on the final char yields. The char yields range between 22% for the heating rate of 10°C/min. and 13% for heating rates close to 1,000°C/min. (Strezov et al., 2006). The decreasing trend of cellulose char yields with increased heating rates was reported previously (Lewellen et al., 1977). Rapid heating rates enhance transfer of volatiles through the biomass particle, reducing the time available for the primary gases and oils to undergo secondary reactions, cracking, and re-polymerization.

Biochar exhibits varying properties based on feedstock and pyrolysis conditions, as indicated in Table 9.7. Physicochemical features of most biochars vary, therefore reacting distinctively due to differences in stability and morphological architecture (Lehmann & Joseph, 2012; Novak & Busscher, 2011; Steinbeiss et al., 2009).

For instance, Steinbeiss et al. (2009) reported the adaptation of soil indigenous microbes and the stability of varying biochar condensation grade and chemical composition as the main drivers for the various production treatments considered. Like most other residues or treatment products, the quality of biochar produced hinges on the type of substrate and process conditions, such as temperature and holding time, used during production (Biagini et al., 2005). Biochar structure and morphology are thus influenced by production temperature. Extreme heat increases the proportion of aromatic carbon and its turbostatical arrangement and graphitic structure, which are responsible for its porosity and surface area (Downie et al., 2009). However, the structural complexities of biochar are observed to be lost during pyrolysis, as posited by Amonette & Joseph (2009) and Haas et al. (2009), respectively.

9.2.2.2 Bio-oil

Bio-oils are dark brown pyrolysis liquids generated when biomass is subjected to heat in the absence of oxygen. This complex mixture of free-flowing organic liquids and water is physically multi-phase with char particles, waxy materials, droplets of different nature and micelles (Oasmaa et al., 2015). Bio-oils are highly oxygenated compounds formed through depolymerization and fragmentation of rapidly heated biomass, especially at fast heating and cooling rates (Evans et al., 2015). Biomass pyrolysis oil reflects elemental composition of the parent feedstock and contains multifunctional compounds, such as aldehydes, ketones, esters, and others, which are prone to further reactions (depending on temperature) at storage to form macro-molecules (Mohan et al., 2006).

TABLE 9.7
Physicochemical Properties of Biochar

Source(s)	Physical Properties				Chemical Properties				Ref.
	pH	EC mS/m	Temp (°C)	Ash	C:N	TP	TK	TC	
Corn residue			350	—	72.6	—	1.04	67.5	(Nguyen & Lehmann, 2009)
			600	—	85.9	—	6.7	790	
Wood	7.0		350	—	144	0.6	—	824	(Rondon et al., 2007)
Poultry litter	9.9		450	—	19	25	22	380	(Chanet al., 2008)
Algae *Ulva flexuousa*	8.0	53	450	—	8.4	7,078 mg/kg	167 mol/kg	22.6%	(Bird et al., 2011)
Sesame			550	36.80	12.50	3.45%	3.38%	86.64	(Volli & Singh, 2012)
Mustard			550	28.10	13.85	2.87%	4.00%	85.43	
Neem			550	24.50	14.30	0.35%	2.38%	82.34	
Wastewater sludge	5.32	4.12	300	52.8	7.71	492.5[a]	<1%	25.6	(Hossain et al., 2011)
	4.87	4.15	400	63.3	8.42	740[a]		20.2	
	7.27	4.7	500	68.2	9.53	567.5[a]		20.3	
	12	2.5	700	72.5	17	527.5[a]		20.4	
Algae *Tetraselmis chui*	12	39[b]	500	30.3	8.70	10[c]	2.1[d]	40	(Grierson et al., 2011)
Food waste digestate	8.39		300	12.6	8.47	2.92	0.87	45.4	(Opatokun et al., 2015)
	9.69		400	49.2	8.27	4.13	1.24	37.3	
	10.1		500	55.1	8.80	4.54	1.39	35.3	
	10.7		700	60.2	17.9	4.78	1.53	34	

[a] Plant-available phosphorus (Colwell phosphorus).
[b] Ds/m.
[c] Phosphorus as P_2O_5.
[d] Potassium as K_2O.

Bio-oil yields are reported to be largely influenced by heating rates, reaction temperatures, vapor residence time and, most importantly, feedstock composition (Kan et al., 2016). For instance, a wood bio-oil yields range of 72 wt%–80 wt% was reported by Mohan et al. (2006), while Isahak et al. (2012) expressed 60 wt%–70 wt% as the average bio-oil yield of fast-pyrolyzed biomass. Biomass often contains active catalysts (potassium and sodium), which promote secondary cracking and consequently impair bio-oil yield and quality (Evans et al., 2015). Sawdust and sugarcane bagasse were reported to produce similar bio-oil yields (approximately 70%), while banana rachis yielded below 30% despite being subjected to the same conditions, indicating substrates' effects on bio-oil output (Montoya et al., 2015).

Moisture content and oxygen concentration of pyrolysis liquids are attributed to their low heating values when compared to hydrocarbon products (see Table 9.8). The higher heating values of 16 MJ/kg–19 MJ/kg in wood-oriented biocrude compared to 40 MJ/kg in conventional heavy fuel were attributed to 15 wt%–30 wt% moisture and 35 wt%–40 wt% of oxygen in the wood against 0.1 wt% moisture and 1 wt% oxygen in the heavy fuel (Czernik & Bridgwater, 2004).

TABLE 9.8

Yields, Heating Values, pH, and Elemental Composition of Bio-oils from Different Biomass

Feedstock	Treatment Process	Yield	HHV	pH	C	H	O	Ref.
Macro-alga	Fast pyrolysis	26.7	28.3	6.1	64.3	7.7	25.3	(Ly et al., 2015)
Food waste	Slow pyrolysis	60.3	11.2a	—	—	—	—	(Opatokun et al., 2015)
Corn stover (multi-pass and single-pass)	Fast pyrolysis	48.7 and 45.0	19.2 and 23.0	—	44.9 and 53.3	14.3 and 17.0	40.1 and 29.0	(Shah et al., 2012)
Paulownia wood	Slow pyrolysis	—	28.6	—	66.1	8.7	25.2	(Yorgun & Yıldız, 2015)
Hardwood shavings	Fast pyrolysis	63.3	22.6	2.7	55.3	6.5	37.6	(Agblevor et al., 2010)
Corn cobs	Fast pyrolysis	61.0	26.2	—	55.1	7.6	36.9	(Mullen et al., 2010)
Soybean	Fast pyrolysis	24.19	33.6	—	67.9	7.8	13.5	(Şensöz & Kaynar, 2006)

Bio-oil's potential as combustion fuel and a source of heat in boilers has been considered (Kan et al., 2016), despite its low heating values and pH. The organic and inorganic constituents of the liquid oil are a viable source of platform chemicals. Although, wood flavor is commercially extracted from bio-oil, essential pharmaceutical and industrial chemicals locked in the oil are expected to be annexed through various upgrading techniques (Evans et al., 2015).

9.2.2.3 Biogas

Biogases or syngases are mixtures of gases produced through biochemical or thermochemical degradation of organic matter (biomass). Gas production from biomass offers renewable and sustainable energy production with a significant potential to contribute to the key economic sectors, such as transportation, electricity, and manufacturing industries. Biogas is produced during AD of organics in various established systems, such as sewage treatment plants, landfills, and digesters. The latter predominately consists of CH_4 (55%–75%) and CO_2 (>40%), with trace components of H_2S, CO, N_2, and volatile organic compounds (VOCs). Meanwhile, syngas is mainly comprised of H_2, CO, hydrocarbons (CH_4, C_2H_2, C_2H_2 and C_2H_6), water, and CO_2, as indicated in Table 9.9. Several endothermic reactions produce H_2 in syngas through cracking of the hydrocarbon at high temperature, while intermediate products are responsible for the light hydrocarbon formation (Rasul & Jahirul, 2012). CO and CO_2 traceable to oxygenated organics decrease with increased temperature, whereas H_2 and CO are posited to increase with the charring temperature. Quantity and constituents of the produced biogas depends on treatment techniques (AD, pyrolysis, or gasification) adopted and feedstock involved (Igoni et al., 2008; Rasi et al., 2011). Temperature, retention time, feed type, stream flow rate, and pre-treatment conditions are generally considered critical parameters for gas yield and quality (Nieves et al., 2011; Yan et al., 2010); substrate efficiency and optimal utilization hinge mostly on process configurations (Opatokun et al., 2015; Vindis et al., 2009). For instance, the H_2:CO ratio of syngas varies with respect to production technology and feedstock (Dayton et al., 2011). Generally, all types of organic wastes, such as putrescible components of MSWs, agricultural waste, sewage sludge, and industrial effluents, are suitable for biogas production.

Biogas is currently a valuable renewable source for electricity production and heating systems, with potential for engine combustion (Holm-Nielsen et al., 2009). Similarly, biomass-oriented syngas is considered as a fuel and feedstock for production of tailored chemicals (Spath & Dayton, 2003).

TABLE 9.9

Composition of Selected Biogas and Syngas and Heating Values

	CH$_4$ (%)	CO$_2$ (%)	O$_2$ (%)	N$_2$ (%)	H$_2$S (ppm)	Benzene (mg/m³)	Toluene (mg/m³)	Heating Value	Ref.
Landfill	47–57	37–41	<1	<1–17	36–115	0.6–2.3	1.7–5.1	—	(Rasi et al., 2007)
Anaerobic sewage digester	61–65	36–38	<1	<2	<0.1	0.1–0.3	2.8–11.8	—	(Rasi et al., 2007)
Landfill	59–68	30–37	—	—	15–428	22–36	83–172	—	(Shin et al., 2002)
Farm biogas plant	55–58	37–38	< 1	<1–2	32–169	0.7–1.3	0.2–0.7	—	(Rasi et al., 2007)

	H$_2$	CO	CO$_2$	CH$_4$	C$_2$H$_2$	C$_2$H$_4$	C$_2$H$_6$	Heating Value	Ref.
Gasified biomass char[a]	52.4	14	27.6	1.7	0.2	4	0	8.3[b]	(Yan et al., 2010)
Pyrolyzed Food waste	5.2	2.2	7	2.6	—	0.14	0.32	15.7	(Opatokun et al., 2015)
Pyrolyzed food waste digestate	9.1	0.7	2.03	1.3	—	0.12	0.3	17.2	(Opatokun et al., 2015)

[a] Measured in v%/dry basis while others are in wt%/min.

[b] Lower heating value in MJ/Nm³, while others are in higher heating values.

9.3 APPLICATIONS OF BIOMASS POST-PROCESSING RESIDUES AND PRODUCTS

9.3.1 ENERGY APPLICATIONS

The focus on biomass for energy cannot be dissociated from its industrial demand as renewable sources. Although the globe requires an additional one-third of its current energy demand (Barnsley et al., 2015), the renewed interest on indigenous energy sources, such as biomass, is driven by its environmental qualities. For instance, approximately 2.73×10^{10} MJ of energy was generated in rural China through the spread of approximately 35 million anaerobic digesters (Chen et al., 2012). Considerable achievement has been recorded in conversion or treatment, often targeted at waste biomass, as sustainable means of primary and secondary energy sources. Recently, a 42% increase in electricity was reported by coupling AD with pyrolysis using agricultural wastes instead of the initial 9,896 KWh$_{el}$ generated on a stand-alone AD plant (Monlau et al., 2015). Constraints related to biomass' initial moisture content and lignocellulose configuration are appropriated through integrated treatment systems, while the environmental challenges of digestate are equally ensured (Opatokun et al., 2015). The latter study reported 96% and 77.3% theoretical efficiency for pyrolyzed food waste and its digestate, respectively, and subsequently produced nutrient-rich biochar with potential for soil applications (Opatokun et al., 2015).

Notwithstanding, AD effluents are equally used as co-substrate or biofertilizer, provided the regulatory standards are fulfilled. The blend of AD liquid effluent (yard and food wastes at 20–30 C/N ratio) were used to increase biogas yield and ensure optimal microbial performance (Brown & Li, 2013) through the blend. Similarly, 60% food wastes were combined with dairy manure to improve throughput (El-Mashad & Zhang, 2010), ensuring zero waste. Co-digestion of feedstock does not only increase biogas yield, but also balances nutrient distribution and dilutes toxic compounds

(Brown & Li, 2013). Dry digestates, like biochar, are also considered as solid fuel, with 85% efficiency achieved when two digestate pellets with combustion power of 44 KW and net calorific values of 15.8 and 15.0 MJ/kg, respectively, were used (Kratzeisen et al., 2010).

9.3.2 METALLURGICAL APPLICATIONS

The metallurgical application of charcoal has traditionally been associated with the reduction of iron oxides in the process of producing metallic iron, although this practice has declined over time, specifically since the progress of the coke-making technology. More recently, charcoal is considered a reductant in the processes of reduction of silica to silicon and other metallic oxides (e.g., nickel, lead) to their corresponding metals. In metallurgy, charcoal supplies the heat and carbon required to maintain the oxide reduction process. Renewed interest in reintroduction of biomass-based metal smelting technology is based on the attempt to improve the sustainability of the metallurgical operations through inclusion of renewable energy sources. The annual energy consumption of the iron and steel industries is equivalent to 5% of the world's total energy consumption and also accounts for 3%– 4% of the global GHG emissions (Chunbao Charles & Cang, 2010). According to Birat (2003), the CO_2 emissions from different ironmaking routes range from 2 tCO_2 per tonne of liquid iron for blast furnace ironmaking, to 0.7 tCO_2/t liquid iron for metal smelting with an electric arc furnace. One alternative approach in reducing the GHG emissions, while maintaining desired iron and steel production levels, is by transforming the existing industries in more-sustainable operations, with biomass as renewable energy and reductant source.

Blast furnace operations require separate coke-making and sintering plants to feed the furnace. Charcoal exhibits high reactivity with CO_2 ($C + CO_2 \rightarrow 2CO$), producing the reductant CO gas that maintains the iron ore reduction process in a stepwise mechanism:

$$3Fe_2O_3 + CO \rightarrow 2Fe_3O_4 + CO_2$$

$$Fe_3O_4 + CO \rightarrow 3FeO + CO_2$$

$$FeO + CO \rightarrow Fe + CO_2$$

Some reports suggest that the iron ore reduction with charcoal is occurring at higher rates when compared to coal and coke (Pandey & Sharma, 2000). The reduction with charcoal in blast furnaces occurs at temperatures up to 250°C lower than coke-blast furnaces, mainly due to differences in carbon reactivity (Meyers & Jennings, 1978). The volatile matter content in the charcoal is expected to additionally contribute to the reduction process, considering charcoal consists 25% by weight of volatile matter, which is mainly hydrogen and carbon monoxide.

Charcoal as an additive to coal blends for coke making was considered in a report published by the NSW SERDF (2001); however, the results showed anti-fissuring effects and decrease in coke strength due to the non-softening properties of charcoal. It appears that unprocessed biomass may have the potential to serve as a blending material in coke production. Das et al. (2002) found that some blends of coal and biomass, in particular molasses, may produce reasonably high swelling ratios, which are required for metallurgical production of coke. Blast furnace technologies can incorporate fuel injection where charcoal and biomass have been considered as potential injectants. NSW SERDF (2001) conducted trials to assess the performance of charcoal as slag foaming injectant, while Takekawa et al. (2003) studied the gasification reactions of waste wood as a blast furnace injectant. The electric arc furnace technology for ironmaking is based on re-melting and recycling of scrap steel. This technology produces lower amounts of CO_2 per liquid iron compared to the blast furnace; however, the production capacity is limited to the availability of scrap steel. Since this technology is essentially an electric-based steelmaking, the potential for inclusion of charcoal in the process is in relation to the electricity generation as a front end of steel production.

The most recent emerging generation of smelting operations consists of direct reduction of iron ores with coal. Direct reduced ironmaking (DRI) processes have several advantages over the conventional blast furnace operations with low pollution effects and low capital-intensive operation, and they can provide successful smelting with low-grade thermal coal. The DRI process consists of carbo-thermic reduction of iron oxides directly with the volatiles liberated during coal devolatilization, carbon monoxide regenerated from coal char, as well as dissolved carbon in iron bath. The DRI technologies offer viable potential for substitution of coal with biomass as a carbon-bearing reductant material. Strezov (2006) found that iron ore can be successfully reduced to predominantly metallic iron using 30 wt% of biomass in a biomass-ore pellet. The shortcomings in potential development of biomass-based metal smelting technology is related to the low density of biomass, requiring larger volumes; hence it potentially can reduce the metal production rates. More realistically, charcoal can potentially provide substitution for coal in the direct reduced ironmaking technologies. Further research will be required to ensure the metallurgical operations maintain the desired levels of energy efficiency, productivity, and process quality.

9.3.3 Agricultural and Carbon Sequestration Applications

The application of post-processing residues, such as digestate, compost, and biochar, as fertilizers or soil enhancements explores the soil-microbe-nutrients interaction to make the available micro- and macro-nutrients in residue to biotic constituent of the system. Anaerobically degraded biomass provides an alternative source of energy and offers an alternative route to synthetic fertilizer, due to mineral richness (Brown et al., 2012) instead of untreated biomass. This mixture of partially degraded organic matter, microbial biomass, and inorganic compounds is considered an inexpensive disposal means and suitable recovery approach to minerals and organic constituents for agricultural use (Tambone et al., 2010). Digestate's ability to eliminate weed seed and impairing pathogen loads provides justification for digestate's use as fertilizer or soil enhancement (Walsh et al., 2012). Meanwhile, heavy metal concentrations, organic loads, odor, and workers' health and safety associated with digestate remains a challenge (Lukehurst et al., 2010). Expectedly, quality standards, national guidelines, and protocols instituted by governments are changing the disposal approach to a recovery process (Al Seadi & Lukehurst, 2012; Alburquerque et al., 2012). Compost remains the most applied means of slow mineralization of stabilized and humified organic materials in soil (Montemurro et al., 2010). Young and mature composts are reported to be influenced by the material of origin, maturity extent, and storage conditions (Fuchs et al., 2008). Similarly, plant-soil disease suppression potential of compost is pivotal to maturity and degree of phytotoxicity. The mineral fertilizer substitution extent of compost depends on its quality and, consequently, nitrogen immobilization.

Unlike digestate and compost, biochar is widely considered as a stable carbon with potential for soil improvement and fertilization. There are several reports on the benefits of biochar in the soil from major effect, such as soil structure enhancement (water retention, mitigation of nutrients leach), increasing soil biological activities, soil remediation, and specific effect on plant and crop growth (Srinivasan et al., 2015). Different animal manures stabilized through pyrolysis indicated the nutrient recycling and management ability of biochar when applied to soil (Cely et al., 2015). The persistence of char in soil for years due to its recalcitrant nature accounts for the carbon sequestration capacity. The O_2 or H_2 to C ratios of biochar are indicators for the stability and carbonization, which equally depends on the parent feedstock elemental constituents (Sohi et al., 2010). Half-lives of less than 100 years are predicted for chars with atomic O:C ratio of greater than 0.6, while at least 1,000 years are attributed to O:C molar ratios lower than 0.2 (Srinivasan et al., 2015). For instance, Opatokun et al. (2015) posited the carbon sequestration potential of food waste, which indicates a molar O:C ratio lower than 0.2, despite the difference in charring temperature as against pyrolyzed food waste digestate with an O:C molar ratio ranges of 0.2–0.6. Therefore, biochar process parameters are fundamental factors that shape char properties, agronomic values, and the large and long-term C sink or sequestration (Sohi et al., 2010).

9.4 LIFECYCLE ANALYSIS OF WASTE PROCESSES AND PRODUCTS

Lifecycle assessment or analysis (LCA) is a management approach to quantify the amount of substrates, energy, and other forms of inputs used over a complete process or product production to identify or evaluate energy cost and social and/or environmental performance of the process or product at all stages. Fundamentally, LCA derives standards from the International Standards Organisation (ISO) 14040 (Principle and framework) and ISO 14044 (Requirement and guidelines) using primary (direct information from facilities or systems) and secondary (e.g., public database, published reports) data sources. Processes and product design are benchmarked and compare to environmental standards through LCA to continuously identify emission and waste during the life cycle of the system or product, while enabling identification of more-sustainable options (Evangelisti et al., 2014).

The major cardinal structures of LCA proposed by the ISO 14040 are comprised of four main phases—namely, the goal and scope definition, inventory analysis, impact evaluation, and interpretation (López-Sabirón et al., 2014), as annotated in Figure 9.3. These iterative phases constitute a framework for holistic assessment of inputs and emissions associated with the stages of a process or product life cycle, from cradle to grave (Ferreira et al., 2015), leading to a more sustainable evaluation. The applicability and uses of the study or assessment are expected to be articulated in the goal and scope, which determine selection of the lifecycle inventory (LCI) framework. Meanwhile, LCI is predicated on a functional unit which is considered the central hub of the entire assessment, since other data in the assessment are referenced and normalized by this unit (Matheys et al., 2007). Functional unit, therefore, quantitatively provides comparability to study or assessment—for instance, 1 kg of food waste indicates the unitary measure of the system. The impact assessment aggregates the inventory waste data through midpoints to endpoints categories, depending on the characterization model used. Impact categories, such as ozone depletion, climate change, terrestrial acidification, and ecotoxicity, are environmental indicators through which environmental burdens are reflected or interpreted, depending on the tool or software (e.g., SimaPro, GaBi, EASEWASTE, ORWARE) and methods (e.g., ReCiPe, Centrum voor Millieukunle Leiden (CML), Environmental Design of Industrial Products (EDIP), EcoIndicator 99) deployed. The choice of impact category, normalization, and weighting needs to be consistent with the study goals. The intended applications of the LCA are related in the interpretation, wherein conclusions of the study are derived, followed by recommendations.

A handful of studies have evaluated waste treatment processes, while some focus on energy and emission related to the products. For instance, the life cycle of biogas and digestate utilization was considered to evaluate the emission mitigation, agricultural benefits, and rural energy needs (heating, illumination, and fuel) of China (Chen et al., 2012). Similarly, the energy, economic, and climate change potential of biochar was estimated using LCA of the product (Roberts et al., 2009), while a comprehensive, cradle-to-grave assessment of AD in terms of energy output was benchmarked with

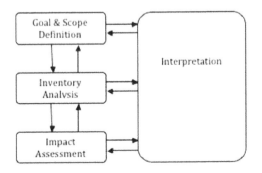

FIGURE 9.3 Structural components of a lifecycle assessment study.

landfill and incineration during waste treatment scenarios (Evangelisti et al., 2014; Moberg et al., 2005). Similar to conventional waste treatment, wherein single treatment techniques are deployed, most LCA studies so far focus on comparing different treatment processes, mainly to evaluate the energy and emission impacts. Moreover, very few studies narrow assessment to specific waste type through which complexity associated with modeling heterogeneity and waste composition of such system (Laurent et al., 2014). Conversely, overburden assumptions and a widening uncertainty threshold, which impair the quality of the report despite the specificities of such LCI (Laurent et al., 2014; Lundie & Peters, 2005).

REFERENCES

Agblevor, F. A., Beis, S., Kim, S. S., Tarrant, R., & Mante, N. O. (2010). Biocrude oils from the fast pyrolysis of poultry litter and hardwood. *Waste Management*, 30(2), 298–307. doi:10.1016/j.wasman.2009.09.042.
Al Seadi, T., & Lukehurst, C. (2012). Quality management of digestate from biogas plants used as fertiliser. *Paper presented at the IEA Bioenergy, Task.*
Alburquerque, J., de la Fuente, C., Campoy, M., Carrasco, L., Nájera, I., Baixauli, C., ... Bernal, M. (2012). Agricultural use of digestate for horticultural crop production and improvement of soil properties. *European Journal of Agronomy*, 43, 119–128.
Alburquerque, J. A., de la Fuente, C., & Bernal, M. P. (2011). Chemical properties of anaerobic digestates affecting C and N dynamics in amended soils. *Agriculture, Ecosystems & Environment*, 160, 15–22.
Alburquerque, J. A., de la Fuente, C., Ferrer-Costa, A., Carrasco, L., Cegarra, J., Abad, M., & Bernal, M. P. (2012). Assessment of the fertiliser potential of digestates from farm and agroindustrial residues. *Biomass and bioenergy*, 40, 181–189.
Amonette, J. E., & Joseph, S. (2009). Characteristics of biochar: Microchemical properties. *Biochar for Environmental Management: Science and Technology*, 33–52.
Antal, M. J., & Grønli, M. (2003). The art, science, and technology of charcoal production. *Industrial & Engineering Chemistry Research*, 42(8), 1619–1640.
Appels, L., Lauwers, J., Degrève, J., Helsen, L., Lievens, B., Willems, K., ... Dewil, R. (2011). Anaerobic digestion in global bio-energy production: Potential and research challenges. *Renewable and Sustainable Energy Reviews*, 15(9), 4295–4301. doi:10.1016/j.rser.2011.07.121.
Barnsley, I., Blank, A., & Brown, A. (2015). Enabling renewable energy and energy efficiency technologies. Opportunities in Eastern Europe, Caucasus, Central Asia, Southern and Eastern Mediterranean, 119. Retrieved from http://www.iea.org/publications/insights/insightpublications/EnablingRenewableEnergyandEnergyEfficiencyTechnologies.pdf
Benito, M., Masaguer, A., Moliner, A., Arrigo, N., & Palma, R. M. (2003). Chemical and microbiological parameters for the characterisation of the stability and maturity of pruning waste compost. *Biology and Fertility of Soils*, 37(3), 184–189.
Biagini, E., Cioni, M., & Tognotti, L. (2005). Development and characterization of a lab-scale entrained flow reactor for testing biomass fuels. *Fuel*, 84(12), 1524–1534.
Bioenergy, I. (2012). Nutrient recovery from digestate and biogas utilisation by up-grading and grid injection. In 37th ed., IEA Bioenergy "Energy from Biogas", Switzerland.
Birat, J.-P. (2003). Recycling and by-products in the steel industry. *Revue de Métallurgie*, 100(4), 339–348.
Bird, M. I., Wurster, C. M., de Paula Silva, P. H., Bass, A. M., & De Nys, R. (2011). Algal biochar–production and properties. *Bioresource Technology*, 102(2), 1886–1891.
Boruff, C. (1947). Recovery of fermentation residues as feeds. *Industrial & Engineering Chemistry*, 39(5), 602–607.
Bossen, D., Mertens, D., & Weisbjerg, M. R. (2008). Influence of fermentation methods on neutral detergent fiber degradation parameters. *Journal of Dairy Science*, 91(4), 1464–1476.
Bouallagui, H., Lahdheb, H., Ben Romdan, E., Rachdi, B., & Hamdi, M. (2009). Improvement of fruit and vegetable waste anaerobic digestion performance and stability with co-substrates addition. *Journal of Environmental Management*, 90(5), 1844–1849.
Bougrier, C., Delgenès, J. P., & Carrere, H. (2006). Combination of thermal treatments and anaerobic digestion to reduce sewage sludge quantity and improve biogas yield. *Process Safety and Environmental Protection*, 84(4), 280–284.
Brewer, L. J., & Sullivan, D. M. (2003). Maturity and stability evaluation of composted yard rimmings. *Compost Science and Utilization*, 11(2), 96–112.

Brown, D., & Li, Y. (2012). Solid state anaerobic co-digestion of yard waste and food waste for biogas production. *Bioresource Technology*.

Brown, D., & Li, Y. (2013). Solid state anaerobic co-digestion of yard waste and food waste for biogas production. *Bioresource Technology*, 127, 275–280.

Brown, D., Shi, J., & Li, Y. (2012). Comparison of solid-state to liquid anaerobic digestion of lignocellulosic feedstocks for biogas production. *Bioresource Technology*, 124, 379–386.

Businelli, D., Massaccesi, L., Said-Pullicino, D., & Gigliotti, G. (2009). Long-term distribution, mobility and plant availability of compost-derived heavy metals in a landfill covering soil. *Science of the Total Environment*, 407(4), 1426–1435.

Bustamante, M., Moral, R., Paredes, C., Vargas-García, M., Suárez-Estrella, F., & Moreno, J. (2008). Evolution of the pathogen content during co-composting of winery and distillery wastes. *Bioresource Technology*, 99(15), 7299–7306.

Bustamante, M., Restrepo, A., Alburquerque, J., Pérez-Murcia, M., Paredes, C., Moral, R., & Bernal, M. (2012). Recycling of anaerobic digestates by composting: Effect of the bulking agent used. *Journal of Cleaner Production*.

Cabaraban, M. T. I., Khire, M. V., & Alocilja, E. C. (2008). Aerobic in-vessel composting versus bioreactor landfilling using life cycle inventory models. *Clean Technologies and Environmental Policy*, 10(1), 39–52.

Cely, P., Gascó, G., Paz-Ferreiro, J., & Méndez, A. (2015). Agronomic properties of biochars from different manure wastes. *Journal of Analytical and Applied Pyrolysis*, 111, 173–182.

Chan, K., Van Zwieten, L., Meszaros, I., Downie, A., & Joseph, S. (2008). Using poultry litter biochars as soil amendments. *Soil Research*, 46(5), 437–444.

Chang, J. I., & Chen, Y. (2010). Effects of bulking agents on food waste composting. *Bioresource Technology*, 101(15), 5917–5924.

Chen, S., Chen, B., & Song, D. (2012). Life-cycle energy production and emissions mitigation by comprehensive biogas–digestate utilization. *Bioresource Technology*, 114, 357–364.

Chen, Y., Cheng, J. J., & Creamer, K. S. (2008). Inhibition of anaerobic digestion process: A review. *Bioresource Technology*, 99(10), 4044–4064. doi:10.1016/j.biortech.2007.01.057.

Chunbao Charles, X., & Cang, D.-q. (2010). A brief overview of low CO_2 emission technologies for iron and steel making. *Journal of Iron and Steel Research, International*, 17(3), 1–7.

Chynoweth, D. P., Owens, J. M., & Legrand, R. (2001). Renewable methane from anaerobic digestion of biomass. *Renewable energy*, 22(1), 1–8.

Clarke, W. P. (2002). Cost-benefit analysis of introducing technology to rapidly degrade municipal solid waste. *Waste Management and Research*, 18(6), 510–524.

Cooney, M., Maynard, N., Cannizzaro, C., & Benemann, J. (2007). Two-phase anaerobic digestion for production of hydrogen–methane mixtures. *Bioresource Technology*, 98(14), 2641–2651. doi:10.1016/j.biortech.2006.09.054.

Czernik, S., & Bridgwater, A. V. (2004). Overview of applications of biomass fast pyrolysis oil. *Energy & Fuels*, 18(2), 590–598. doi:10.1021/ef034067u.

Das, S., Sharma, S., & Choudhury, R. (2002). Non-coking coal to coke: Use of biomass based blending material. *Energy*, 27(4), 405–414.

Datar, R. P., Shenkman, R. M., Cateni, B. G., Huhnke, R. L., & Lewis, R. S. (2004). Fermentation of biomass-generated producer gas to ethanol. *Biotechnology and bioengineering*, 86(5), 587–594.

Dayton, D. C., Turk, B., & Gupta, R. (2011). Syngas cleanup, conditioning, and utilization. *Thermochemical Processing of Biomass: Conversion into Fuels, Chemicals and Power*, 12, 78.

Demattê, J., Gama, M., Cooper, M., Araújo, J., Nanni, M., & Fiorio, P. (2004). Effect of fermentation residue on the spectral reflectance properties of soils. *Geoderma*, 120(3), 187–200.

Demirbas, A. (2001). Biomass to charcoal, liquid, and gaseous products via carbonization process. *Energy Sources*, 23(6), 579–588.

Dennis, W. (1945). Method of treating fermentation residues: Google patents.

Dogan, E., Sengorur, B., & Koklu, R. (2009). Modeling biological oxygen demand of the melen river in Turkey using an artificial neural network technique. *Journal of Environmental Management*, 90(2), 1229–1235.

Downie, A., Crosky, A., & Munroe, P. (2009). Physical properties of biochar. *Biochar for Environmental Management: Science and Technology*, 13–32.

El-Mashad, H. M., & Zhang, R. (2010). Biogas production from co-digestion of dairy manure and food waste. *Bioresource Technology*, 101(11), 4021–4028.

Evangelisti, S., Lettieri, P., Borello, D., & Clift, R. (2014). Life cycle assessment of energy from waste via anaerobic digestion: A UK case study. *Waste Management*, 34(1), 226–237.

Evans, A., Strezov, V., & Evans, T. J. (2015). Bio-oil applications and processing.

Fenández, L. E. M., Sørensen, H. R., Jørgensen, C., Pedersen, S., Meyer, A. S., & Roepstorff, P. (2007). Characterization of oligosaccharides from industrial fermentation residues by matrix-assisted laser desorption/ionization, electro spray mass spectrometry, and gas chromatography mass spectrometry. *Molecular Biotechnology*, 35(2), 149–160.

Ferreira, V. J., López-Sabirón, A. M., Royo, P., Aranda-Usón, A., & Ferreira, G. (2015). Integration of environmental indicators in the optimization of industrial energy management using phase change materials. *Energy Conversion and Management*, 104, 67–77.

Ferreras, L., Gómez, E., Toresani, S., Firpo, I., & Rotondo, R. (2006). Effect of organic amendments on some physical, chemical and biological properties in a horticultural soil. *Bioresource Technology*, 97(4), 635–640.

Forster-Carneiro, T., Pérez, M., & Romero, L. (2008). Influence of total solid and inoculum contents on performance of anaerobic reactors treating food waste. *Bioresource Technology*, 99(15), 6994–7002.

Fuchs, J. G., Berner, A., Mayer, J., Smidt, E., & Schleiss, K. (2008). Influence of compost and digestates on plant growth and health: Potentials and limits. *Paper Presented at the Proceedings of the International Congress CODIS 2008.*

Fukuda, H., Kondo, A., & Noda, H. (2001). Biodiesel fuel production by transesterification of oils. *Journal of Bioscience and Bioengineering*, 92(5), 405–416.

Fuwape, J. A. (1996). Effects of carbonisation temperature on charcoal from some tropical trees. *Bioresource Technology*, 57(1), 91–94.

Gómez, X., Cuetos, M., García, A., & Morán, A. (2007). An evaluation of stability by thermogravimetric analysis of digestate obtained from different biowastes. *Journal of Hazardous Materials*, 149(1), 97–105.

Goyal, S., Dhull, S., & Kapoor, K. (2005). Chemical and biological changes during composting of different organic wastes and assessment of compost maturity. *Bioresource Technology*, 96(14), 1584–1591.

Grierson, S. (2012). *A Systems Approach to Thermochemical Conversion and Carbon Sequestration from Microalgae.* Sydney, Australia: Macquarie University Sydney.

Grierson, S., Strezov, V., & Shah, P. (2011). Properties of oil and char derived from slow pyrolysis of tetraselmis chui. *Bioresource Technology*, 102(17), 8232–8240.

Haas, T. J., Nimlos, M. R., & Donohoe, B. S. (2009). Real-time and post-reaction microscopic structural analysis of biomass undergoing pyrolysis. *Energy & Fuels*, 23(7), 3810–3817.

Hao, X., Chang, C., Larney, F. J., & Travis, G. R. (2001). Greenhouse gas emissions during cattle feedlot manure composting. *Journal of Environmental Quality*, 30(2), 376–386.

Holm-Nielsen, J. B., Al Seadi, T., & Oleskowicz-Popiel, P. (2009). The future of anaerobic digestion and biogas utilization. *Bioresource Technology*, 100(22), 5478–5484.

Hossain, M. K., Strezov, V., Chan, K. Y., Ziolkowski, A., & Nelson, P. F. (2011). Influence of pyrolysis temperature on production and nutrient properties of wastewater sludge biochar. *Journal of Environmental Management*, 92(1), 223–228.

Igoni, A. H., Ayotamuno, M., Eze, C., Ogaji, S., & Probert, S. (2008). Designs of anaerobic digesters for producing biogas from municipal solid-waste. *Applied Energy*, 85(6), 430–438.

Isahak, W. N. R. W., Hisham, M. W. M., Yarmo, M. A., & Yun Hin, T.-Y. (2012). A review on bio-oil production from biomass by using pyrolysis method. *Renewable and Sustainable Energy Reviews*, 16(8), 5910–5923.

Ishii, K., & Takii, S. (2003). Comparison of microbial communities in four different composting processes as evaluated by denaturing gradient gel electrophoresis analysis. *Journal of Applied Microbiology*, 95(1), 109–119.

Izumi, K., Okishio, Y.-k., Nagao, N., Niwa, C., Yamamoto, S., & Toda, T. (2010). Effects of particle size on anaerobic digestion of food waste. *International Biodeterioration & Biodegradation*, 64(7), 601–608.

Juang, C.-P., Whang, L.-M., & Cheng, H.-H. (2011). Evaluation of bioenergy recovery processes treating organic residues from ethanol fermentation process. *Bioresource Technology*, 102(9), 5394–5399.

Kan, T., Strezov, V., & Evans, T. J. (2016). Lignocellulosic biomass pyrolysis: A review of product properties and effects of pyrolysis parameters. *Renewable and Sustainable Energy Reviews*, 57, 1126–1140.

Kaparaju, P., & Rintala, J. (2005). Anaerobic co-digestion of potato tuber and its industrial by-products with pig manure. *Resources, Conservation and Recycling*, 43(2), 175–188. doi:10.1016/j.resconrec.2004.06.001.

Khan, A. W., Rahman, M. S., & Takashi, A. (2009). Application of malt residue in submerged fermentation of bacillus subtilis. *Journal of Environmental Sciences*, 21, S33–S35.

Klass, D. L. (2004). Biomass for renewable energy and fuels. *Encyclopedia of Energy*, 1(1), 193–212.

Koido, K., Hanaoka, T., & Sakanishi, K. (2013). Pressurised gasification of wet ethanol fermentation residue for synthesis gas production. *Bioresource Technology*, 131, 341–348.

Kratzeisen, M., Starcevic, N., Martinov, M., Maurer, C., & Müller, J. (2010). Applicability of biogas digestate as solid fuel. *Fuel*, 89(9), 2544–2548.

Lasaridi, K. E., & Stentiford, E. I. (1998). A simple respirometric technique for assessing compost stability. *Water Research*, 32(12), 3717–3723.

Laurent, A., Clavreul, J., Bernstad, A., Bakas, I., Niero, M., Gentil, E., ... Hauschild, M. Z. (2014). Review of LCA studies of solid waste management systems–Part II: Methodological guidance for a better practice. *Waste Management*, 34(3), 589–606.

Lehmann, J., & Joseph, S. (2012). *Biochar for Environmental Management: Science and Technology*. London, UK: Routledge.

Lehtomäki, A., & Björnsson, L. (2006). Two-stage anaerobic digestion of energy crops: Methane production, nitrogen mineralisation and heavy metal mobilisation. *Environmental Technology*, 27(2), 209–218.

Lewellen, P., Peters, W., & Howard, J. (1977). Cellulose pyrolysis kinetics and char formation mechanism. *Paper presented at the Symposium (International) on Combustion*.

Li, Y., Park, S. Y., & Zhu, J. (2011). Solid-state anaerobic digestion for methane production from organic waste. *Renewable and Sustainable Energy Reviews*, 15(1), 821–826.

Liew, L. N., Shi, J., & Li, Y. (2012). Methane production from solid-state anaerobic digestion of lignocellulosic biomass. *Biomass and Bioenergy*, 46, 125–132.

López-Sabirón, A. M., Royo, P., Ferreira, V. J., Aranda-Usón, A., & Ferreira, G. (2014). Carbon footprint of a thermal energy storage system using phase change materials for industrial energy recovery to reduce the fossil fuel consumption. *Applied Energy*, 135, 616–624.

Lukehurst, C. T., Frost, P., & Al Seadi, T. (2010). Utilisation of digestate from biogas plants as biofertiliser. *IEA Bioenergy*, 1–36.

Lundie, S., & Peters, G. M. (2005). Life cycle assessment of food waste management options. *Journal of Cleaner Production*, 13(3), 275–286.

Ly, Kim, S.-S., Woo, H. C., Choi, J. H., Suh, D. J., & Kim, J. (2015). Fast pyrolysis of macroalga saccharina japonica in a bubbling fluidized-bed reactor for bio-oil production. *Energy*, 93, Part 2, 1436–1446. doi:10.1016/j.energy.2015.10.011.

Macias-Corral, M., Samani, Z., Hanson, A., Smith, G., Funk, P., Yu, H., & Longworth, J. (2008). Anaerobic digestion of municipal solid waste and agricultural waste and the effect of co-digestion with dairy cow manure. *Bioresource Technology*, 99(17), 8288–8293.

Manungufala, T., Chimuka, L., & Maswanganyi, B. (2008). Evaluating the quality of communities made compost manure in South Africa: A case study of content and sources of metals in compost manure from Thulamela Municipality, Limpopo province. *Bioresource Technology*, 99(5), 1491–1496.

Martin, J. H. (2005). *An Evaluation of a Mesophilic, Modified Plug Flow Anaerobic Digester For Dairy Cattle Manure*. Chantilly VA: Eastern Research Group, Inc. EPA Contract No. GS 10F-0036K Work Assignment/Task Order (9).

Matheys, J., Van Autenboer, W., Timmermans, J.-M., Van Mierlo, J., Van den Bossche, P., & Maggetto, G. (2007). Influence of functional unit on the life cycle assessment of traction batteries. *The International Journal of Life Cycle Assessment*, 12(3), 191–196.

McKendry, P. (2002). Energy production from biomass (part 2): Conversion technologies. *Bioresource Technology*, 83(1), 47–54.

Meyers, H., & Jennings, R. F. (1978). Charcoal ironmaking: A technical and economic review of Brazilian experience. UNIDO/IOD.228; 1978.b

Moberg, Å., Finnveden, G., Johansson, J., & Lind, P. (2005). Life cycle assessment of energy from solid waste—part 2: Landfilling compared to other treatment methods. *Journal of Cleaner Production*, 13(3), 231–240.

Mochidzuki, K., Soutric, F., Tadokoro, K., Antal, M. J., Tóth, M., Zelei, B., & Várhegyi, G. (2003). Electrical and physical properties of carbonized charcoals. *Industrial & Engineering Chemistry Research*, 42(21), 5140–5151.

Mohan, D., Pittman, C. U., & Steele, P. H. (2006). Pyrolysis of wood/biomass for bio-oil: A critical review. *Energy & Fuels*, 20(3), 848–889.

Monlau, F., Sambusiti, C., Antoniou, N., Barakat, A., & Zabaniotou, A. (2015). A new concept for enhancing energy recovery from agricultural residues by coupling anaerobic digestion and pyrolysis process. *Applied Energy*, 148, 32–38.

Montemurro, F., Ferri, D., Tittarelli, F., Canali, S., & Vitti, C. (2010). Anaerobic digestate and on-farm compost application: Effects on lettuce (lactuca sativa L.) crop production and soil properties. *Compost Science & Utilization*, 18(3), 184–193. doi:10.1080/1065657x.2010.10736954.

Montoya, J., Valdés, C., Chejne, F., Gómez, C., Blanco, A., Marrugo, G., ... Acero, J. (2015). Bio-oil production from Colombian bagasse by fast pyrolysis in a fluidized bed: An experimental study. *Journal of Analytical and Applied Pyrolysis*, 112, 379–387.

Mullen, C. A., Boateng, A. A., Goldberg, N. M., Lima, I. M., Laird, D. A., & Hicks, K. B. (2010). Bio-oil and bio-char production from corn cobs and stover by fast pyrolysis. *Biomass and Bioenergy*, 34(1), 67–74. doi:10.1016/j.biombioe.2009.09.012.

Nallathambi Gunaseelan, V. (1997). Anaerobic digestion of biomass for methane production: A review. *Biomass and Bioenergy*, 13(1), 83–114.

Nguyen, B. T., & Lehmann, J. (2009). Black carbon decomposition under varying water regimes. *Organic Geochemistry*, 40(8), 846–853.

Ni, M., Leung, D. Y. C., Leung, M. K. H., & Sumathy, K. (2006). An overview of hydrogen production from biomass. *Fuel Processing Technology*, 87(5), 461–472.

Nieves, D. C., Karimi, K., & Horváth, I. S. (2011). Improvement of biogas production from oil palm empty fruit bunches (OPEFB). *Industrial Crops and Products*, 34(1), 1097–1101.

Novak, J., & Busscher, W. (2011). Selection and use of designer biochars to improve characteristics of southeastern USA Coastal Plain degraded soils. *Advanced Biofuels and Bioproducts*, 69–96.

NSW SERDF (2001). Sustainable steelmaking using renewable forest energy, NSW sustainable energy research and development fund, ACARP, Brisbane, Australia.

Oasmaa, A., Sundqvist, T., Kuoppala, E., Garcia-Perez, M., Solantausta, Y., Lindfors, C., & Paasikallio, V. (2015). Controlling the phase stability of biomass fast pyrolysis bio-oils. *Energy & Fuels*, 29(7), 4373–4381.

Opatokun, S. A., Kan, T., Al Shoaibi, A. S., Srinivasakannan, C., & Strezov, V. (2015). Characterisation of food waste and its digestate as feedstock for thermochemical processing. *Energy & Fuels*, 30(3), 1589–1597.

Opatokun, S. A., Strezov, V., & Kan, T. (2015). Product based evaluation of pyrolysis of food waste and its digestate. *Energy*, 92, 349–354.

Paavola, T., & Rintala, J. (2008). Effects of storage on characteristics and hygienic quality of digestates from four co-digestion concepts of manure and biowaste. *Bioresource Technology*, 99(15), 7041–7050. doi:10.1016/j.biortech.2008.01.005.

Pandey, B., & Sharma, T. (2000). Reducing agents and double-layered iron ore pellets. *International Journal of Mineral Processing*, 59(4), 295–304.

Parawira, W., Read, J. S., Mattiasson, B., & Björnsson, L. (2008). Energy production from agricultural residues: High methane yields in pilot-scale two-stage anaerobic digestion. *Biomass and Bioenergy*, 32(1), 44–50. doi:10.1016/j.biombioe.2007.06.003

Parikka, M. (2004). Global biomass fuel resources. *Biomass and Bioenergy*, 27(6), 613–620.

Peigne, J., & Girardin, P. (2004). Environmental impacts of farm-scale composting practices. *Water, Air, & Soil Pollution*, 153(1), 45–68.

Perez, M., Romero, L., & Sales, D. (2001). Organic matter degradation kinetics in an anaerobic thermophilic fluidised bed bioreactor. *Anaerobe*, 7(1), 25–35.

Pötsch, E., Pfundtner, E., Much, P., Lüscher, A., Jeangros, B., Kessler, W., ... Suter, D. (2004). Nutrient content and hygienic properties of fermentation residues from agricultural biogas plants. Paper presented at the Land use systems in grassland dominated regions. *Proceedings of the 20th General Meeting of the European Grassland Federation*, Luzern, Switzerland, June 21–24, 2004.

Raposo, F., Banks, C., Siegert, I., Heaven, S., & Borja, R. (2006). Influence of inoculum to substrate ratio on the biochemical methane potential of maize in batch tests. *Process Biochemistry*, 41(6), 1444–1450.

Rasi, S., Läntelä, J., & Rintala, J. (2011). Trace compounds affecting biogas energy utilisation–A review. *Energy Conversion and Management*, 52(12), 3369–3375.

Rasi, S., Veijanen, A., & Rintala, J. (2007). Trace compounds of biogas from different biogas production plants. *Energy*, 32(8), 1375–1380.

Rasul, M., & Jahirul, M. (2012). *Recent Developments in Biomass Pyrolysis for Bio-fuel Production: Its Potential for Commercial Applications*. Australia: Central Queensland University, Centre for Plant and Water Science, Faculty of Sciences, Engineering and Health.

Roberts, K. G., Gloy, B. A., Joseph, S., Scott, N. R., & Lehmann, J. (2009). Life cycle assessment of biochar systems: Estimating the energetic, economic, and climate change potential. *Environmental Science & Technology*, 44(2), 827–833.

Rondon, M. A., Lehmann, J., Ramírez, J., & Hurtado, M. (2007). Biological nitrogen fixation by common beans (*Phaseolus vulgaris* L.) increases with bio-char additions. *Biology and Fertility of Soils*, 43(6), 699–708.

Said-Pullicino, D., Erriquens, F. G., & Gigliotti, G. (2007). Changes in the chemical characteristics of water-extractable organic matter during composting and their influence on compost stability and maturity. *Bioresource Technology*, 98(9), 1822–1831.

Salminen, E. A., & Rintala, J. A. (2002). Semi-continuous anaerobic digestion of solid poultry slaughterhouse waste: Effect of hydraulic retention time and loading. *Water Research*, 36(13), 3175–3182.

Santibáñez, C., Varnero, M. T., & Bustamante, M. (2011). Residual glycerol from biodiesel manufacturing, waste or potential source of bioenergy: A review. *Chilean Journal of Agricultural Research*, 71, 3.

Selling, R., Hakansson, T., & Bjornsson, L. (2008). Two-stage anaerobic digestion enables heavy metal removal. *Water Science and Technology*, 57(4), 553–558.

Şensöz, S., & Kaynar, İ. (2006). Bio-oil production from soybean (Glycine max L.); fuel properties of Bio-oil. *Industrial Crops and Products*, 23(1), 99–105. doi:10.1016/j.indcrop.2005.04.005.

Shah, A., Darr, M. J., Dalluge, D., Medic, D., Webster, K., & Brown, R. C. (2012). Physicochemical properties of bio-oil and biochar produced by fast pyrolysis of stored single-pass corn stover and cobs. *Bioresource Technology*, 125, 348–352. doi:10.1016/j.biortech.2012.09.061.

Shin, H.-C., Park, J.-W., Park, K., & Song, H.-C. (2002). Removal characteristics of trace compounds of landfill gas by activated carbon adsorption. *Environmental Pollution*, 119(2), 227–236.

Singh, A., Nizami, A. S., Korres, N. E., & Murphy, J. D. (2011). The effect of reactor design on the sustainability of grass biomethane. *Renewable and Sustainable Energy Reviews*, 15(3), 1567–1574.

Smith, D. C., & Hughes, J. C. (2004). Changes in maturity indicators during the degradation of organic wastes subjected to simple composting procedures. *Biology and Fertility of Soils*, 39(4), 280–286.

Smith, K. A., Grylls, J., Jeffrey, B., & Sinclair, A. (2007). Nutrient value of digestate from farm-based biogas plants in Scotland. In *Report for Scottish Executive Environment and Rural Affair Department* (Ed.), (pp. 44). ADAS UK Ltd., and SAC Commercial Ltd.

Smith, K. A., Jeffrey, W. A., Metcalfe, J. P., Sinclair, A. H., & Williams, J. R. (2010). Nutrient value of digestate from farm-based biogas plants. *Paper Presented at the Proceedings of the 14th RAMIRAN International Conference*, Lisbon, Portugal.

Smith, S. R. (2009). A critical review of the bioavailability and impacts of heavy metals in municipal solid waste composts compared to sewage sludge. *Environment International*, 35(1), 142.

Sohi, S., Krull, E., Lopez-Capel, E., & Bol, R. (2010). A review of biochar and its use and function in soil. *Advances in Agronomy*, 105, 47–82.

Sonesson, U., Björklund, A., Carlsson, M., & Dalemo, M. (2000). Environmental and economic analysis of management systems for biodegradable waste. *Resources, Conservation and Recycling*, 28(1), 29–53.

Spath, P. L., & Dayton, D. C. (2003). Preliminary screening-technical and economic assessment of synthesis gas to fuels and chemicals with emphasis on the potential for biomass-derived syngas: DTIC Document.

Srinivasan, P., Sarmah, A. K., Smernik, R., Das, O., Farid, M., & Gao, W. (2015). A feasibility study of agricultural and sewage biomass as biochar, bioenergy and biocomposite feedstock: Production, characterization and potential applications. *Science of the Total Environment*, 512, 495–505.

Steinbeiss, S., Gleixner, G., & Antonietti, M. (2009). Effect of biochar amendment on soil carbon balance and soil microbial activity. *Soil Biology and Biochemistry*, 41(6), 1301–1310.

Strezov, V. (2006). Iron ore reduction using sawdust: Experimental analysis and kinetic modelling. *Renewable Energy*, 31(12), 1892–1905.

Strezov, V., Evans, T. J. & Nelson, P. F. (2006) Carbonization of biomass fuels, in *Biomass and Bioenergy: New Research*, M. D. Brenes (Ed.), New York: Nova Science Publishers, pp. 91–123.

Strezov, V., Patterson, M., Zymla, V., Fisher, K., Evans, T. J., & Nelson, P. F. (2007). Fundamental aspects of biomass carbonisation. *Journal of Analytical and Applied Pyrolysis*, 79(1), 91–100.

Stroot, P. G., McMahon, K. D., Mackie, R. I., & Raskin, L. (2001). Anaerobic codigestion of municipal solid waste and biosolids under various mixing conditions—I. Digester performance. *Water Research*, 35(7), 1804–1816.

Su, W., Ma, H., Wang, Q., Li, J., & Ma, J. (2012). Thermal behavior and gaseous emission analysis during co-combustion of ethanol fermentation residue from food waste and coal using TG-FTIR. *Journal of Analytical and Applied Pyrolysis*.

Sung, S., & Liu, T. (2003). Ammonia inhibition on thermophilic anaerobic digestion. *Chemosphere*, 53(1), 43–52. doi:10.1016/s0045-6535(03)00434-x.

Takekawa, M., Wakimoto, K., Matsu-Ura, M., Hasegawa, M., Iwase, M., & McLean, A. (2003). Investigation of waste wood as a blast furnace injectant. *Steel Research International*, 74(6), 347–350.

Tambone, F., Scaglia, B., D'Imporzano, G., Schievano, A., Orzi, V., Salati, S., & Adani, F. (2010). Assessing amendment and fertilizing properties of digestates from anaerobic digestion through a comparative study with digested sludge and compost. *Chemosphere*, 81(5), 577–583.

Tang, J. C., Maie, N., Tada, Y., & Katayama, A. (2006). Characterization of the maturing process of cattle manure compost. *Process Biochemistry*, 41(2), 380–389.

Tao, W., Hall, K. J., & Duff, S. J. B. (2006). Performance evaluation and effects of hydraulic retention time and mass loading rate on treatment of woodwaste leachate in surface-flow constructed wetlands. *Ecological Engineering*, 26(3), 252–265.

Thommes, K., & Strezov, V. (2015). Fermentation of biomass.

Tiquia, S. (2002). Microbial transformation of nitrogen during composting. In *Microbiology of Composting*, pp. 237–245. Heidelberg, Germany: Springer Verlag.

Treusch, O., Hofenauer, A., Tröger, F., Fromm, J., & Wegener, G. (2004). Basic properties of specific wood-based materials carbonised in a nitrogen atmosphere. *Wood Science and Technology*, 38(5), 323–333.

UNEP. (2011). Waste investing in energy and resource efficiency. Towards a green economy, 285–327. Retrieved from http://www.unep.org/greeneconomy/Portals/88/documents/ger/GER_8_Waste.pdf

Vavilin, V., Fernandez, B., Palatsi, J., & Flotats, X. (2008). Hydrolysis kinetics in anaerobic degradation of particulate organic material: An overview. *Waste Management*, 28(6), 939–951.

Vindis, P., Mursec, B., Janzekovic, M., & Cus, F. (2009). The impact of mesophilic and thermophilic anaerobic digestion on biogas production. *Journal of Achievements in Materials and Manufacturing Engineering*, 36(2), 192–198.

Viotti, P., Di Genova, P., & Falcioli, F. (2004). Numerical analysis of the anaerobic co-digestion of the organic fraction from municipal solid waste and wastewater: Prediction of the possible performances at Olmeto plant in Perugia (Italy). *Waste Management & Research*, 22(2), 115–128.

Volli, V., & Singh, R. (2012). Production of bio-oil from de-oiled cakes by thermal pyrolysis. *Fuel*, 96, 579–585.

Walsh, J. J., Rousk, J., Edwards-Jones, G., Jones, D. L., & Williams, A. P. (2012). Fungal and bacterial growth following the application of slurry and anaerobic digestate of livestock manure to temperate pasture soils. *Biology and Fertility of Soils*, 48(8), 889–897.

Wang, Z., Lin, W., Song, W., & Wu, X. (2012). Pyrolysis of the lignocellulose fermentation residue by fixed-bed micro reactor. *Energy*.

Ward, A. J., Hobbs, P. J., Holliman, P. J., & Jones, D. L. (2008). Optimisation of the anaerobic digestion of agricultural resources. *Bioresource Technology*, 99(17), 7928–7940.

Weiland, P. (2010). Biogas production: Current state and perspectives. *Applied Microbiology and Biotechnology*, 85(4), 849–860.

Xu, F., Shi, J., Lv, W., Yu, Z., & Li, Y. (2012). Comparison of different liquid anaerobic digestion effluents as inocula and nitrogen sources for solid-state batch anaerobic digestion of corn stover. *Waste Management*.

Yan, F., Luo, S.-y., Hu, Z.-q., Xiao, B., & Cheng, G. (2010). Hydrogen-rich gas production by steam gasification of char from biomass fast pyrolysis in a fixed-bed reactor: Influence of temperature and steam on hydrogen yield and syngas composition. *Bioresource Technology*, 101(14), 5633–5637.

Yorgun, S., & Yıldız, D. (2015). Slow pyrolysis of paulownia wood: Effects of pyrolysis parameters on product yields and bio-oil characterization. *Journal of Analytical and Applied Pyrolysis*, 114, 68–78. doi:10.1016/j.jaap.2015.05.003.

Zheljazkov, V. D., & Warman, P. R. (2002). Comparison of three digestion methods for the recovery of 17 plant essential nutrients and trace elements from six composts. *Compost Science & Utilization*, 10(3), 197–203.

Zheng, Y., Li, Z., Feng, S., Lucas, M., Wu, G., Li, Y., ... Jiang, G. (2010). Biomass energy utilization in rural areas may contribute to alleviating energy crisis and global warming: A case study in a typical agro-village of Shandong, China. *Renewable and Sustainable Energy Reviews*, 14(9), 3132–3139.

10 Application of Biochar for Carbon Sequestration in Soils

Yani Kendra, Vladimir Strezov, and Hossain M. Anawar

CONTENTS

10.1 INTRODUCTION

Climate change is one of the major environmental challenges at present. Since the 1950s, it has been observed that the atmospheric greenhouse gas concentrations have increased, impacting the global climate (IPCC, 2013). It is predicted that if the climate continues to change at its current rate, it is likely to result in a number of adverse impacts, such as increases in the duration, intensity, and spatial extent of heat waves; an increased frequency of heavy precipitation events; and a possible increase in the intensity and frequency of tropical cyclones (Kirtman et al., 2013). However, through the introduction of effective mitigation strategies, there is the opportunity to decrease the rate at which climate change is occurring and, as a result, reduce the possibility of adverse climate change impacts.

A number of natural and anthropogenic substances and processes have been identified as the drivers of climate change. The largest contribution to the increase in surface warming has been attributed to the increase in atmospheric CO_2 concentrations (IPCC, 2013). For this reason, designing mitigation strategies and technologies aimed at reducing anthropogenic contributions to atmospheric CO_2 concentrations may be the most promising solution for reducing the adverse climate change impacts. One possible mitigation strategy for reduction of atmospheric CO_2 concentrations is the use of biochar for carbon sequestration in soils.

Biochar is a charcoal substance that is produced through pyrolysis of biomass. During the pyrolysis process, biomass undergoes thermal decomposition in the absence of oxygen, producing solid (biochar), liquid (bio-oil), and gas products (Bhattacharya et al., 2015). Biochar consists of highly recalcitrant carbon structures, preventing its degradation (Woolf et al., 2010; Bhattacharya et al., 2015), which allows for long-term storage and sequestration of carbon in the soils. In addition to its carbon-sequestration potential, the addition of biochar to soils improves the quality of low-fertile and degraded soils and improves agricultural productivity (Woolf et al., 2010) by increasing nutrient and soil moisture availability, ameliorating acidic soils, and stimulating soil microbial activity (Anawar et al., 2015). For these reasons, the use of biochar to sequester carbon in soils may be a promising and highly beneficial climate change mitigation strategy.

For biochar to be effective in the sequestration of carbon in soils, the properties relating to its stability in soil must be fully understood. A number of studies have been conducted related to factors that may affect the stability of biochar in soil. These have included studies assessing various production conditions and feedstock types, as well as a number of soil properties that may affect the stability of biochar. A variety of analytical methods have been utilized, and the scale of research has varied from laboratory research to larger-scale field incubation trials. This chapter aims to first discuss selected methods used for the analysis of biochar stability in soils and synthesize information gained from research undertaken in the past five years to determine the production conditions, feedstock types, and soil conditions under which biochar is most stable and therefore will be most effective in sequestering carbon in soil to mitigate climate change.

10.2 METHODS FOR TESTING STABILITY OF BIOCHARS IN SOILS

10.2.1 FIELD STUDIES OF BIOCHAR STABILITY IN SOILS

For the analysis of biochar stability in soils, studies in which biochar is emplaced and monitored at a field scale can provide insight into biochar stability under realistic environmental conditions (Lehmann et al., 2015). Through the comparison of results yielded from a variety of field studies undertaken in different locations, the stability of biochar can be assessed in a variety of natural ecosystems and naturally occurring environmental conditions, including temperature, moisture, and native microbial communities (Gurwick et al., 2013).

The randomized complete block design method is a typical method of choice for studying biochar stability in the field (Major et al., 2010; Haefele et al., 2011; Rasse et al., 2017; Lanza et al., 2018). This method sets out a number of plots, arranged in longitudinal blocks with a buffer zone between each plot (Figure 10.1).

Plots may range in size, from as small as 4 m × 5 m (Major et al., 2010) to larger plots, such as 10 m × 4.5 m (Lanza et al., 2018); however, the selected size should be consistent throughout the experiment. Within each block, the biochar treatments to be compared are applied at random with the application of three to four replicate treatments throughout the blocks. For example, four replicate blocks may be laid out with each of the four treatments applied once in each block (Figure 10.1). The main advantage of this method in the field is that it accounts for the possibility of

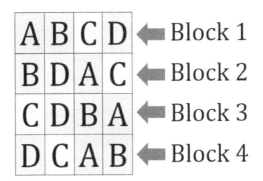

FIGURE 10.1 An example of four replications of four different treatments (A, B, C, and D) laid out using the randomized complete block design method. In the field, separation between each plot within and between the longitudinal blocks would exist.

soil inhomogeneity and uneven surface shadowing (Lanza et al., 2018) that may otherwise produce unaccountable variances between plots that have received the same treatment.

The biochar is first applied by hand, then raked over the plots for an even distribution. This is then either ploughed (Rasse et al., 2017; Lanza et al., 2018) or tilled (Major et al., 2010; Haefele et al., 2011; Dong et al., 2017) to a predetermined depth. This may range from as shallow as 5 cm (De la Rosa et al., 2018) to as deep as 23 cm (Rasse et al., 2017), with this depth kept consistent throughout the experiment. A number of analytical methods are used to determine the recalcitrance of biochar in soil in combination with the field incubation. For example, respiration can be measured in the field using chamber collars (Rasse et al., 2017; Lanza et al., 2018), or soils can be sampled with a core sampler or drilling (Haefele et al., 2011; Lanza et al., 2018). Biochar properties, such as aromaticity, can be analyzed using nuclear magnetic resonance spectroscopy (NMR), and the alteration of biochar surface topography can be assessed using field emissions scanning electron microscopy (De la Rosa et al., 2018).

Despite the advantages of the field studies to assess the stability of biochar in soil, constraints at this scale do exist. Extrapolation of the results obtained from the two- to three-year field studies to assess the recalcitration of biochar over larger time periods may only produce conservative estimates due to the possibility of later-emerging, long-term effects that may not be evident within a shorter time period (Lanza et al., 2018). Unlike controlled laboratory experiments, exposure to environmental conditions may also result in the loss of biochar through runoff, as a result of heavy rain events (Major et al., 2010). Furthermore, due to the scale of field studies, there are often limitations to the number of replications that may be used, as well as the number of biochar varieties that may be compared (Lehmann et al., 2015). The availability of land, along with the necessary amount of time and material resources that are required for field trials, can restrict the use of field study methods in the analysis of biochar stability (Brewer et al., 2011).

10.2.2 Laboratory Incubation Studies

Laboratory studies are conducted on a smaller scale and allow for greater control over experimental conditions, as well as the convenience of smaller soil and biochar sample sizes, allowing for analysis of a number of variables and the use of additional replicates. Laboratory studies can investigate the effect of specifically selected environmental conditions, soil types, varying biochar feedstock types, and pyrolysis conditions on the stability of biochar in soils (Lehmann et al., 2015). The preparation and setup of laboratory incubation studies is largely dependent on the methods selected to analyze biochar degradation at the completion or throughout the incubation period. A common method is the use of an aerobic incubation setup with the inclusion of a CO_2 trap used to quantify the respiration of CO_2 over the course of the incubation period, as a measure of biochar mineralization and stability (Knoblauch et al., 2011; Singh et al., 2012; Fang et al., 2014a; Kuzyakov et al., 2014; Wu et al., 2016).

Prior to the incubation, soil and biochar mixtures are prepared by air or oven drying and uniformly mixing with a weight for weight (w/w) percentage of approximately 2.0%–2.5% biochar in soils (Fang et al., 2014a; Wu et al., 2016). Depending on the tested variables, a nutrient solution and microbial inoculum may be added to the mixture at this stage (Singh et al., 2012; Fang et al., 2014a). A known volume of the soil and biochar mixture is packed into a plastic jar (Singh et al., 2012; Kuzyakov et al., 2014; Hansen et al., 2015; Wu et al., 2016) and water added to a known water-holding capacity, most commonly 70% (Fang et al., 2014a; Herath et al., 2014; Kuzyakov et al., 2014a). This is maintained throughout the experiment through regular weighing and watering (Hansen et al., 2015). The jars are then placed inside a sealed incubation vessel or bucket and kept in the dark at a constant temperature, unless temperature is the variable being tested. As with the field trials, a control sample without the addition of biochar and a minimum of three to four replicates of each treatment should be tested (Knoblauch et al., 2011; Singh et al., 2012; Fang et al., 2014a; Hansen et al., 2015).

Along with the soil mixtures, a CO_2 trap must be placed inside each incubation vessel. The CO_2 traps commonly consist of a glass cap or jar containing a known volume of 1.0–2.5 M solution of NaOH (Knoblauch et al., 2011; Singh et al., 2012; Fang et al., 2014a; Kuzyakov et al., 2014; Wu et al., 2016). The CO_2 is trapped by the NaOH solution via the following reaction:

$$2NaOH_{(aq)} + CO_{2(g)} \rightarrow Na_2CO_{3(aq)} + H_2O_{(l)}$$

As NaOH is the limiting reactant, the reaction reaches an equilibrium, whereby further CO_2 cannot be absorbed (Yoo et al., 2013). For this reason, the CO_2 trap must be replaced periodically throughout the incubation period. This should occur more frequently at the beginning of the incubation period, when CO_2 is expected to be mineralized at a faster rate, and less frequently with time (Fang et al., 2014a; Wu et al., 2016).

CO_2 lost from the soil through respiration is quantified by first precipitating the $Na_2CO_{3(aq)} + H_2O_{(l)}$ solution through the addition of $BaCl_2$, and then the remaining NaOH titrated with HCl using phenolphthalein as the indicator (Knoblauch et al., 2011; Singh et al., 2012; Fang et al., 2014a; Kuzyakov et al., 2014; Wu et al., 2016). Alternative methods of CO_2 quantification include CO_2 capture in KOH solution and quantifying the changes in electrolyte electrical conductivity (Bamminger et al., 2014) and measuring CO_2 emissions with a gas analyzer (Hansen et al., 2015).

There are some limitations to the use of laboratory incubation studies to measure biochar stability in soils. Results obtained from laboratory incubations cannot provide a realistic representation of biochar stability in the natural environment, as they occur in the absence of naturally occurring processes, such as litter input, and varying temperature and water dynamics (Lehmann et al., 2015). In addition, extrapolated results to assess biochar recalcitrance over longer time periods are conservative and do not consider the impact of possible long-term effects on the stability of biochar (Lanza et al., 2018).

10.3 INFLUENCE OF PRODUCTION CONDITIONS AND FEEDSTOCK TYPE ON CARBON-SEQUESTRATION POTENTIAL OF BIOCHAR

The carbon-sequestration potential of biochar is reliant on the long-term stability of biochars in soils, which is dependent on a number of factors. One factor is the intrinsic physical and chemical properties of the biochar (Wu et al., 2016), which depend on the pyrolysis conditions, such as the treatment temperature, and the properties of the original feedstock from which the biochar was produced (Jindo et al., 2014; Purakayastha et al., 2015). A number of recent studies have investigated the impacts of the pyrolysis conditions on biochar stability using a variety of techniques, which are summarized in the following section.

10.3.1 PYROLYSIS TEMPERATURE

Pyrolysis temperature has been shown to influence the chemical and physical properties of biochar, including the hydrogen-to-carbon and oxygen-to-carbon ratios (H:C and O:C) (Purakayastha et al., 2016). Decreased H:C is associated with an increase in the aromaticity of biochar, which is related to increased biochar stability (Xiao et al., 2016), while a decrease in O:C is associated with the loss of highly reactive oxygen, containing functional groups that decrease the stability of biochar (Chen et al., 2016). It is assumed that biochars with low H:C and O:C ratios could be more effective in sequestering carbon in soils.

Elemental analysis of biochar produced at varying pyrolysis temperatures, ranging from 300°C to 800°C, show an obvious trend in decreasing H:C ratio with increasing pyrolysis temperatures (Figure 10.2) (Jindo et al., 2014; Chen et al., 2016; Purakayastha et al., 2016). This trend remains strong and consistent for a variety of feedstock types, suggesting that higher pyrolysis temperatures produce biochars with lower H:C ratios. Similarly, the O:C ratio decreases with increasing

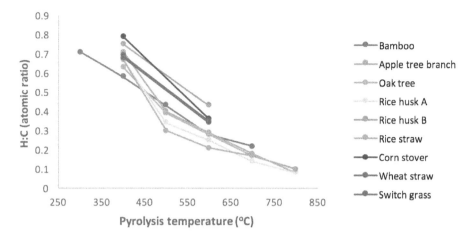

FIGURE 10.2 Effect of pyrolysis temperature on the H:C atomic ratio for a variety of feedstocks. (Data was obtained from Chen, D. et al. 2016. *Bioresour Technol* 218: 1303–1306; Jindo, K. et al. 2014. *Biogeosciences* 11: 6613–6621; and Purakayastha, T.J. et al. 2016. *Soil Tillage Res* 155: 107–115.)

pyrolysis temperatures; however, this trend is less consistent (Jindo et al., 2014; Chen et al., 2016). O:C ratios for some feedstock types increase between 500°C and 700°C. When biochars are produced at temperatures greater than 700°C, O:C ratios are lower than those for biochars produced at 400°C (Figure 10.3). This indicates that biochar produced at temperatures greater than 700°C consist of the least amount of oxygen containing functional groups and therefore are more stable than those produced at temperatures less than 400°C.

The presence of aromatic carbon structures is a defining biochar property and is a main contributor to its stability. It is believed that these aromatic structures exist in two phases within biochar, an amorphous phase and a crystalline phase consisting of a number of condensed polyaromatic sheets (Wiedemeier et al., 2015). The aromaticity of biochar refers to the total aromatic carbon in both phases, while the degree of aromatic condensation refers to the condensed aromatic carbon in the crystalline phase. These aromatic carbon structures are less available for microbial degradation and therefore are a major contributing factor to the carbon-sequestration potential of biochar in soil (Rasse et al., 2017).

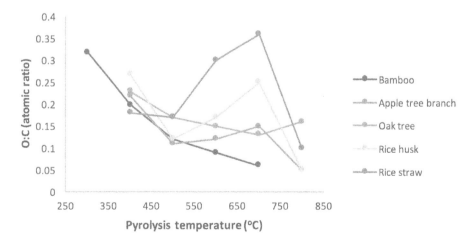

FIGURE 10.3 The effect of pyrolysis temperature on the O:C atomic ratio for a variety of feedstocks. (Data was obtained from Chen, D. et al. 2016. *Bioresour Technol* 218: 1303–1306; and Jindo, K. et al. 2014. *Biogeosciences* 11: 6613–6621.)

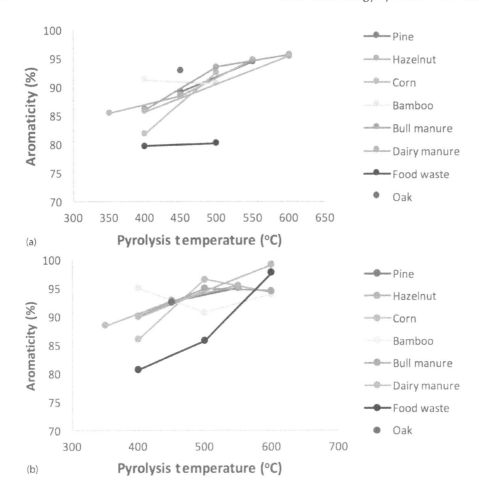

FIGURE 10.4 The effect of pyrolysis temperature on the aromaticity of biochars produced from a variety of feedstocks. (a) Results obtained using cross-polarisation spectra. (b) Results obtained using direct-polarization spectra. (Data from Mcbeath, A.V. et al. 2014. *Biomass Bioenergy* 60: 121–129.)

In order to determine the effect of pyrolysis temperature on the aromaticity and presence of aromatic carbon structures in biochar, studies have utilized nuclear NMR (Jindo et al., 2014; Mcbeath et al., 2014) and Fourier-transform infrared (FTIR) (Jindo et al., 2014; Zheng et al., 2018) spectroscopy to assess the aromatic structures present in biochars produced at different pyrolysis temperatures. NMR and FTIR results reveal that the aromaticity of biochar produced at pyrolysis temperatures ranging between 350°C and 600°C increased with the pyrolysis temperature (Figure 10.4) (Jindo et al., 2014; Mcbeath et al., 2014; Zheng et al., 2018); however, at temperatures beyond 600°C, aromaticity was observed to decrease with the decreasing intensity of peaks for aromatic groups (Mcbeath et al., 2014). This suggests that biochar produced at 600°C may contain the most aromatic structure and therefore have the most carbon-sequestering potential.

Biochar degradation is often evaluated as carbon lost through soil respiration or mineralization (Bird et al., 2017). The soil priming effect is related to mineralization, as it refers to the increase (positive priming) or decrease (negative priming) of microbial-induced carbon degradation and mineralization as a result of the addition of soil amendments, such as biochar (Cely et al., 2014). Therefore, biochar with a higher carbon-sequestering potential should exhibit lower mineralization and more negative or less positive priming effects when added to soil.

TABLE 10.1
Summary of Biochar Properties Relevant to Its Carbon-Sequestration Potential

Desired Biochar Characteristic	Optimal Pyrolysis Temperature	References
Low H:C	High pyrolysis temperatures	Chen et al. (2016); Jindo et al. (2014); Purakayastha et al. (2016)
Low O:C	>700°C	Chen et al. (2016); Jindo et al. (2014)
High aromaticity and aromatic condensation	600°C	Jindo et al. (2014); Mcbeath et al. (2014); Zheng et al. (2018)
Negative or less-positive soil priming effect	High pyrolysis temperatures	Fang et al. (2015); Sheng et al. (2016)
Lower carbon mineralization	High pyrolysis temperatures	Fang et al. (2014b, 2015)

It has been observed that, as the pyrolysis temperature increases, the soil priming effect resulting from the addition of biochar to soil becomes less positive or more negative (Fang et al., 2015; Sheng et al., 2016). However the degree of priming varies between different soil types (Sheng et al., 2016) and incubation temperatures (Fang et al., 2015). The rate of mineralization is initially higher for biochars produced at higher pyrolysis temperatures; however, this decreases rapidly, and overall total carbon and biochar carbon mineralization are found to be lower for biochars produced at higher temperatures (Fang et al., 2014b, 2015).

It is difficult to define a single pyrolysis temperature that would produce a biochar with optimal physical and chemical properties. However, from the observations listed in Table 10.1, it can be deduced that high pyrolysis temperatures within the range of 600°C–700°C may produce the most stable biochars with the highest carbon-sequestration potential.

10.3.2 FEEDSTOCK TYPE

There are a variety of potential biomass feedstock types, such as agricultural waste, biomass from forestry, municipal waste, and animal manure (Windeatt et al., 2014). Different feedstock types vary in composition, which in turn affects the properties of the resulting biochar (Purakayastha et al., 2015). Different plant-based feedstocks vary in their lignin, cellulose, and hemicellulose content, which has been shown to affect the fixed carbon content, aromaticity, and degree of aromatic condensation of biochar (Windeatt et al., 2014; Wiedemeier et al., 2015). Woody feedstocks have high lignin content, while some agricultural residues, such as sugarcane bagasse, have high cellulose content (Jindo et al., 2014; Windeatt et al., 2014). A study by Windeatt et al. (2014) found that feedstock types with higher lignin contents produced biochars with a greater fixed carbon, while those that were high in cellulose produced biochars with a lower fixed carbon content (Figure 10.5).

In addition to producing biochar with high fixed carbon contents, biochar produced from woody feedstock materials, such as oak and pine, are also shown to exhibit greater aromaticity and degrees of aromatic condensation when compared to mineral-rich agricultural residue and manure biochars (Figure 10.4) (Mcbeath et al., 2014). With higher fixed-carbon contents, aromaticity, and degrees of aromatic condensation, lignin-rich feedstocks should subsequently produce a more recalcitrant biochar. Using a recalcitrance index, in which the recalcitrance of biochars are estimated in relation to that of graphite, Windeatt et al. (2014) found a correlation between recalcitrance and the lignin content of feedstocks. Biochar produced from palm shell, which was the most lignin rich, was found to be the most recalcitrant, while the biochar produced from wheat straw, which contained the lowest lignin content, was found to be the least recalcitrant (Windeatt et al., 2014).

The ash content in the biomass has also been shown to influence the properties of biochar. Ash may shield some of the organic material during pyrolysis, affecting the carbonization process and resulting

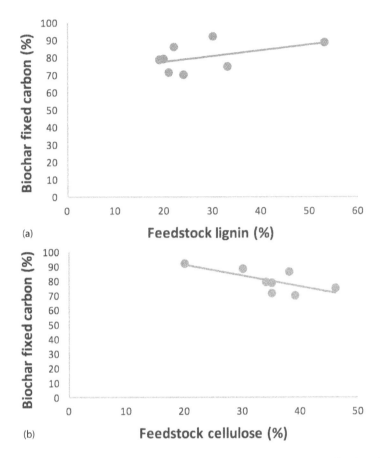

FIGURE 10.5 The effect of cellulose content (a) and lignin content (b) on the fixed carbon content of biochar produced from a variety of feedstocks. (Data from Windeatt, J.H. et al. 2014. *J Environ Manage* 146: 189–197.)

in a biochar with a lower fixed carbon content (Jindo et al., 2014; McBeath et al., 2015). In addition to reducing the fixed carbon content of biochar, there is also the possibility that significant amounts of ash in feedstocks may result in the inhibition of the formation of aromatic structures, reducing cross-links between the components and, as a result, reducing the overall stability of the biochar (McBeath et al., 2015). Although the ash content of biochar increases with pyrolysis temperature, the ash content is mostly dependent on the type of feedstock used (Figure 10.6) (Herath et al., 2014; Bhattacharya et al., 2015; Chen et al., 2016; Purakayastha et al., 2016; Bird et al., 2017). In general, it is found that woody feedstocks have low ash content, while mineral-rich feedstocks, such as crop residues and manures, have a much higher ash content (Figure 10.7) (Wang et al., 2013; McBeath et al., 2015). When comparing the aromaticity and aromatic condensation of biochars produced from different feedstock material, it is found that the woody feedstocks with a lower ash content, such as oak and pine, have greater aromaticity and degrees of aromatic condensation compared to the more mineral- and ash-rich feedstocks, such as crop residues and manures (McBeath et al., 2015).

Presence of amorphous silicon in leaf and grass feedstocks may also influence biochar properties. Amorphous silicon is found in leaf and grass tissues, typically in the form of phytoliths that encapsulate and protect carbon in plant tissues from degradation (Prabha et al., 2017). The effect of this on biochar stability and carbon-sequestering potential is relatively unknown. Jindo et al. (2014) found that the Si rice husk biochar produced at 500°C encapsulated carbon, forming a dense carbon

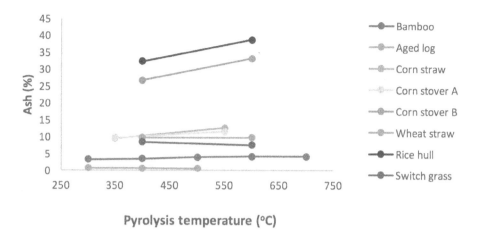

FIGURE 10.6 The effect of different pyrolysis temperatures and different feedstock types on the ash content of biochars. (Data from Bird, M.I. et al. 2017. *Soil Biol Biochem* 106: 80–89; Chen, D. et al. 2016. *Bioresour Technol* 218: 1303–1306; Herath, H.M.S.K. et al. 2014. *Org Geochem* 73: 35–46; Purakayastha, T.J. et al. 2016. *Soil Tillage Res* 155: 107–115; Zheng, H. et al. 2018. *Sci Total Environ* 610–611: 951–960.)

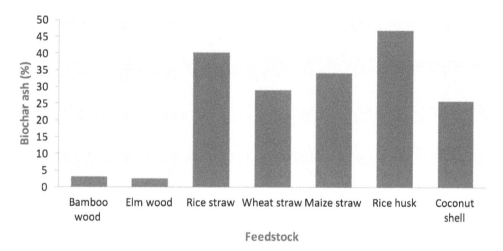

FIGURE 10.7 Influence of different feedstock types on the ash content of biochars. (Data obtained from Wang, Y. et al. 2013. *Energy Fuels* 27: 5890–5899.)

structure, where at 700°C this did not occur. Conversely McBeath et al. (2015) suggest that Silicon in Silicon-rich feedstocks, such as rice husks, may inhibit the formation of aromatic carbon structures and decrease the stability of the biochar.

10.4 INFLUENCE OF SOIL PROPERTIES ON THE CARBON-SEQUESTRATION POTENTIAL OF BIOCHARS

There are a variety of soil properties and characteristics that differ between soils of different origins. In order to establish scenarios in which biochar will have the greatest potential for sequestering carbon in soil, understanding of the influence of the soil properties on the stability of biochar is required. A number of studies have attempted to address this topic in recent years, and relationships between some soil properties and biochar stability have been established.

10.4.1 Soil Mineral Content

The mineral composition of soils is a parameter of importance for the biochar stability in soils. Due to the heterogenous nature of the biochars, minerals present in the soil, and the dissolved organic and inorganic matter of soils, the reactions between soil minerals and biochar are complex (Yang et al., 2016). However, studies comparing the CO_2 mineralized from biochar-amended soils of different properties have found that the mineral composition of soils has a great influence over the rate and cumulative mineralization of biochar carbon (Fang et al., 2014a; Nguyen et al., 2014). It has been observed that biochars incubated in mineral-rich clayey soils exhibit decreased CO_2 mineralization when compared to those incubated in mineral-poor and sandy soils (Fang et al., 2014a; Nguyen et al., 2014).

It has been observed that in Aluminium and Iron-rich soil, biochars react with the minerals, forming crystalline structures on the surface of the biochar, increasing its resistance to microbial degradation (Yang et al., 2016). Decreased CO_2 mineralization has also been attributed to biochar stabilization via calcium bridging and ligand exchange reactions that occur between biochar functional groups, phyllosilicates, carboxyl and phenolic groups of organic matter, and hydroxyl groups on the surface of Fe and Al oxides that exist in mineral-rich soils (Fang et al., 2014a). It was also found that greater stabilization and encapsulation of biochar samples occur in the presence of variably charged minerals, such as kaolinite, as opposed to permanently charged minerals, such as smectite (Fang et al., 2014a).

Minerals also incorporate into the biochar pores, thereby limiting the access for microbial degradation and subsequently decreasing the mineralization of biochar carbon (Fang et al., 2014a; Yang et al., 2016; Sheng and Zhu, 2018). The fine silica alumina minerals in clay have the ability to fill the pore spaces of biochar via adsorption and ligand exchange, providing protection against microbial degradation (Fang et al., 2014a; Sheng and Zhu, 2018). From these observations, it can be established that biochar stability is the greatest when biochar is incubated in clayey soils rich in Al, Fe, and variably charged minerals as opposed to mineral-poor soils with a high content of permanently charged minerals.

10.4.2 Soil pH

Another soil property to influence the mineralization of biochar is the soil pH, which has been shown to influence biochar degradation and priming effects, affect interactions between biochar and soil minerals, and influence soil microbiomes (Sheng and Zhu, 2018). Studies assessing the stability of biochars in acidic soils have found a number of trends in biochar mineralization and degradation as a result of soil pH.

Addition of biochar to acidic soils causes a significant increase in CO_2 emissions during the initial stages of incubation, when compared to soils with a neutral or alkaline pH (Sheng et al., 2016; Sheng and Zhu, 2018). The increase in CO_2 emissions and initial positive priming effect that have been recorded in acidic soils are attributed to an abiotic release of biochar carbonates, a liming effect, and the co-metabolism that occurs due to the addition of biochar of a higher pH to lower-pH soils (Sheng et al., 2016; Sheng and Zhu, 2018). This occurs as acidic soils that contain a low proportion of alphatic hydrocarbon and a greater portion of aromatic compounds have limited bioavailable organic carbon; therefore, upon the addition of the biochar, the labile carbon fraction provides a bioavailable substrate, facilitating an increase in microbial diversity and in turn the increased biochar carbon mineralization (Sheng and Zhu, 2018). Cumulative biochar carbon mineralization was observed to be higher in acidic soils for both studies conducted by Sheng and Zhu (2017) and Sheng et al. (2016). In contrast, neutral and alkaline soils have been shown to exhibit negative priming effects sooner after the biochar has been added, in addition to total CO_2 mineralization's being suppressed (Sheng et al., 2016).

Although positive priming occurs in the initial stages after biochar has been added to acidic soil, this effect was observed to decrease to either a negative or less-positive priming effect after a period of time, as a result of the biochar and soil-mineral interactions' being enhanced in acidic soils

(Fang et al., 2014a; Sheng and Zhu, 2018). It was also found that there is an increase in hydrogen- and oxygen-containing groups that form hydrophilic functional groups on the surface of biochars in acidic soil. This increase in hydrophilic carboxyl and hydroxyl functional groups subsequently promotes ligand exchange between biochar and soil minerals, limiting biochar carbon accessibility to soil microorganisms (Fang et al., 2014a; Sheng and Zhu, 2018).

10.4.3 SOIL ORGANIC MATTER, MICROBIAL ACTIVITY, AND INCUBATION TEMPERATURE

Wu et al. (2016) investigated the effect of soil organic matter content on the stability of biochar in paddy soils. It was found that the native soil organic carbon may significantly influence the rate of biochar carbon mineralization in the later stages of incubation. Cumulative biochar carbon mineralization was found to be significantly positively correlated with the native organic carbon content, this being attributed to soil organic carbon providing a substrate for microbial activity and the subsequent co-metabolism of biochar carbon (Wu et al., 2016). The greater soil organic carbon content and subsequent increased microbial biomass were also believed to be the cause of biochar carbon mineralization in alkaline soils (Sheng et al., 2016). Substantial differences in biochar carbon mineralization between different soil types was also attributed to variations in soil microbial activity, supported by differing soil organic carbon contents (Nguyen et al., 2014).

Mineralization of soil organic carbon and biochar carbon increased with increasing incubation temperatures; however, the temperature sensitivities of the biochar are significantly dependent on soil type (Fang et al., 2014b, 2015). Mineralization rates of the biochar carbon increase with increasing incubation temperatures, with the extent varying between soils with different properties. The temperature sensitivity was observed to be greater for the range of 20°C–40°C than for the range of 40°C–60°C. The priming of native soil organic carbon appeared to be more negative or less positive as incubation temperatures increased; however, after 450 days of incubation, the priming effect for all soils became more positive or less negative in all soil types but at varying rates (Fang et al., 2015). At lower temperatures, increases in biochar carbon mineralization are attributed mainly to increased microbial activity, whereas the influence of higher temperatures on mineralization is more likely to be a result of abiotic oxidation reactions.

10.5   CONCLUSION

The stability of biochar in soils for carbon sequestration is typically evaluated using field methods and laboratory measurements. Field methods provide valuable information for the behavior of biochar in natural environments, while laboratory incubations allow for evaluation of the impacts that a variety of variables may have on the stability of biochars under controlled conditions. Biochars produced at temperatures ranging between 600°C and 700°C from feedstocks of high lignin and low ash contents in combination with clayey soils of neutral or alkaline pH, with low soil organic matter and a high variably charged mineral content, and of a fine texture, may produce conditions favorable for long-term carbon sequestration when biochars are added to soils.

REFERENCES

Anawar, H.M., Akter, F., Solaiman, Z.M. and Strezov, V. 2015, Biochar: An emerging Panacea for remediation of soil contaminants from mining, industry and sewage wastes, *Pedosphere*, vol. 25, no. 5, pp. 654–665.
Bamminger, C., Marshcner, B. and Juschke, E. 2014, An incubation study on the stability and biological effects of pyrogenic and hydrothermal biochar in two soils, *European Journal of Soil Science*, vol. 65, pp. 72–82.
Bhattacharya, I., Yadav, J.S.S., More, T.T., Yan, S., Tyagi, R.D., Surampalli, R.Y. and Zhang, T.C. 2015, Biochar, in R.Y. Surampalli, T.C. Zhang, R.D. Tyagi, R. Naidu, B.R. Gurjar, C.S.P. Ojha, S. Yan, S.K. Brar, A. Ramakrishnan and C.M. Kao (Eds.), *Carbon Capture and Storage - Physical, Chemical, and Biological Methods.* American Society of Civil Engineers (ASCE), Reston, VA, pp. 421–454.

Bird, M.I., McBeath, A.V., Ascough, P.L., Levchenko, V.A., Wurster, C.M., Munksgaard, N.C., Smernik, R.J. and Williams, A. 2017, Loss and gain of carbon during char degradation, *Soil Biology and Biochemistry*, vol. 106, pp. 80–89.

Brewer, C.E., Unger, R., Schmidt-Rohr, K. and Brown, R.C. 2011, Criteria to select biochars for field studies based on biochar chemical properties, *Bioenergy Research*, vol. 4, no. 4, pp. 312–323.

Cely, P., Tarquis, A.M., Paz-Ferreiro, J., Méndez, A. and Gascó, G. 2014, Factors driving the carbon mineralization priming effect in a sandy loam soil amended with different types of biochar, *Solid Earth*, vol. 5, pp. 585–594.

Chen, D., Yu, X., Song, C., Pang, X., Huang, J. and Li, Y. 2016, Effect of pyrolysis temperature on the chemical oxidation stability of bamboo biochar, *Bioresource Technology*, vol. 218, pp. 1303–1306.

De la Rosa, J.M., Rosado, M., Paneque, M., Miller, A.Z. and Knicker, H. 2018, Effects of aging under field conditions on biochar structure and composition: Implications for biochar stability in soils, *Science of the Total Environment*, vol. 613–614, pp. 969–976.

Dong, X., Li, G., Lin, Q. and Zhao, X. 2017, Quantity amd quality changes of biochar aged for 5 years in soil under field conditions, *Catena*, vol. 159, pp. 136–143.

Fang, Y., Singh, B. and Singh, B.P. 2015, Effect of temperature on biochar priming effects and its stability in soils, *Soil Biology & Biochemistry*, vol. 80, pp. 136–145.

Fang, Y., Singh, B., Singh, B.P. and Krull, E. 2014a, Biochar carbon stability in four contrasting soils, *European Journal of Soil Science*, vol. 65, no. 1, pp. 60–71.

Fang, Y., Singh, B.P. and Singh, B. 2014b, Temperature sensitivity of biochar and native carbon mineralisation in biochar-amended soils, *Agriculture, Ecosystems and Environment*, vol. 191, pp. 158–167.

Gurwick, N.P., Moore, L.A., Kelly, C. and Elias, P. 2013, A systematic review of biochar research, with a focus on its stability in situ and its promise as a climate mitigation strategy, *PLOS one*, vol. 8, no. 9, p. e75932.

Haefele, S., Konboon, Y., Wongboon, W., Amarante, S., Maarifat, A.A., Pfeiffer, E.M. and Knoblauch, C. 2011, Effects and fate of biochar from rice residues in rice-based systems, *Field Crops Research*, vol. 121, pp. 430–440.

Hansen, V., Muller-Stover, D., Ahrenfeldt, J., Holm, J.K., Henriksen, U.B. and Hauggaard-Nielsen, H. 2015, Gasification biochar as a valuable by-product for carbon sequestration and soil amendment, *Biomass and Bioenergy*, vol. 72, pp. 300–208.

Herath, H.M.S.K., Camps-Arbestain, M., Hedley, M., Van Hale, R. and Kaal, J. 2014, Fate of biochar in chemically- and physically-defined soil organic carbon pools, *Organic Geochemistry*, vol. 73, pp. 35–46.

IPCC. 2013, Summary for policymakers, in T.F. Stocker, D. Qin, G.-K. Plattner, M. Tignor, S.K. Allen, J. Boschung, A. Nauels, Y. Xia, V. Bex and P.M. Midgley (Eds.), *Climate Change 2013: The Physical Science Basis. Contribution of Working Group I to the Fifth Assessment Report of the Intergovernmental Panel on Climate Change*. Cambridge University Press, Cambridge, UK.

Jindo, K., Mizumoto, H., Sawada, Y., Sanchez-Monedero, M.A. and Sonoki, T. 2014, Physical and chemical characterization of biochars derived from different agricultural residues, *Biogeosciences*, vol. 11, pp. 6613–6621.

Kirtman, B., Power, S.B., Adedoyin, J.A., Boer, G.J., Bojariu, R., Camilloni, I., Doblas-Reyes, F.J. et al. 2013, Near-term climate change: Projections and predictibility, in T.F. Stocker, D. Qin, G.-K. Plattner, M. Tignor, S.K. Allen, J. Boschung, A. Nauels, Y. Xia, V. Bex and P.M. Midgley (Eds.), *Climate Change 2013: The Physical Science Basis. Contribution of Working Group I to the Fifth Assessment Report of the Intergovernmental Panel on Climate Change*. Cambridge University Press, Cambridge, UK.

Knoblauch, C., Maarifat, A.-A., Pfeiffer, E.-M. and Haefele, S.M. 2011, Degradanility of black carbon and its impact on trace gas fluxes and carbon turnover in paddy soils, *Soil Biology & Biochemistry*, vol. 43, pp. 1768–1778.

Kuzyakov, Y., Bogomolova, I. and Glaser, B. 2014, Biochar stability in soil: decomposition during eight years and transformation as assessed by compound-specific 14C analysis, *Soil Biology & Biochemistry*, vol. 70, pp. 229–236.

Lanza, G., Stang, A., Kern, J., Wirth, S. and Gessler, A. 2018, Degradability of raw and post-processed chars in two-year field experiment, *Science of the Total Environment*, vol. 628–629, pp. 1600–1608.

Lehmann, J., Abiven, S., Kleber, M., Pan, G., Singh, B.P., Sohi, S.P. and Zimmerman, A.R. 2015, Persistence of biochar in soil, in J. Lehmann and S. Joseph (Eds.), *Biochar for Environmental Management: Science, Technology and Implementation*, 2nd ed. Routledge, New York, pp. 235–282.

Major, J., Lehmann, J., Rondon, M. and Goodale, C. 2010, Fate of soil-applied black carbon: Downward migration leaching and soil respiration, *Global Change Biology*, vol. 16, no. 4, pp. 1366–1379.

McBeath, A.V., Wurster, C.M. and Bird, M.I. 2015, Influence of feedstock properties and pyrolysis conditions on biochar carbon stability as determined by hydrogen pyrolysis, *Biomass and Bioenergy*, vol. 73, pp. 155–173.

Mcbeath, A.V., Smernik, R.J., Krull, E.S. and Lehmann, J. 2014, The influence of feedstock and production temperature on biochar carbon chemistry: A solid-state 13 C NMR study, *Biomass and Bioenergy*, vol. 60, pp. 121–129.

Nguyen, B.T., Koide, R.T., Dell, C., Drohan, P., Skinner, H., Adler, P.R. and Nord, A. 2014, Turnover of soil carbon following addition of switchgrass-derived biochar to four soils, *Soil Science Society of America Journal*, vol. 78, no. 2, pp. 531–537.

Prabha, S., Sreekanth, N.P., Padmakumar, B. and Thomas, A.P. 2017, Characterization of selected biochars to determine their suitability as a soil amendment from a climate change mitigation perspective, *Octa Journal of Environmental Research*, vol. 5, no. 1, pp. 53–63.

Purakayastha, T.J., Das, K.C., Gaskin, J., Harris, K., Smith, J.L. and Kumari, S. 2016, Effect of pyrolysis temperatures on stability and priming effects of C3 and C4 biochars applied to two different soils, *Soil & Tillage Research*, vol. 155, pp. 107–115.

Purakayastha, T.J., Kumari, S. and Pathak, H. 2015, Characterisation, stability, and microbial effects of four biochars produced from crop residues, *Geoderma*, vol. 239–240, pp. 293–303.

Rasse, D.P., Budai, A., O'Toole, A., Ma, X., Rumpel, C. and Abiven, S. 2017, Persistence in soil of Miscanthus biochar in laboratory and field conditions, J. Paz-Ferreiro (Ed.), *PLoS One*, vol. 12, no. 9, p. e0184383.

Sheng, Y., Zhan, Y. and Zhu, L. 2016, Reduced carbon sequestration potential of biochar in acidic soil, *Science of the Total Environment*, vol. 572, pp. 129–137.

Sheng, Y. and Zhu, L. 2018, Biochar alters microbial community and carbon sequestration potential across different soil pH, *Science of the Total Environment*, vol. 622–623, pp. 1391–1399.

Singh, B.P., Cowie, A.L. and Smernik, R.J. 2012, Biochar carbon stability in a clayey soil as a function of feedstock and pyrolysis temperature, *Environmental Science & Technology*, vol. 46, pp. 11770–11778.

Wang, Y., Hu, Y., Zhao, X., Wang, S. and Xing, G. 2013, Comparisons of biochar properties from wood material and crop residues at different temperatures and residence times, *Energy & Fuels*, vol. 27, no. 10, pp. 5890–5899.

Wiedemeier, D.B., Abiven, S., Hockaday, W.C., Keiluweit, M., Kleber, M., Masiello, C.A., Mcbeath, A.V. et al. 2015, Aromaticity and degree of aromatic condensation of char, *Organic Geochemistry*, vol. 78, pp. 135–43.

Windeatt, J.H., Ross, A.B., Williams, P.T., Forster, P.M., Nahil, M.A. and Singh, S. 2014, Characteristics of biochars from crop residues: Potential for carbon sequestration and soil amendment, *Journal of Environmental Management*, vol. 146, pp. 189–197.

Woolf, D., Amonette, J.E., Alayne Street-Perrott, F., Lehmann, J. and Joseph, S. 2010, Sustainable biochar to mitigate global climate change, *Nature Communications*, vol. 1, p. 56.

Wu, M., Han, X., Zhong, T., Yuan, M. and Wu, W. 2016, Soil organic carbon content affects the stability of biochar in paddy soil, *Agriculture, Ecosystems and Environment*, vol. 223, pp. 59–66.

Xiao, X., Chen, Z. and Chen, B. 2016, H/C atomic ratio as a smart linkage between pyrolytic temperatures, aromatic clusters and sorption properties of biochars derived from diverse precursory materials, *Scientific Reports*, vol. 6, no. 1, p. 13.

Yang, F., Zhao, L., Gao, B., Xu, X. and Cao, X. 2016, The interfacial behavior between biochar and soil minerals and its effect on biochar stability, *Environmental Science & Technology*, vol. 50, no. 5, pp. 2264–2271.

Yoo, M., Han, S.-J. and Wee, J.-H. 2013, Carbon dioxide capture capacity of sodium hydroxide aqueous solution, *Journal of Environmental Management*, vol. 114, pp. 512–519.

Zheng, H., Wang, X., Luo, X., Wang, Z. and Xing, B. 2018, Biochar-induced negative carbon mineralization priming effects in a coastal wetland soil: Roles of soil aggregation and microbial modulation, *Science of The Total Environment*, vol. 610–611, pp. 951–960.

11 Integration of Biomass, Solar, Wind, and Hydro-energy Systems and Contribution to Agricultural Production in the Rural Areas

Hossain M. Anawar and Vladimir Strezov

CONTENTS

11.1 INTRODUCTION

Renewable energy provides sustainable energy access, protects the environment, and augments the economic benefits of farmers in rural communities. Renewable energy technologies, such as solar energy, wind energy, geothermal energy, and biomass, use unlimited amounts of primary energy resources; however, hydropower that is affected by the annual hydrologic cycle belongs to the mix of renewable technologies (Frey and Linke, 2002). The photovoltaic cell converts sunlight directly into electricity indefinitely, due to the sun's energy supply that is 10,000 times that of current human use. Renewable energy is recognized as a sustainable form of energy and environmentally friendly power-generation technologies, which have minimal negative impacts on human health and global environment and reduction in greenhouse gas (GHG) emissions, as well as the economic consideration of fossil fuel use (Omer, 2008). The renewable energy sources produce nearly one-fifth of all global power (IEA, 2003). The renewables are the second-largest power source after coal (39%) with respect to power generation, status, and prospects, but ahead of nuclear (17%), natural gas (17%), and oil (8%) (Omer, 2008). From 1973 to 2000, renewables grew at 9.3% per year (Omer, 2008).

There are no electricity and water networks in the rural and desert parts of the arid area that occupies more than 50% of the world's area. People living in this area receive pumped water from borehole wells powered by diesel engines, causing maintenance problems, high running costs, and environmental pollution problems. Renewable energy systems, such as solar and wind, can provide a sustainable energy source for water pumps, crushing grain, and other activities in the rural life and agricultural activities of these regions after they are integrated in the regional development plans (Omer, 2008). Hydropower energy, derived from the movement of water in rivers, oceans, and waterfalls, can be used to generate electricity, using turbines for useful work. Geothermal power, which directly harnesses the natural flow of heat from the ground, is approximately equal to the incoming solar energy, especially during the day. Bioenergy crops, such as corn, sugar cane, switch grass, and others, are also a renewable source of energy, along with alcohol, biofuels, biogas, oils, hydrogen, et cetera. Methane, if produced from anaerobic digestion of biomass, can be considered as a renewable source of energy. Depletion of fossil fuels is another concern, but it may be prevented through integration of various sources of energy, such as solar, wind, biomass, and biofuels, and combined heat and power.

11.2 NEXUS BETWEEN RENEWABLE ENERGY, ELECTRICITY PRODUCTION, AGRICULTURE, AND GREENHOUSE GAS EMISSION

Environment, economy, and energy (EEE) contribute significantly to the development of the world (Azad et al., 2015), where energy plays a significant role for human, economic, and social improvements, as well as sustainable development (Ibrahiem, 2015). The 56% increase in energy consumption will occur from 2010 to 2040 (Azad et al., 2015) and result in increased carbon dioxide (CO_2) emission, which is the main component with approximately 61.4% of total GHG emissions (Sadorsky, 2010). The forest and vegetable areas contribute to carbon sequestration. Different types of renewable energy, termed *clean energy*, such as solar, wind, geothermal, biofuel, and hydropower, improve the environment, level of employment, output, and income with the provision of 3.5 million jobs (Al-Mulali et al., 2013; Ibrahiem, 2015). The hydropower reservoirs can emit biogenic GHGs, which are sometimes higher than the emission rate of thermal power plants (Scherer and Pfister, 2016). However, renewable energy has a potential of 50% decrease in CO_2 emissions by 2050 if the increase in temperature of the world is limited to 2.0°C–2.4°C in the long run (Apergis et al., 2010). The consumption of renewable energy resources contributes to the social, economic, and environmental improvements of farmers (Jebli and Youssef, 2016b, 2017b). Table 11.1 shows that the energy consumption causes pollution, and the associated problems can be bidirectional between energy use and environmental pollutants (Akhmat et al., 2014a).

Based on the annual data from 1981 to 2015, the nexus between agriculture value added, coal electricity, hydroelectricity, renewable energy, forest area, vegetable area, and GHG emission in Pakistan was investigated by Khan et al. (2018). The unidirectional and bidirectional causality was observed among the variables using the Toda and Yamamoto approach. The long-run causality of GHG emission, agriculture value added, and forest area was observed based on the vector error correction model (VECM) results. The annual convergence from short- to long-run equilibrium was 36.8%. GHG emission reductions were observed due to increases in agriculture (0.124%), renewable energy (1.086%), vegetable area (0.153%) and forest area (0.240%).

TABLE 11.1

Relationship of Energy Consumption with Pollution and Causality Effects

Author(s)	Variables	Method	Countries	Time Period	Causality Results
Apergis et al. (2010)	CO_2, renewable energy, economic growth, and nuclear energy	VECM	19 developing and developed countries	1984–2007	1. Nuclear energy decreases CO_2 2. Renewable energy increases CO_2 3. Nuclear energy $\rightarrow CO_2$
Ahmed et al. (2014)	Deforestation, trade openness, energy consumption, growth in the economy, and population	ARDL, VECM	Pakistan	1980–2013	1. The decreasing negative effect of growth in the economy on deforestation 2. Growth in the economy $\rightarrow$ deforestation 3. Consumption of energy $\rightarrow$ deforestation 4. Growth in the economy $\leftrightarrow$ consumption of energy 5. Trade openness $\rightarrow$ deforestation
Akhmat et al. (2014a)	Energy consumption and environmental pollutants	Engle and Granger Causality test	Pakistan, Bangladesh, Nepal, India, Sri Lanka	1975–2011	1. Energy consumption increases pollutants 2. Energy consumption $\rightarrow$ environmental pollutants 3. Energy consumption $\leftrightarrow CO_2$ (Nepal) 4. CO_2 exerts changes in electric power consumption (Bangladesh and Nepal)
Akhmat et al. (2014b)	GHG emissions, agricultural CH_4 emissions, industrial N_2O emission, and energy sources (coal, gas, oil energy; fossil fuel; nuclear)	FMOLS, DOLS	35 developing countries	1975–2012	1. Nuclear energy decreases GHGs 2. Electricity from oil, gas, and coal increases GHGs
Cowan et al. (2014)	CO_2, growth in the economy and consumption of electricity	Panel bootstrap causality approach	BRICS	1990–2010	1. GDP $\leftrightarrow CO_2$ (Russia) 2. GDP $\rightarrow CO_2$ (South Africa) 3. $CO_2 \rightarrow$ GDP (Brazil) 4. Electricity consumption $\rightarrow CO_2$ (India)
Jebli and Youssef (2016a)	CO_2, trade openness, GDP, renewable energy, nonrenewable energy, and agriculture value added	VECM	Tunisia	1980–2011	1. Agriculture value added $\leftrightarrow CO_2$ 2. Agriculture value added $\leftrightarrow$ trade openness 3. Renewable energy $\rightarrow$ GDP 4. $CO_2 \rightarrow$ renewable energy 5. Agriculture value added, non-renewable energy, and trade increase CO_2 6. Renewable energy decreases CO_2

(Continued)

TABLE 11.1 (*Continued*)
Relationship of Energy Consumption with Pollution and Causality Effects

Author(s)	Variables	Method	Countries	Time Period	Causality Results
Jebli and Youssef (2016b)	CO_2, agricultural value added, GDP, and combustible renewables and waste	VECM	Brazil	1980–2011	1. Agriculture value added → CO_2 2. Agriculture value added → GDP 3. Combustible renewables and waste, and agriculture value added increases growth in the economy 4. Agriculture value added and combustible renewables and waste decrease CO_2
Lau et al. (2016)	CO_2, hydroelectricity consumption, and economic growth	VECM	Malaysia	1965–2010	1. Hydroelectricity consumption → CO_2 2. GDP → CO_2
Asumadu-Sarkodie and Owusu (2017)	CO_2, biomass-burned crop residues, agriculture value added, agricultural machinery, cereal production, livestock, and rice area	VECM	Ghana	1961–2012	1. Rice production increases CO_2 2. Cereal production increases CO_2 3. Agricultural machinery decreases CO_2 4. Cereal production ↔ CO_2 5. Biomass-burned crop residues ↔ CO_2
Jebli and Youssef (2017a)	CO_2, renewable energy, agriculture value added, GDP	VECM	5 North African countries	1980–2011	1. Agriculture value added ↔ CO_2 2. Agriculture value added → GDP 3. GDP/renewable energy 4. Agriculture value added → renewable energy 5. Renewable energy → agriculture value added 6. Renewable energy → CO_2 7. Increase in GDP and renewable energy increase CO_2 8. Increase in agricultural value added decreases CO_2
Jebli and Youssef (2017b)	CO_2, renewable energy, arable land, agriculture value added, and GDP	VECM	Morocco	1980–2013	1. Renewable energy increases due to increase in economic growth, arable land, and agricultural production 2. Renewable energy decreases CO_2 3. Agriculture value added ↔ renewable energy 4. Arable land use/renewable energy

(Continued)

TABLE 11.1 (*Continued*)
Relationship of Energy Consumption with Pollution and Causality Effects

Author(s)	Variables	Method	Countries	Time Period	Causality Results
Khan et al. (2017)	GHG emission, financial development, energy use, renewable energy, trade, and urbanization	VECM, FMOLS, GMM	34 upper-middle-income countries	2001–2014	1. The decrease in GHG emission was observed due to increase in renewable energy in Europe

Source: Reprinted from *Renew. Energ.*, 118, Khan, M.T.I. et al., The nexus between greenhouse gas emission, electricity production, renewable energy and agriculture in Pakistan, 437–451, Copyright 2018, with permission from Elsevier.
FMOLS: fully modified ordinary least square; GDP: gross domestic product; GHG: greenhouse gas.

11.3 ENERGY AND THE FOOD SYSTEM

Modern agriculture is heavily dependent on fossil energy resources, where energy has been used for crop management, fertilizers, pesticides, and machinery production since the 1960s. Low-energy inputs can contribute to lower yields, while higher energy use contributes to higher output. But this relationship is not always linear, indicating that increasing energy inputs can lead to ever-smaller yield gains. The large amounts of natural gas and some coal are used in nitrogen fertilizer production that can account for more than 50% of the total energy use in commercial agriculture. Depending on the cropping system, the United Kingdom's agriculture uses oil between 30% and 75% of the energy inputs. Food prices will rise with increase of fossil energy prices, and food production contributes significantly to anthropogenic GHG emissions due to dependence of agriculture on fossil sources of energy. Utilization of renewable energy with technological developments and changes in crop management may contribute to improved energy efficiency of agriculture and reduce the reliance of this sector from fossil resources (Woods et al., 2010).

11.4 CLIMATE-FRIENDLY AGRICULTURE AND RENEWABLE ENERGY

Agricultural activities contribute between 14% and 30% of human-caused GHG emissions from their direct and indirect sources, such as running fuel-powered farm equipment, pumping water for irrigation, raising dense populations of livestock in indoor facilities, and applying nitrogen-rich fertilizers (Reynolds and Wenzlau, 2012). Livestock production alone contributes around 18% of the global emissions, including 9% of CO_2, 35% of methane, and 65% of nitrous oxide. However, the climate-friendly agricultural practices have "significant" potential to reduce emissions, including removing 80%–88% of the CO_2 of its current production (UN Food and Agriculture Organization, FAO) through replacement of fossil fuels by renewable energy, including biofuels and annual crops, by perennial crops that require considerably less fertilizer, pesticides, and herbicides. The climate-friendly food production needs further research and investment to contribute to climate change mitigation. Many farmers already produce renewable energy by growing corn to make ethanol. An increasing number of farmers and ranchers are using wind power to produce electricity on their land (Azad et al., 2014). Wind, solar, and biomass energy can be harvested forever, providing farmers with a long-term source of income (Ali et al., 2012). Renewable energy can be used on the farm to replace other fuels or sold as a "cash crop."

A wide range of government policy interventions to obtain the climate, energy, and environment-related objectives affect the agriculture sector (Troost et al., 2015). Climate-friendly, renewable energy production in agriculture contributes to land use extensification and landscape preservation, resulting in possible potential conflicts with agri-environmental policies. This policy implementation, reflecting potential trade-offs and inconsistencies, is not clearly understood, because it does not fully include the farmers' responses when a choice is made between investments in biogas production and participation in agri-environmental policy schemes. The incentives set by the German Renewable Energy Act (EEG) and the agri-environmental policy scheme Compensation Scheme for Market Easing and Landscape Protection (Marktentlastungs-und Kultur-landschaftsausgleich, MEKA), with their impact on the farming population in southwest Germany, were investigated by a farm-level model. A potentially large decrease of MEKA participation was reported by simulations due to biogas production supported under EEG. The conflicts between the expansion of renewable energy and environmental considerations were not alleviated by a 2012 EEG revision; rather, priorities shifted from the former to the latter.

11.5   RENEWABLE ENERGY FOR SUSTAINABLE AGRICULTURE

Based on annual data from 1980 to 2012, Paramati et al. (2018) studied the impact of renewable and non-renewable energy consumption on agriculture, industry, services, and overall economic activities gross domestic product (GDP) across a panel of G20 nations by applying several econometric models. The empirical findings indicate that both renewable and non-renewable energy consumptions positively affect the economy across multiple sectors and contribute positively to overall economic output. The results found greater positive economic effects of renewable energy compared to non-renewable energy (Apergis and Payne, 2010, 2011; Bhattacharya et al., 2016). Therefore, effective policies are necessary to turn domestic and foreign investments into renewable energy projects that can ensure low carbon emissions and sustainable economic development across the G20 nations (Apergis and Payne, 2012; Paramati et al. 2016, 2017). During the period of 1990–2012, renewable energy generation internationally was increased from 16.6% to 18.1% of total energy consumption (IEA, 2014a, 2014b). Furthermore, it is predicted that the average growth of renewable energy will increase at a rate of 0.17% per year, contributing to 21% of total final energy consumption by 2030.

Sustainable agriculture can be achieved by minimizing the adverse environmental impacts of the sector and reduction in the use of finite natural resources, while achieving maximized crop productivity and economic stability (Corwin et al., 1999). There are two important burdens that will need to be overcome to achieve sustainable agriculture, such as physical, which include changes in soil properties due to long-term effects of agriculture and processes considered important for crop productivity, and socio-economic, which involve resource management of labor and energy (Chel and Kaushik, 2011). The renewable energy technologies are suitable for agriculture in respect of cost for any location, with the additional benefit of earning carbon credits, as compared with conventional fossil fuel–based technologies. However, governmental support is essential to implement these technologies in the agricultural sector due to high capital cost investments.

It is necessary to promote the energy efficient, low-carbon technologies and greater penetration of renewable energy to achieve greater environmental sustainability of the agriculture. Studies have estimated various farming energy requirements, and it was found that on-farm electricity consumption can vary from 330 kWh/cow/year to 566 kWh/cow/year (Nacer et al., 2016). Technical feasibility and economic viability of managing this electricity demand can be modelled by, as an example, the hybrid optimization model for electric renewable (HOMER) software, developed by the U.S. National Renewable Energy Laboratory, which can assist to propose systems that will improve reliability of the utility grid during peak load periods.

11.6 RENEWABLE ENERGY DEVELOPMENT AND IMPLICATIONS TO AGRICULTURAL VIABILITY AND CONFLICT

The development of renewable energy sources has occupied the land resources (Pimentel et al., 1994). For instance, biofuels for transportation fuel supply from ethanol cultivate corn and soybeans on land. Similarly, electricity production from wind energy requires a significant amount of land. There is a conflicting pressure between the need to meet the energy demands and the significant need for land resources to cultivate energy crops. The global increase of food prices due to cultivation of biofuel crops on agricultural land creates critical conflicts regarding the tradeoff between long-term energy security and food security. Therefore, the interface between agriculture and energy has a potential contribution to agricultural viability (Smith et al., 2008).

The links between wind energy development on agricultural fields instead of wind farm siting in high-density areas and implications to the agricultural sector have been identified. The 30 states of the United States have introduced renewable portfolio standards (RPS) legislation to promote renewable energy projects, including wind energy (Steve et al., 2007; US DOE, 2008). The RPS policy mandates that a fraction of the electricity supplied by the utilities is generated from renewable energy sources, which include solar, wind, hydro-, biomass, and wave/tidal electricity (Huang et al., 2008). The interest of the policy is how wind energy developments on agricultural land impact the economic viability of farming practices (Berton, 2015). The payments from leasing land for wind turbine co-location are projected up to $50 million per year in the United States, which may have an effect on agricultural viability. The potential cross-sectoral impacts should be carefully investigated, because renewable energy development occupies large land areas (Adelaja and Hailu, 2008). Social impacts, which may include creating new monopolies and injustices due to renewable energy developments, should be combatted.

11.7 RENEWABLE ENERGY CONTRIBUTION TO AGRICULTURE

Mosher and Corscadden (2012) reviewed different feed-in tariff (FIT) policies in some of the countries (Canada, Denmark, Germany, Netherlands, and USA) and their associated impacts on promoting renewable energy developments and the pressure on agricultural land. The effect of organic agriculture on the economic and social development of the communities was also identified (Ramaraj and Dussadee, 2015), where the development of organic agriculture has significant potential in the production and consumption of renewable energy. Furthermore, organic agriculture can reduce the application of expensive and energy-intensive synthetic fertilizers. Organic agricultural practices obtain equal or even higher productivity than conventional agricultural activities in the developing countries. Therefore, it can potentially improve the food security and sustainable livelihoods of the rural population to mitigate the adverse effect of climate change (El-Hage Scialabbaa and Muller-Lindenlaufa, 2010).

The major source of fuel in some of the African countries, such as Uganda, is biomass energy, used mainly for cooking and space heating, which can supply up to 90% of the national energy demand (Turyareeba, 2001). Investments in conventional energy technologies are expected have low social impacts in this country, mainly due to high and unaffordable costs and limited access to electricity. The livelihood of the rural population in Uganda can be improved by development of modern and accessible renewable energy systems. The major source of income and employment in this country is driven by old technologies and techniques used for agricultural farming, harvesting, and processing. Therefore, development and adaptation of renewable energy technologies is expected to improve the livelihood of rural Ugandans and to modernize the agriculture in this country, but also other countries with similar economies (Turyareeba, 2001). Table 11.2 presents some of the energy options available to modernize traditional agricultural practices.

TABLE 11.2

Conventional and Renewable Energy Options for Modernizing Agriculture

Activity	Energy Source	Conventional Alternative	Renewable Alternative
Tilling, planting, and weeding	Manual labor using the hand hoe	Tractors using diesel	Hand ploughs/planters, ox ploughs, and tractor using biofuels
Irrigation	Manual labor using head loads	Canals, irrigation using diesel-powered water pumps	Canals, irrigation using solar, wind and biofuels
Application of fertilizer	Rarely used, applied manually	Inorganic fertilizer	Compost and slurry from biogas, mulching
Harvesting	Manual labor using pangas, axes, hoes	Combine harvesters and power saws using petroleum fuels	Ox-drawn harvesters, combine harvesters and power saws using biofuels
Transport	Manual labor using head loads	Tractors, trucks/lorries	Wheelbarrows, hand pushed carts, ox/donkey/horse drawn cart, biofueled vehicles
Preservation	Direct sun drying, smoking	Electric or kerosene-powered refrigerators, electric-powered dryers (electricity generated using petroleum)	Solar and biomass kilns and dryers, solar/biomass/wind/micro-hydro-powered refrigerators
Processing	Manual labor using pestle and mortar or grinding stones	Mechanized grinders using diesel generators	Paddle grater, mata mills, biofueled generators

Source: Reprinted from *Renew. Energ.*, 24, Turyareeba, P.J., Renewable energy: Its contribution to improved standards of living and modernisation of agriculture in Uganda, 453–457, Copyright 2001, with permission from Elsevier.

11.8 SOLAR ENERGY FOR AGRICULTURE

The abundance of solar energy can be used to increase application of renewable energy in agriculture, specifically in remote or rural areas. Sunlight can be used for drying of the agricultural products. There is a range of solar devices developed to aid sustainable agriculture, and they may include water and space heating, lighting, battery charging, driving small motors, water pumping, powering electric fences, and power generation. Solar energy use in agriculture can save in electricity and heating costs, achieve self-sustainable agriculture, and contribute to pollution reduction. Solar heat collectors are the most advanced technologies that can be used for drying crops, to provide heating in livestock buildings and greenhouses. In addition, hot water for dairy operations, pen cleaning, and homes can be supplied with solar water heaters.

Low capital investment is required in photovoltaic projects that can be developed at small-to-medium scales. Photovoltaics (solar electric panels) and off-grid photovoltaic systems can power farm operations and remote water pumps, low-pressure drip irrigation at low cost, lights, and electric fences (Jamea, 2013). Pump spy technologies, which are pumps connected remotely to an operator, can be installed in boreholes, tanks, cisterns, or rivers and used for movement of water for drinking or to operate irrigation systems. The buildings and barns can be modified or designed to maximize the natural daylight and reduce the need for lighting. In cases when extended power lines are required, solar power is a less-expensive solution (Azad et al., 2014). Solar pumps have led the agricultural community towards constant high-yield farming and tripled their income (Smart World: Solar Pumps, 2018) in India by replacing conventional diesel-driven pumps with solar pumps subsidized by the government and private funding. Depending on the type and depth of the water source, a pump is selected for use. If a 1,000 Wp pump is installed to draw from a 10-meter-deep water source, two acres of land can be irrigated using 40,000 L of water in a day (Sources from Tata Power Solar).

The development of new technologies is expected to increase solar cell efficiency and reduce costs. For instance, new "quantum dot" materials are expected to increase the efficiencies by almost double, providing 65% of the sun's energy conversion into electricity. For comparison, the most efficient commercially available solar cells achieve conversion efficiencies of up to 30%. The first-generation solar cells made of bulk polycrystalline Si wafers produce solar conversion efficiencies between 12% and 16%.

11.9 WIND ENERGY FOR AGRICULTURE

Installation of wind turbines in agricultural land offers farmers the opportunity to lease their land to wind developers, use wind power as the primary energy source in their farms, or decide to invest in wind power and become wind power producers. A typical farm would require only a small wind generator of between 400 W and 40 kW to meet the in-farm electricity needs, or to supply electricity to different appliances and utilities. An example is the wind-powered water pumping for cattle, which is more efficient than the fan-bladed windmills, and cheaper and environmentally more appropriate than the diesel-powered water pumps. The typical investment required for installation of each turbine on agricultural land with strong winds ranges between $2,000 and $5,000 per year. Farmers can still utilize the land for agricultural purposes and plant crops and graze livestock around the turbine base. Individual farmers' investment in wind farms and forming wind power cooperatives are considered the two main initiatives by the farmers in these developments (Azad et al., 2014). Diversification in the agricultural sector is important for providing enough spaces to build wind turbines in countries with a lack of available land (Jamea, 2013). For instance, in Denmark and Germany, farmer cooperatives are formed to diversify their incomes through wind farm investments.

Although utilization of wind energy in agriculture is still challenging for farmers, its use for water pumping is the most profitable that is conducted by windmills and does not require high wind speeds, has simple construction, and is cheaper by approximately 50%. It has capacity of 50,000 L/day to irrigate approximately 8–10 ha of land using a drop-by-drop irrigation system. Wind turbines used to power specific utilities can range in size from 750 kW to 5 MW. A number of turbines are typically grouped into wind farms to provide bulk power to the electrical grid (AWEA, 2006).

11.10 GEOTHERMAL, POWER PLANT, SEA, AND RIVER WATER HEAT SOURCES IN HORTICULTURE FACILITIES

Modelling studies have been conducted to investigate the use of waste heat as an energy source to drive a heat pump in a large-scale horticulture facility (Hyun et al., 2014). The study recommended use of power plant waste heat as the major source of heat and highest efficiency for utilization in horticulture. The costs for heating in a horticulture facility can range about 19%–58% of the total horticulture management cost, depending on the cultured crop.

Use of a horizontal type of geothermal heat pump in horticulture has been shown to reduce the cost of heating by approximately 67.8% (Park, 2007; Ryu et al., 2012). Groundwater as a heat source has also been used to drive a thermal heat pump in horticulture (Kim, 2013) and found the lifecycle costs are reduced by 42% compared to natural gas, by 62% compared to a vertical closed type, and by 72% compared to the Standing Column Well (SCW) type. As a result, CO_2 emissions were reduced by 24%, 71%, and 82%, respectively. Baek et al. (2012) compared the performance of heat pumps using sea water versus air as a heat source and concluded that, when sea water is used as a heat source, the heating performance was improved by about 8.5% over a year, compared to the use of air as heat source. This study considered apartment buildings as the end-user.

Geothermal power has different applications in sustainable agricultural practices. High-temperature geothermal resources with temperatures above 150°C are used in organic agriculture. Geothermal energy is used to supply heat in greenhouse operations to grow vegetables, flowers, ornamentals, and tree seedlings. Geothermal heat can also be applied in aquaculture operations to raise different types of fish and other aquatic species. Geothermal heat finds use in dehydration and drying of food and grains (Dickson and Fanelli, 2003; Lund et al., 2011).

11.11 HYBRID RENEWABLE ENERGY SOURCE: WATER PUMPING SYSTEMS FOR AGRICULTURE

Integration of renewable energy sources with water pumps has been a subject of intensive investigations in the past (Gobal et al., 2013). Use of solar power as a source of energy for water pumps is the most commercially advanced technology, with solar water pumps available in high capacities of 10 kW. However, the pumps used in remote areas are smaller in scale and are less than 1,500 W. There are two major groups of solar water pump systems, solar photovoltaic and solar thermal water pumping systems. In addition to these two, the other renewable energy source water pumping systems can include wind energy, biomass driven, or a hybrid renewable energy water pumping system, which includes two or more different renewable energy sources. A typical solar water pump system is presented in Figure 11.1. Compared with conventional fuel, a solar water pumping system has different advantages, which include no need for fuel, lower noise, and better environmental performance (Jafar, 2000; Cakir et al., 2013; Mekhilef et al., 2013).

11.12 RENEWABLE ENERGY SOURCE AS FERTILIZER FOR ORGANIC AGRICULTURE

The most widely used and advanced technology to treat organic waste is through anaerobic digestion, which produces methane as a biogas and digestate residue. The residue contains the nutrients used for cultivation of the crop and can be used as an organic fertilizer to improve crop yield and soil fertility, and close the energy and nutrient cycles (Arthurson, 2009). The digestate by-product is the result of mineralization; has a high nitrogen-to-carbon ratio, suitable as a fertilizer; enhances nutrient penetration into the soil; and reduces up to 80% of odors from the feedstock (Weiland, 2010; Rodriguez-Navas et al., 2013). This high content is advantageous to the agricultural crops, as they are primarily capable of utilizing ammonium nitrogen.

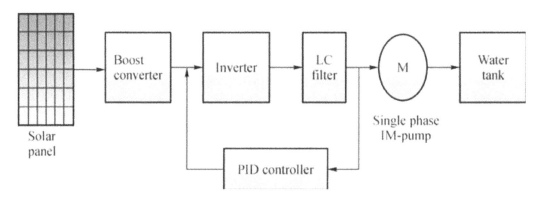

FIGURE 11.1 Photovoltaic water pumping system for irrigation in agriculture. IM: induction motor. (Reprinted from Binshad, T.A. et al., *Front. Energ.*, 10, 319–328, 2016, with permission from Springer Nature.)

11.13   AGRICULTURE-BASED RENEWABLE ENERGY PRODUCTION

Both federal and state levels of US policymakers have introduced a range of incentives, regulations, and programs to increase production and use of agriculture-based renewable energy since the late 1970s (Schnepf, 2006). The rapid growth in use of bioenergy can adversely affect the other uses of biomass feedstocks, such as food, animal feed, and industrial processing.

11.14   BIOENERGY PRODUCTION

Plant biomass is the main source of renewable materials, including a potential source of renewable energy and bio-based products (Guo et al., 2015; Wannapokin et al., 2017). Energy production from anaerobic digestion of the agricultural waste residues from the agricultural sector, agriculture industry, and grassland biomass is widely applied in different countries. Furthermore, animal manures have been used as feedstock in anaerobic digestion either on their own or in mixtures with agricultural waste. For instance, buffalo grass, which is used as a feedstock for animal feeding, was shown to enhance biogas production when co-digested with buffalo dung (Figure 11.2, Chuanchai and Ramaraj, 2018). The addition of grass can help raise the C:N ratio of the feedstock that is suitable for metabolic activities in anaerobic digestion systems.

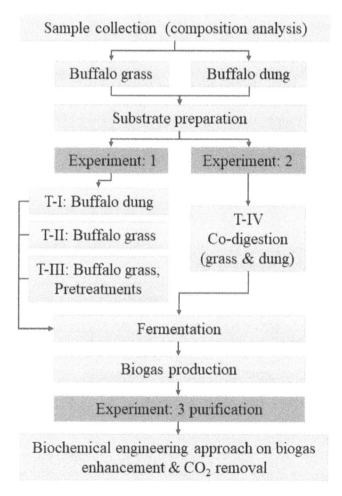

FIGURE 11.2   The flowchart of biogas generation methodology using Buffalo grass and Buffalo dung. (Reprinted from Chuanchai, A., and Ramaraj, R., *Biotechnology*, 8, 151, 2018.)

The various methods of pre-treatment can modify the physical structure and chemical composition of lignocellulosic materials, which are otherwise not suitable for anaerobic digestion (Wannapokin et al., 2018). Therefore, pre-treatments of lignocellulosic waste can be applied so that they can be anaerobically digested and to increase the methane content in the biogas. Boiling with different retention time was shown to increase the biodegradability of the grass substrate (Chuanchai and Ramaraj, 2018), while the best condition was obtained at 100°C with two-hour retention time and a 1:1 ratio of co-digestion mixture with manure.

Biomass can be used in different ways to produce electricity, either directly through gasification, or by co-firing with coal at an existing power plant. Biomass can be also used to produce cellulosic ethanol. Although corn is currently used as the most prevailing energy crop, native prairie grasses, such as switchgrass or fast-growing trees like poplar and willow, can be used as well (Sadaka et al., 2014). Biochar produced from torrefaction or pyrolysis of energy crops, such as switchgrass, are high-quality solid fuels based on their energy content. These perennial crops need lower maintenance than the annual row crops, making them cheaper and more sustainable to produce (Azad et al., 2014). The agricultural biomass resources can be used to produce a variety of fuels, such as liquid and gaseous fuels (Jamea, 2013). According to the US Department of Energy, tripling US use of biomass energy could contribute to $20 billion in new income for farmers and rural communities and reduce greenhouse emissions equivalent to removal of 70 million cars off the road.

11.15   EDUCATION, ENERGY PLANNING, AND RESEARCH

The Agricultural Engineering has broad scope to contribute to the renewable energy sector, in particular through three critical aspects: teaching, energy planning, and research (Pedretti et al., 2010). Education is essential to raise public awareness, spread the energy culture, and introduce energy production systems based on non-fossil substrates, such as biomass, because absence of in-depth knowledge of the matter can prevent success and turn even valuable initiatives into failures. Therefore, Agricultural Engineering must devise and develop innovative teaching pathways to involve and attract students, as well as provide them with sound knowledge and a critical approach that can help in the production and use of renewable sources. Agricultural Engineering as a discipline has a critical role in education and traineeship of professionals and scholars who will participate in the actions directed at improving efficiency and energy planning in various sectors. An alternative to the future solutions is interconnection between agronomy and biology to establish connection to biotechnologies, which would innovate the traditional distribution of competencies. Furthermore, putting biotechnology research into use and processing biofuels can contribute to making Agricultural Engineering a discipline of Biosystems Engineering.

11.16   CASE STUDIES OF DIFFERENT COUNTRIES

The large availability of unexploited lands and high abundance of sunlight make solar energy systems, especially photovoltaics and wind energy, attractive opportunities in the eastern and southern countries of the Mediterranean region (Jamea, 2013). The study by Mexico Renewable Energy in Agriculture Project (MREAP 2006) shows that the renewable energy in the agricultural sector can increase the productivity and income of the estimated 600,000 unelectrified livestock farmers in Mexico by supporting adoption of investments in production and improved farming practices and improving the executing agency, Fideicomiso de Riesgo Compartido (FIRCO's), ability to catalyze the penetration of renewable energy technologies in the agriculture sector.

European Union (EU) agriculture and forestry play an increasing role in supplying renewable energy. However, the production of renewable energy has increased more rapidly in the agricultural sector than in the forestry sector. The production of renewable energy from agriculture increased by almost sevenfold between 2004 and 2010, while the production from forestry increased by 54% in the period 2000–2010, at an average annual growth rate of 4.4%. The energy efficiency and renewable energy sources benefit for the agricultural sector in Macedonia (Petrovska et al., 2012). The proper utilization of energy and environmentally friendly practices can increase the efficiency of production and provide larger benefits. Mbzibain et al. (2015) examined how national institutional structures affect farmers' intentions to invest in renewable energy enterprises in the UK agricultural sector. Cognitive institutions were positively related to intentions, and they can play a more-important role in determining farm entrepreneurship in the renewable energy sector.

According to Natarajan (2015), at least 37% electricity connection goes for farming in Vellore taluk of India, where farmers are severely affected by the power shortage and water supply. Therefore, it is necessary to develop low-cost models so that people can use renewable energy in this area. An enterprise named Solargao is trying to integrate electrification in rural areas with mechanization and modernization in the agricultural sector for effective utilization of renewable energy, such as solar power generation, in Bangladesh (Binshad et al., 2016). The solar irrigation system plays a crucial role in the operation of the eco-village (Figure 11.1). Energy scarcity in the agricultural sector is one of the major challenges facing sustainable food security in Nigeria (Tijjani et al., 2013), where renewable energy is the next alternative option (Kainth, 2010).

11.17 CONCLUSIONS

There are enormous opportunities in the application of renewable energy technologies for the socio-economic development in the agricultural production of rural areas. However, further technological improvement is needed to upgrade their equipment for maximum energy use efficiency (for example, cold stores, irrigation pumps), and to install renewable energy to run intensive operations to provide the energy for high-energy-intensity activities, such as irrigation systems, packaging and processing equipment, water heating, and sterilization. Through government and private energy incentives, the farmers will increase their energy self-sufficiency through energy efficiency, and renewable energy will lower their energy bills and costs, strengthen the food supply chain, lower GHG emissions, and provide clean produce.

REFERENCES

Adelaja, S., and Y. G. Hailu. 2008. *Renewable Energy Development and Implications to Agricultural Viability. American Agricultural Economics Association Annual Meeting*, Orlando, FL.

Ahmed, K., M. Shahbaz, A. Qasim, and W. Long. 2014. The linkages between deforestation, energy and growth for environmental degradation in Pakistan. *Ecological Indicators* 49: 95–103.

Akhmat, G., K. Zaman, T. Shukui, D. Irfan, and M. M. Khan. 2014a. Does energy consumption contribute to environmental pollutants? Evidence from SAARC countries. *Environmental Science and Pollution Research* 21: 5940–5951.

Akhmat, G., K. Zaman, T. Shukui, F. Sajjad, M. A. Khan, and M. Z. Khan. 2014b. The challenges of reducing greenhouse gas emissions and air pollution through energy sources: Evidence from a panel of developed countries. *Environmental Science and Pollution Research* 21: 7425–7435.

Al-Mulali, U., H. G. Fereidouni, J. Y. Lee, and C. N. B. C. Sab. 2013. Examining the bi directional long run relationship between renewable energy consumption and GDP growth. *Renewable and Sustainable Energy Reviews* 22: 209–222.

Ali, S. M., N. Dash, and A. Pradhan. 2012. Role of renewable energy on agriculture. *International Journal of Engineering Sciences & Emerging Technologies* 4(1): 51–57.

American Wind Energy Association. 2006. U.S. wind industry ends most productive year, sustained growth expected for at least next two years, 2006. Available: http://www.awea.org/news/US_Wind_Industry_Ends_Most_Productive_Year_012406.html.

Apergis, N., and J. E. Payne. 2010. Renewable energy consumption and economic growth: Evidence from a panel of OECD countries. *Energy Policy* 38(1): 656–660.

Apergis, N., and J. E. Payne. 2011. The renewable energy consumption–growth nexus in Central America. *Applied Energy* 88(1): 343–347.

Apergis, N., and J. E. Payne. 2012. Renewable and non-renewable energy consumption-growth nexus: Evidence from a panel error correction model. *Energy Economics* 34(3): 733–738.

Apergis, N., J. E. Payne, K. Menyah, and Y. Wolde-Rufael. 2010. On the causal dynamics between emissions, nuclear energy, renewable energy, and economic growth. *Ecological Economics* 69: 2255–2260.

Arthurson, V. 2009. Closing the global energy and nutrient cycles through application of biogas residue to agricultural land—Potential benefits and drawback. *Energies* 2: 226–242.

Asumadu-Sarkodie, S., and P. A. Owusu. 2017. The causal nexus between carbon dioxide emissions and agricultural ecosystem—An econometric approach. *Environmental Science and Pollution Research* 24(2): 1608–1618. doi:10.1007/s11356-016-7908-2.

Azad, A. K., M. G. Rasul, and T. Yusaf. 2014. Statistical diagnosis of the best Weibull methods for wind power assessment for agricultural applications. *Energies* 7: 3056–3085.

Azad, A. K., M. G. Rasul, M. M. K. Khan, S. C. Sharma, and M. M. K. Bhuiya. 2015. Study on Australian energy policy, socio-economic, and environment issues. *Journal of Renewable and Sustainable Energy* 7: 63131–63220.

Baek, Y. J., S. H. Lee, M. S. Kim, Y. S. Lee, G. C. Jang, and H. S. Na. 2012. A simulation study on the annual heating performance of a seawater-source heat pump and air-source heat pump. In *Proceedings of the Autumn Conference of the Korean Solar Energy Society (KSES)*, Seoul, Korea.

Berton, M. 2015. President of AIEL (Associazione Italiana Energie Agroforestali) - Confederazione Italiana Agricoltori Renewable Energy and Agriculture: A Conflict or an Opportunity? Italy. Available: http://www.wfo-oma.org/climate-change/articles/renewable-energy-and-agriculture-a-conflict-or-an-opportunity.html.

Bhattacharya, M., S. R. Paramati, I. Ozturk, and S. Bhattacharya. 2016. The effect of renewable energy consumption on economic growth: evidence from top 38 countries. *Applied Energy* 162: 733–741.

Binshad, T. A., K. Vijayakumar, and M. Kaleeswari. 2016. PV based water pumping system for agricultural irrigation. *Frontiers in Energy* 10: 319–328.

Cakir, U., K. Comaklı, O. Comakl, and S. Karsl. 2013. An experimental exergetic comparison of four different heat pump systems working at same conditions: As air to air, air to water, water to water and water to air. *Energy* 58: 210–219.

Chel, A., and G. Kaushik. 2011. Renewable energy for sustainable agriculture. *Agronomy for Sustainable Development* 31(1): 91–118.

Chuanchai, A., and R. Ramaraj. 2018. Sustainability assessment of biogas production from buffalo grass and dung: Biogas purification and bio-fertilizer. *Biotechnology* 8: 151. doi:10.1007/s13205-018-1170-x.

Corwin, D. L., K. Loague, and T. R. Ellsworth. 1999. Assessing non-point source pollution in the vadose zone with advanced information technologies. In Corwin D. L., Loague K., and Ellsworth T. R. (Eds.), *Assessment of Non-point Source Pollution in the Vadose Zone*. Geophysical Monograph 108, pp. 1–20. Washington, DC: American Geophysical Union.

Cowan, W. N., T. Chang, R. Inglesi-Lotz, and R. Gupta. 2014. The nexus of electricity consumption, economic growth and CO_2 emissions in the BRICS countries. *Energy Policies* 66: 359–368.

Dickson, M. H., and M. Fanelli (Eds.). 2003. *Geothermal Energy: Utilization and Technology*. Renewable Energy Series. Paris, France: UNESCO.

El-Hage Scialabbaa, N., and M. Muller-Lindenlaufa. 2010. Organic agriculture and climate change. *Renewable Agriculture and Food Systems* 25: 158–169.

Gobal, C., M. Mohanraj, P. Chandramohan, and P. Chandrasekar. 2013. Renewable energy source water pumping systems—A literature review. *Renewable and Sustainable Energy Reviews* 25: 351–370.

Guo, M., W. Song, and J. Buhain. 2015. Bioenergy and biofuels: History, status, and perspective. *Renewable and Sustainable Energy Reviews* 42: 712–725. doi:10.1016/j.rser.2014.10.013.

Huang, M., J. Alavalapat, D. Carter, and M. Langholtz. 2008. Is the choice of renewable portfolio standards random? *Energy Policy* 35: 5571–5575. doi:10.1016/j.enpol.2007.06.010.

Hyun, I. T., J. H. Lee, Y. B. Yoon, K. H. Lee, and Y. Nam. 2014. The potential and utilization of unused energy sources for large-scale horticulture facility applications under Korean climatic conditions. *Energies* 7: 4781–4801.

Ibrahiem, D. M. 2015. Renewable electricity consumption, foreign direct investment and economic growth in Egypt: An ARDL approach. *Procedia Economics and Finance* 30: 313–323.

International Energy Administration (IEA). 2003. *Renewables for Power Generation: Status and Prospects*. Paris, France: International Energy Agency.

International Energy Administration (IEA). 2014a. *World Energy Statistics and Balances (2014)*. Paris, France: UN Energy Statistics.

International Energy Administration (IEA). 2014b. *World Energy Statistics and Balances*. Paris, France: UN Energy Statistics.

Frey, G. W., and D. M. Linke. 2002. Hydropower as a renewable and sustainable energy resource meeting global energy challenges in a reasonable way. *Energy Policy* 30: 141261–141265.

Jafar, M. 2000. A model for small-scale photovoltaic solar water pumping. *Renewable Energy* 19: 85–89.

Jamea, E. M. 2013. Role of Agricultural Sector in Harnessing Renewable Energy, December 25, 2013 - 11:15 am | Biomass Energy, Solar Energy, Wind Energy. MENA, Mediterranean region.

Jebli, M. B., and S. B. Youssef. 2016a. Renewable energy consumption and agriculture: Evidence for co-integration and Granger causality for Tunisian economy. *International Journal of Sustainable Development and World Ecology* 24(2): 149–158. doi:10.1080/13504509.2016.1196467.

Jebli, M. B., and S. B. Youssef. 2016b. Combustible renewables and waste consumption, agriculture, CO_2 emissions and economic growth in Brazil. *MPRA* Paper No. 69694. Available: https://mpra.ub.uni-muenchen.de/69694/.

Jebli, M. B., and S. B. Youssef. 2017a. The role of renewable energy and agriculture in reducing CO_2 emissions: Evidence for North Africa countries. *Ecological Indicators* 74: 295–301.

Jebli, M. B., and S. B. Youssef. 2017b. Renewable energy, Arable Land, agriculture, CO_2 emissions, and economic growth in Morocco. *MPRA* Paper No. 76798. Available: https://mpra.ub.uni-muenchen.de/76798/.

Kainth, G. S. 2010. Food security and sustainability in India. Available: http://www.merinews.com/article/; accessed October 12, 2011.

Khan, M. T. I., Q. Ali, and M. Ashfaq. 2018. The nexus between greenhouse gas emission, electricity production, renewable energy and agriculture in Pakistan. *Renewable Energy* 118: 437–451.

Khan, M. T. I., M. R. Yaseen, and Q. Ali. 2017. Dynamic relationship between financial development, energy consumption, trade and greenhouse gas: Comparison of upper middle-income countries from Asia, Europe, Africa and America. *Journal of Cleaner Production* 161: 567–580.

Kim, J. S. 2013. Efficiency analysis of geothermal heating and cooling system using groundwater for controlled horticulture. PhD thesis, Andong University, Andong, Korea.

Lau, E., C. Tan, and C. Tang. 2016. Dynamic linkages among hydroelectricity consumption, economic growth, and carbon dioxide emission in Malaysia. *Energy Sources, Part B: Economics, Planning and Policy* 11(11): 1042–1049.

Lund, J. W., D. H. Freeston, and T. L. Boyd. 2011. Direct utilization of geothermal energy 2010 worldwide review. *Geothermics* 40: 159–180.

Mexico Renewable Energy in Agriculture Project (MREAP). 2006. World Bank – Case Study (1999–2006). Mexico.

Mbzibain, A., G. Tate, and A. Shaukat. 2015. The adoption of renewable energy enterprises in the UK. *Journal of Small Business and Enterprise Development* 22(2): 249–272.

Mekhilef, S., S. Z. Faramarzi, R. Saidur, and Z. Salam. 2013. The application of solar technologies for sustainable development of agricultural sector. *Renewable and Sustainable Energy Reviews* 18: 583–594.

Mosher, J. N., and K. W. Corscadden. 2012. Agriculture's contribution to the renewable energy sector: Policy and economics—Do they add up? *Renewable & Sustainable Energy Reviews* 16(6): 4157–4164.

Nacer, T., A. Hamidat, and O. Nadjemi. 2016. A comprehensive method to assess the feasibility of renewable energy on Algerian dairy farms. *Journal of Cleaner Production* 112: 3631–3642.

Natarajan, R. 2015. director, Carbon Dioxide Research and Green Technology Centre, VIT University, Chennai, India.

Omer, A. M. 2008. Green energies and the environment. *Renewable & Sustainable Energy Reviews* 12: 1789–1821.

Paramati, S. R., N. Apergis, and M. Ummalla. 2017. Financing clean energy projects through domestic and foreign capital: The role of political cooperation among the EU, the G20 and OECD countries. *Energy Economics* 61: 62–71.

Paramati, S. R., N. Apergis, and M. Ummalla. 2018. Dynamics of renewable energy consumption and economic activities across the agriculture, industry, and service sectors: Evidence in the perspective of sustainable development. *Environmental Science Pollution Research* 25: 1375–1387.

Paramati, S. R., M. Ummalla, and N. Apergis. 2016. The effect of foreign direct investment and stock market growth on clean energy use across a panel of emerging market economies. *Energy Economics* 56: 29–41.

Park, Y. J. 2007. *A Study on Field Test of the Horizontal Ground Source Heat Pump System for Greenhouse*; 2005-N-GE-P-01; Ministry of Commerce, Industry and Energy (MOTIE): Sejong, Korea.

Pedretti, E. F., G. Riva, G. Toscano, and D. Duca. 2010. Considerations on renewable energy sources and their related perspectives of agricultural engineering. *Journal of Agricultural Engineering* 2: 35–45.

Pimentel, D., G. Rodrigues, T. Wane, R. Abrams, K. Goldberg, H. Staecker, E. Ma et al. 1994. Renewable energy: Economic issues. *BioScience* 44(8).

Petrovska, M., D. Filiposki, and S. Petrovska. 2012. Energy efficiency and renewable energy sources—Benefit for the agricultural sector in the Republic of Macedonia. *International Journal of Ecosystems and Ecology Science* 2(2): 65–68.

Ramaraj, R., and N. Dussadee. 2015. Renewable energy application for organic agriculture: A review. *International Journal of Sustainable and Green Energy* 4(1–1): 33–38. Special Issue: New Approaches to Renewable and Sustainable Energy.

Reynolds, L., and S. Wenzlau. 2012. *Climate-Friendly Agriculture and Renewable Energy*: Working Hand-in-Hand toward Climate Mitigation. Washington, DC: Worldwatch Institute.

Rodriguez-Navas, C., E. Bjorklund, B. Halling-Sorensen, and M. Hansen. 2013. Biogas final digestive by-product applied to croplands as fertilizer contains high levels of steroid hormones. *Environmental Pollution* 180: 368–371.

Ryu, Y. S., H. J. Joo, J. W. Kim, and M. R. Park. 2012. Economic analysis of cooling-heating system using ground source heat in horticultural greenhouse. *Journal of the Korean Solar Energy Society* 32: 60–67.

Sadaka, S., M. A. Sharara, A. Ashworth, P. Keyser, F. Allen, and A. Wright. 2014. Characterization of Biochar from Switchgrass Carbonization. *Energies* 7: 548–567. doi:10.3390/en7020548.

Sadorsky, P. 2010. The impact of financial development on energy consumption in emerging economies. *Energy Policy* 38: 2528–2535.

Scherer, L., and S. Pfister. 2016. Hydropower's biogenic carbon footprint. *PLoS One* 11(9): e0161947. doi:10.1371/journal.pone.0161947.

Schnepf, R. 2006. *Agriculture-Based Renewable Energy Production*. Washington, DC: CRS Report for Congress.

Smart World. 2018. *Solar Pumps: Renewable Energy for Better Agriculture*. New Delhi, India: Electronics for You.

Smith, J. R., W. Richards, D. Acker, B. Flinchbaugh, R. Hahn, R. Heck, B. Horan et al. 2008. 25 by 25 agriculture's role in ensuring U.S. energy independence. Available: http://www.bio.org/ind/25×25.pdf; accessed April 2008.

Steve, J., A. Severn, and B. Raum. 2007. Renewable Portfolio Standard (RPS). California, USA. Available: http://www.awea.org/legislative/pdf/RPS%20factsheet%20Dec%202007.pdf; accessed April 2008.

Tijjani, N., B. Alhassan, A. I. Saddik, I. Muhammad, A. M. Lawal, and S. A. Maje. 2013. Renewable energy and sustainable food security in Nigeria. *Journal of Energy Technologies and Policy* 3(4): 1–6.

Troost, C., T. Walter, and T. Berger. 2015. Climate, energy and environmental policies in agriculture: Simulating likely farmer responses in Southwest Germany. *Land Use Policy* 46: 50–64.

Turyareeba, P. J. 2001. Renewable energy: Its contribution to improved standards of living and modernisation of agriculture in Uganda. *Renewable Energy* 24: 453–457.

U.S. Department of Energy (USDOE). 2008. States with renewable portfolio standards. Washington, DC. Available: http://www.eere.energy.gov/states/maps/renewable_portfolio_states.cfm; accessed April 2008.

Wannapokin, A., R. Ramaraj, and Y. Unpaprom. 2017. An investigation of biogas production potential from fallen teak leaves (*Tectona grandis*). *Emer Life Science Research* 3: 1–10. doi:10.7324/ELSR.2017.31110.

Wannapokin, A., R. Ramaraj, K. Whangchai, and Y. Unpaprom. 2018. Potential improvement of biogas production from fallen teak leaves with co-digestion of microalgae. *3 Biotech* 8(2): 1–18. doi:10.1007/s1320 5-018-1084-7.

Weiland, P. 2010. Biogas production: Current state and perspectives. *Applied Microbiology and Biotechnology* 85: 849–860.

Woods, J., A. Williams, J. K. Hughes, M. Black, and R. Murphy. 2010. Energy and the food system. *Philosophical Transactions of the Royal Society B* 365: 2991–3006.

12 Solar Energy for Biofuel Extraction

Haftom Weldekidan, Vladimir Strezov, and Graham Town

CONTENTS

12.1 INTRODUCTION

Due to the rapid global population growth and rising living standards, there has been a significant increase in energy demand and consumption over the last several decades (Chen et al., 2015). By 2040, the total energy use is expected to grow by about 40% of the current use. Even though the share of fossil fuels in the entire energy mix is expected to fall, it will still remain the dominant source of energy, with oil, coal, and gas each expected to account for over 25% of the global energy needs (Cronshaw, 2015). It is also estimated that the world population will reach 9.3 billion by 2050 (Morales et al., 2014). This rapid population growth will increase the energy demand, while fossil fuels, being dominant energy sources, are estimated to significantly deplete after 70 years (Metzger and Hüttermann, 2009). It is inevitable that sustainability and environmental challenges will continue, unless an alternative source of energy is put in place ahead of time. The existing pattern of energy supply cannot be sustained in the near future because of the depletion of fossil fuel reserves, as well as environmental impacts from their use (Rahman et al., 2014). According to Morales et al. (2014), one of the most complex challenges faced today is managing and halting climate changes produced by the over-exploitation of natural resources.

    Biomass is seen as the most promising energy source to mitigate greenhouse gas emissions. Substantial adoption of this ubiquitous energy source could alleviate the environmental, social, and economic problems faced by the modern society (Khan et al., 2009). Many researchers have shown the possibility of a substantial contribution of biomass to our energy demand for the years to come.

Until 2012, global biomass use was 8%–14% of the world's final energy consumption. The annual availability of biomass is estimated to reach as high as 108 Gtoe, which is almost 10 times the world's current energy requirement (Hoogwijk et al., 2005, Demirbas, 2007, Williams et al., 2012, Kan et al., 2016).

There should be efficient utilization of biomass through the adoption of improved energy technologies. There are many existing processes that convert raw biomass to usable forms of energy and chemicals. These include combustion, pyrolysis, gasification, torrefaction, liquefaction, esterification and fermentation (Elliott et al., 1991, Strezov and Evans, 2015). These processes are considered as critical biomass utilization alternatives, offering economic benefits through the production of high-value fuel gasses and liquids, char and chemicals (Rapagnà et al., 1998, Han and Kim, 2008, Bulushev and Ross, 2011, Zhang et al., 2016). These processes are highly endothermic requiring large heat input, generally supplied from non-renewable sources of energy (Morales et al., 2014). Solar energy can be captured and stored in chemicals or fuels, also known as solar fuels, for later use and easy transportation. Utilization of solar energy for assisting the biomass conversion through distillation or thermochemical processing is expected to significantly improve the overall biofuel life cycle performance. Recently biofuel extraction technologies using concentrated solar energy have been tested in solar reactors with real sun (Zeng et al., 2017). Current technologies consist concentrating part with polished aluminium or glass mirror as reflecting surface, biomass reactors mostly made of quartz or borosilicate glasses and different types of metals such as copper and steel, and controllers for temperatures, heating rates, pressure and tracking units (Weldekidan et al., 2017).

The objective of this work is to review solar-based technologies and their applications for solar assisted biomass utilization and conversion technologies. The first part of the paper describes the fundamental conversion mechanisms of biomass to biofuels, with emphasis on the thermochemical conversion mechanisms. Different types of solar concentrating technologies with potential to capture the solar heat to drive the thermochemical conversion process are further discussed. Integration of the prospective solar collectors with biomass reactors are additionally elaborated. Finally, review of the solar assisted pyrolysis, gasification and status of solar-assisted distillation process, together with characterization of the different product fractions obtained from the processes, are presented.

12.2 OVERVIEW OF BIOMASS TO BIOFUEL CONVERSION MECHANISMS

Biofuel is a type of energy derived from biomass such as plants, agricultural, animal, domestic, and industrial wastes. Biomass can be converted into higher value biofuels either through biochemical, thermochemical or physico-chemical processes.

Biochemical conversion process involves fermentation of the sugars into alcohols, such as ethanol. This includes biomass pre-treatment followed by fermentation of the sugars to ethanol then separation and purification to produce pure ethanol (Ullah et al., 2015). Figure 12.1 shows the recent trends for the second generation of biofuel production through biochemical process from lignocellulosic biomass.

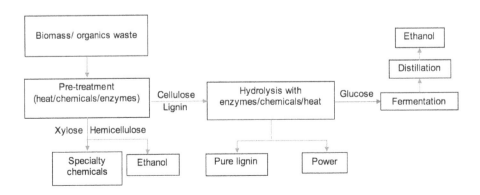

FIGURE 12.1 Biochemical process of biofuel production.

The efficiency of the biochemical conversion process is between 35%wt and 50%wt (Singh et al., 2010). This process can also be used to transform biomass into any type of petrochemical product compounds, such as olefins and aromatics, which are made from petroleum or fossil fuels.

Distillation is typically used to produce, separate, and distil ethanol into usable fuels. The energy is typically supplied through either an external heat source, such as gas or electricity from the grid. In both cases, this practice reduces the environmental benefits of the biomass conversion processes on a full-lifecycle basis.

Thermochemical processes of converting biomass into biofuels involve application of heat energy to treat the biomass in the conversion process with conversion efficiencies in the range between 41%wt and 77%wt (Singh et al., 2010). The treatment processes include combustion, gasification, and pyrolysis.

Combustion is the direct burning of biomass in the air for the purpose of heating and power generation, practiced since mankind has started using fire. Gasification is a process that converts organic or fossil fuel–based carbonaceous materials into carbon monoxide, carbon dioxide, and hydrogen, known as syngas. The syngas can be processed to produce different types of gaseous biofuels and liquids. Gasification is achieved by reacting the material at high temperatures (>700°C), without combustion, but with a controlled amount of oxygen and/or steam. The Fisher-Tropsch process with chemical catalytic conversion is an advanced engineering process developed to optimize the production of syngas for biofuel production. Gasification is a highly endothermic process. The heat required to maintain this process is supplied from non-renewable sources of energy by burning a significant portion (at least 35%) of the feedstock or using the electric grid, which lowers the final efficiency of the process (Piatkowski and Steinfeld, 2008, Kenarsari and Zheng, 2014, Pozzobon et al., 2016).

Pyrolysis is another thermochemical process that uses heat in the near absence of oxygen to destruct and distill biomass to produce biofuels, bio-oils (biocrude), biogas, and char. Pyrolysis has high flexibility in that it can be used to produce heavy fuel oil for heat and power applications or upgraded for conventional refinery operations, or it can be gasified for syngas, which can be converted to hydrogen. Pyrolysis can convert over 60 wt% of the biomass into liquid bio-oil (Hertwich and Zhang, 2009). Pyrolysis requires moderate temperature, in the range of 400°C–600°C, to depolymerize biomass to a mixture of oxygenates (or bio-oil) that are liquid at room temperature (Mettler et al., 2012). This external heat energy generally comes from burning part of the biomass or fossil fuels, or using grid electricity (Morales et al., 2014).

12.3 ENERGY FROM THE SUN

Approximately 885 million terawatt hours reach the earth's surface in a year, which is 4,200 times the energy that mankind would consume in 2035, according to the International Energy Agency (IEA's) Current Policies Scenario (Philibert, 2011). In just three hours, the earth collects enough solar radiation to meet the world's energy needs for one year. If one-tenth of one percent of the solar energy is captured and distributed, then the energy supply problem disappears (Philibert, 2011).

Biomass captures and converts solar radiation into energy ($C_xH_{2x}O_x$) through photosynthesis. Agrawal and Singh (2010) provided a review on the fraction of solar energy that can be recovered as biofuels, mainly liquid fuels for transportation purpose, via the cultivation and then conversion of the biomass using different methods (pyrolysis, gasification, fermentation, H_2bioil B, and H_2CAR processes). The highest sunlight conversion efficiency for a full-season growing biomass can be achieved up to 3.7% (Zhu et al., 2008). Fast pyrolysis of biomass can generate 524–627 L of liquid fuel per year, which corresponds to recovery of 65%–77% of the absorbed solar energy by the biomass. Estimates of the solar energy recovery as liquid fuels from fermentation, gasification, H_2bioil-B, and H_2CAR processing of biomass were found to be 41%–50%, 35%–50%, 59%–69%, and 58%, respectively (Agrawal and Singh, 2010). Using supplementary energy, such as H_2 or electricity that is recovered from solar energy at higher efficiencies than the biomass, can increase the fuel yield by a factor of 1.5–3.0 (Agrawal and Singh, 2010).

These estimates can be considerably improved if the heat energy for biomass conversion can be supplied from the sun using solar concentrating technologies. The following section describes the solar concentrating technologies available to integrate with biomass conversion for production of biofuels from biomass.

12.3.1 Solar Concentrating Technologies

Solar concentrators are devices that focus the solar energy incident over a large surface onto a smaller surface. There are several types of solar concentrating technologies that can capture the energy from the sun and make it available for application in the biomass-to-biofuel conversion processes. These are parabolic troughs (Hotz et al., 2010), heliostat fields, linear Fresnel reflectors, parabolic dishes, compound parabolic concentrators, flat plates (Kraemer et al., 2011), box type, and linear Fresnel lenses (Hotz et al., 2010, Lee et al., 2010, Kraemer et al., 2011, Bernardo et al., 2012, Baral et al., 2015). Each of these technologies differs in their temperature achievable, focal type, reflective material, operation characteristics, design, and application, and therefore each has its own advantages and drawbacks for use in biofuel production. It is important to make an appropriate technology choice for use in converting biomass to higher-value biofuels.

Flat-plate solar collectors are suitable to produce hot water up to 80°C, but flat-plate solar collectors integrated with evacuated tubes can reach a temperature of 125°C. The ability to boil water under ambient sunlight without optical concentration was demonstrated by Ni et al. (2016). Ni et al. (2016) used graphite as absorber material. To reduce heat losses, the bottom of the absorber was insulated using thermal foam, and a sheet of transparent bubble wrap was placed on top of it. The arrangement was able to generate saturated steam at 100°C, at efficiency as high as 64%.

Higher temperatures can be obtained with solar parabolic trough collectors, which achieve temperatures higher than 400°C (Morales et al., 2014). This type of arrangement is used to produce steam in industrial operations, whereby thermal energy collected in the receiver part is transported by a heat transfer medium to the intended place. Higher temperature can be achieved with central receiver systems or dish concentrators, which can achieve temperatures of up to 2,000°C.

12.3.2 Parabolic Trough Concentrator

A parabolic trough is a type of solar thermal device that is straight in one dimension and curved as a parabola (two-dimensional, U-shaped symmetrical curve) in the other two. The side that faces the sun is lined with a highly reflective material; thus, solar radiation is reflected onto a linear receiver placed at the focal line of the parabola (Blanco et al., 1986, Feuermann and Gordon, 1991, Lovejoy et al., 1993, Kaygusuz, 2001, Nixon et al., 2010, Duffie and Beckman, 2013, Abid et al., 2016). The reactor can be placed at this linear focal point of the parabolic trough concentrator. Typically, a reactor made from metal of high thermal conductivity or evacuated glass tube can reach working temperatures of over 400°C and concentration ratios (ratio between the concentrator opening area and the aperture area that receives) of 30–100 (Duffie and Beckman, 2013). Copper or bimetallic copper-steel are good options for this purpose, but stratification is unavoidable (Flores and Almanza, 2004). An optical efficiency of 80% has been recorded in California, providing 354 W/m^2 and a stagnation temperature of 600°C (Lovejoy et al., 1993). A biomass reactor system, if used with parabolic troughs, can be either heated directly (Alonso and Romero, 2015) in the focal line or by a heat-transferring medium to heat the reactor placed out of the focal line. There are technical challenges that should be taken into consideration while using parabolic trough for biomass process-ing, including the risk of overheating tubes and instability of the bioreactor. Also, complex control mechanisms are required if an indirect heating system of the bioreactor is to be employed (Nixon et al., 2010).

12.3.3   LINEAR COMPOUND PARABOLIC COLLECTOR

A linear compound parabolic collector reflector is used as a non-tracking system to concentrate solar energy to a double-sided flat receiver/reactor, which is normal to the compound parabolic col-lector axis (Gu et al., 2014). The common configuration has a bottom section resembling a circle, while the upper section is a parabola; thus, the focus is a line stretching from edge to edge. This is a non-imaging linear concentrator that can also work as a stationary collector without tracking system (Blanco et al., 1986). The maximum achievable temperature is only 200°C for a concentration ratio of 3. Thus, unless augmented by an additional heat source, the compound parabolic collector alone cannot be used for the purpose of biomass thermal processing.

12.3.4   LINEAR FRESNEL REFLECTORS

Linear Fresnel reflectors use long and thin segments of mirrors to focus sunlight onto a fixed absorber/reactor located at a common focal point of the reflectors. The reflectors are made from cheap flat mirrors and can concentrate the sun's rays 30 times. The operation temperature at the focal line is 150°C, but, with the use of a secondary concentrator, temperatures of up to 300°C can be reached. Additionally, if a compound parabolic collector is integrated with linear Fresnels, the optical and capture efficiencies can be improved to 60% and 76%, respectively (Feuermann and Gordon, 1991, Nixon et al., 2010). The reactor used with linear Fresnel is separated from the reflec-tor field and is stationary. The capital and maintenance costs are much lower than the other types of solar collectors (Feuermann and Gordon, 1991, Nixon et al., 2010).

12.3.5   PARABOLIC DISH REFLECTOR

A parabolic dish is a surface generated by a parabola revolving around its axis. It can be used to concentrate the solar rays and achieve reactor temperatures as high as 2,000°C. Depending on the size, a solar parabolic dish can have a concentration ratio (Tesfay et al., 2014) in the range of 500–2,000. The parabolic dish is mainly used to concentrate solar radiation for low-, medium-, and high-temperature applications (Tsoutsos et al., 2003, Abu Bakar et al., 2015) but requires two-dimensional continuous tracking, as the concentrated solar rays should be focused onto a single focal point (Nixon et al., 2010). Parabolic dish has the highest capture of solar energy, achieving optical efficiency (ratio of energy reaching the absorber to the irradiance falling on the collector surface) of up to 94%.

   With the correct reflective filming of the collector and black reactor coating, a reactor or receiver placed at the focal point can reach a temperature well over 1,000°C (Kaygusuz, 2001). For solar thermochemical processes, particularly with solar parabolic dish, the best candidate reflective coat-ings are silver-coated glass and silvered polymer films. Polymer reflectors are lighter in weight, offer greater system design flexibility, and have the potential for a lower cost than glass reflectors (Schissel et al., 1995, Kennedy and Terwilliger, 2005). The current research trends in the assessment of efficient solar reflective materials for long-term outdoor application range from various silvered glass mirrors, silvered polymer films, and anodized sheet aluminium with additional protective polymer coating (Fend et al., 2000, DiGrazia et al., 2009, Auti et al., 2015). Plain aluminium, with reflectivity of 85%, is the other reflector coating that is of interest for its low cost (Nostell et al., 1997, Kumar et al., 2015).

TABLE 12.1
Solar Collectors, Focal Type, and Achievable Temperature Levels

Type of Concentrator	Focus Type	Temperature (°C)	Reference
Parabolic dish reflector	Focal point	>1,500	(Kaygusuz, 2001, Tsoutsos et al., 2003, Nixon et al., 2010, Abu Bakar et al., 2015)
Parabolic trough	Focal line	400	(Lovejoy et al., 1993, Duffie and Beckman, 2013)
Linear compound parabolic	Focal line	200	(Blanco et al., 1986)
Linear Fresnel	Focal line	300	(Feuermann and Gordon, 1991, Nixon et al., 2010)

A number of studies are available on black coatings for solar thermal applications. Black coating materials should be low-cost, easy to manufacture, chemically stable, and able to withstand high temperatures (Kennedy, 2002). Moon et al. (2015) performed several experiments on black oxide nanoparticles as solar-absorbing material for a high-temperature concentrating solar system. Accordingly, a cobalt oxide (Co_3O_4) black-coated layer exhibited high-temperature durability and hardly degraded in structure after long working hours. The light-absorbing performance of Co_3O_4 was found to be 88.2%, making it a promising candidate for solar absorption in the next-generation, high-temperature solar concentrating systems.

Higher temperature can also be achieved if the receiver or reactor is enveloped in a glass tube. This gives the parabolic dishes the potential to eventually become one of the important devices for solar thermochemical conversion processes. Moreover, parabolic dish systems are typically designed for small-to-moderate capacity applications on the order of 10 kWh, which are suitable for remote power needs in rural areas and the places far from the national electricity grid. Another advantage of the parabolic dish is that, unlike other solar thermal systems, such as parabolic trough, Fresnel mirrors, and compound parabolic, leveled ground is not a requirement for its installation or operation (Nixon et al., 2010). Despite all these benefits, the drawback with the parabolic dish is its high cost and manufacturing difficulties. The reflector, in many cases the mirror, is the major contributor to the high cost, although there are alternatives, such as stretched-aluminium silvered polymer, which considerably reduces the cost from $80/m²–$150/m² to $40/m²–$80/m² and has longer life span than reflective materials used to make mirrors (Nixon et al., 2010). Table 12.1 summarizes potential solar collectors for biomass thermochemical conversion process.

12.4 SOLAR-ASSISTED THERMOCHEMICAL CONVERSIONS

12.4.1 SOLAR-ASSISTED THERMOCHEMICAL REACTORS

In the solar thermochemical production of biofuels, the feedstock is placed in a reactor and heated by the solar collector, either directly or indirectly. In a directly heated reactor, Figure 12.2a, the substrate is exposed to the concentrated solar radiation through a transparent container made of borosilicate glass or fused quartz, providing efficient energy transfer to the reaction site by direct radiation. In the directly heated reactor (Alonso and Romero, 2015), the solar reactor walls should be clean at all times, so as not to hinder the passage of the concentrated rays to the feedstock (Melchior et al., 2009, Kodama et al., 2010, Piatkowski and Steinfeld, 2011). The challenge to keep reactor windows clean can be overcome by using indirectly irradiated reactors, but at the expense of heat transfer efficiency. In indirect reactors, the solar energy is first absorbed by opaque wall reactor, then transferred to the biomass by conduction (Tesfay et al., 2014) or convection with heat-transferring fluid (Asmelash et al., 2014a, 2014b). Heat transmission may be limited, depending on the absorber and conductive material used. Maximum operating temperature, thermal conductivity,

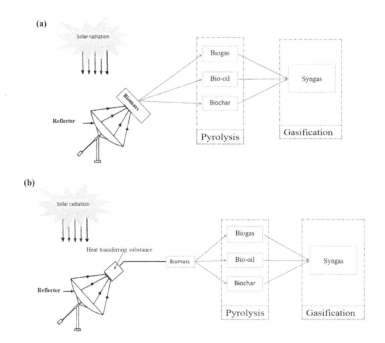

FIGURE 12.2   Schematic diagram of directly (a) and indirectly (b) irradiated solar reactors.

inertness to chemical reaction, resistance to thermal shocks, and radiative absorbance are a few of the drawbacks (Piatkowski and Steinfeld, 2011). Moreover, packing density of the biomass, ability to move through the reactor, the size of the particles, and physical properties of the reactants can affect the heat transfer conditions (Adinberg et al., 2004). Adinberg et al. (2004) suggested use of intermediate fluids, such as gases, liquid metals, or molten salts, to improve the heat transfer conditions from the reactor to the feedstock. Indirect reactors can be made from metals with high thermal conductivity. Assuming thermal conductivity, cost, manufacturing ability, durability, and weight of the metals, the candidate reactor materials are commercial copper (K = 423 W/m.K), aluminium (K = 215 W/m.K), and pure silver (K = 418 W/m.K). Enveloping the reactor by evacuated tube prevents the radiative heat loss and hence improves reactor performance.

12.4.2   SOLAR-ASSISTED PYROLYSIS

Application of solar energy for biomass thermochemical conversion dates back to the 1980s, starting with solar simulators, called *furnace images*, as sources of radiation, and parabolic or elliptic mirrors as concentrators (Zeng et al., 2017). Solar simulators were produced from powerful light sources, such as carbon arcs, xenon lamps, and mercury-xenon arc lamps. Some of the pioneering works that use solar simulators for biomass pyrolysis were Hopkins et al. (1984a, 1984b), Tabatabaie-Raissi and Antal (1986), Hunjan et al. (1989), Boutin et al. (1998, 2002), and Authier et al. (2009).

Solar-assisted pyrolysis is currently an emerging technology (Authier et al., 2009, Zeng et al., 2015a) for biomass conversion to biofuels and biochar that is attracting considerable research interest. Table 12.2 summarizes the performance parameters of the solar-assisted pyrolysis performed for production of biogas, liquid biofuels, and biochar.

Zeng et al. (2015a) conducted a laboratory-scale solar pyrolysis experiment on pellets of wood and investigated the influence of temperature, using different flow rates of argon as a sweep gas. The experiments were conducted using a downward-facing, 1.5 kW parabolic solar dish concentrator focusing the solar rays into a pellet placed at the focus of the parabolic dish inside a

TABLE 12.2
Summary of Solar-Assisted Biomass Pyrolysis

Reference	Heat Source	Power (KW)	Temperatures (°C)	Heating Rates (°C/s)	Samples Studied	Yield Summary
(Zeng et al., 2015a)	Solar simulator	1.5	600–2,000	50	Pellet wood	Gases (CO, CH_4) 15.3%–37.1% and liquids 70.7%–51.6%
(Zeng et al., 2015b)	Solar dish	1.5	600–2,000	5–450	Beechwood	28% liquid bio oil, 10% char and 62% gas (H_2, CH_4, CO, CO_2 and C_2H_6)
(Zeng et al., 2015c)	Solar dish	1.5	800–2,000	50–450	Beechwood	Gases (H_2, CO, CO_2, CH_4, and C_2H_6) heating value increased about five times (3527–14,589 kJ/kg)
(Zeng et al., 2015d)	Solar simulator	1.5	600–2,000	5–450	Wood	Char yield decreased with temperature and heating rate in the temperature range
(Authier et al., 2009)	Xenon lamp	//	907–487	//	Oak wood	Gas and char
(Morales et al., 2014)	Parabolic trough	//	465	//	Orange peel	Liquid (77.64 wt%), a non-condensable gas (1.43 wt%) and char 20.93 wt%
(Li et al., 2016)	Sun simulator	//	800–2,000	50	Sawdust, peach pit, grape stalk, and grape marc	Gas (63%wt), tar, and char (37%wt)
(Joardder et al., 2014)	//	//	500	5	Date seed	Liquid (50%wt), solid char, and gas (50%)
(Ramos and Pérez-Márquez, 2014)	Parabolic trough	//	>270	0.5	Wood	Charcoal (39%)
(Zeaiter et al., 2015)	Fresnel lens	//	850	//	Scrap tires	Oil and gas
(Hans et al., 2015)	Linear mirror	//	500	//	Agricultural wastes	Charcoal
(Soria et al., 2017)	Solar dish	1.5–2.0	600–2,000	10 & 50	Beechwood	Char, tar, and gases
(Arribas et al., 2017)	7 Kw xenon short-arc lamp	5,800 kW/m²	//	//	Algae, wheat straw, and sludge	63–90 vol% syngas

//: Data not available.

transparent and insulated graphite crucible. A series of heliostat mirrors were used to reflect the solar rays and continuously redirect them to the parabolic dish for exact concentration at the focal point. The temperature and heating rate of the sample were controlled by a shutter, which applies a proportional-integral-derivative controller that modulates the incident radiation. The reactor was made to accommodate gas inlets and outlets to permit entrance and exit of argon gas and reaction products, respectively. It was demonstrated that the temperature greatly affects products of pyrolysis compared to the sweep gas. The gas yields (mainly CO and CH_4) were found to increase from 15.3%wt to 37.7%wt when the temperature increased from 600°C to 2,000°C at 50°C/s heating rate. The higher the temperature, the higher was the gas yield, but the liquid yield decreased from 70.7%wt to 51.6%wt, with temperature showing most of the tar was decomposed at lower temperature ranges.

Zeng et al. (2015b) further performed pyrolysis of wood using solar energy. The solar concentrator was made from a downward-facing parabolic mirror 2 m in diameter and 0.85 m focal length. It was equipped with a sensor that detects the sun and adjusts the system for maximum concentration at the focal point where the substrate was placed and directly heated by the solar radiation. The maximum power was 1.5 kW and 15,000 W/m² flux density. A sweeping argon gas was used to wash walls of the transparent Pyrex reactor wall, so as to be clean and pass the radiation. The main objective of the experiment was to determine the optimal parameters to maximize lower heating values of wood gas products of pyrolysis during solar pyrolysis processes. Heating rates (5°C/s–450°C/s) and pressure (0.44–1.14 bar) were the investigated parameters in a temperature range of 600°C–2,000°C. Of the products, 62% were gases (H_2, CH_4, CO, and CO_2), and the remaining 28 and 10% were liquid bio-oil and char, respectively. The effect of temperature was found to be the most significant parameter in determining the characteristics of solar pyrolysis products and gas composition. More hydrogen and carbon monoxide were obtained at 1,200°C, 50°C/s heating rate, and atmospheric pressure. The lower heating value (LHV) increased five times when the temperature increased from 600°C to 1,200°C, and the heating rate increased from 5°C/s to 50°C/s. Moreover, the heating rate had substantial influence, but the effect of pressure was not significant on the product distribution of the solar pyrolysis process. The highest LHV, which was 10,376 kJ/kg, was found at 1,200°C, heating rate 50°C/s, and 0.85 bar.

Zeng et al. (2015c) performed a solar pyrolysis experiment to identify the effects of process parameters (temperature and heating rate) to optimize the solar pyrolysis process to produce combustible gases from sawdust, using a solar dish, which can generate flux intensity of 15,000 kW/m² for 1,000 W/m² of direct normal irradiance. The setup was equipped with a solar "blind optical" pyrometer, for measuring the sample temperature and solar tracker for adjusting the system to achieve maximum solar radiation during the day. A sample of wood (diameter of 10 mm and height of 5 mm) placed in a 6 L transparent Pyrex balloon reactor was directly heated from a solar dish with a 1.5 kW power. A temperature of 800°C–2,000°C and heating rate of 50°C/s–450°C/s were the operating variables. Box-Behnken design experiments were performed to optimize the process. It was shown that temperature and heating rate were the most influencing factors for the LHV, gas composition, and product distributions, but the effect of argon flow rate was found to be minimal. The LHV of the produced gas (H_2, CO, CO_2, CH_4, and C_2H_6) increased with temperature and heating rates. Particularly, the LHV increased four times (from 3,527 to 14, 589 kJ/kg) using solar pyrolysis of the wood.

A similar experimental setup was used by Zeng et al. (2015d), who conducted solar pyrolysis in a laboratory scale reactor, producing char from wood. The biomass samples were prepared into cylinders 10 mm in diameter and 5 mm in thickness. The char produced increased its surface area and pore volume until the temperature reached 1,200°C, then decreased significantly at 2,000°C. In a similar way, the pore volume and surface area increased with the heating rate until 150°C/s but slightly decreased afterward.

Theoretical and experimental studies of oak wood fast pyrolysis were conducted under controlled heat flux densities (0.3–0.8 MW/m²) and a temperature range of 907°C–487°C, generated from a xenon lamp in elliptical solar mirrors (Authier et al., 2009). The study was conducted to investigate the effect of heat flux density on the pyrolysis zone thickness of the wood. It was shown that the pyrolysis zone thickness of the wood decreased with the increasing flux density and influenced the char yields and gas composition for the lower flux density; the thickness varied from 180×10^{-6} to 750×10^{-6} m, but for higher heat flux, it was significantly smaller and varied from 100×10^{-6} to 225×10^{-6} m.

Morales et al. (2014) investigated orange-peel pyrolysis using solar energy from a parabolic trough solar concentrator. The trough was able to generate 27,088 W/m^2, which is equivalent to 31 times of the available solar energy in the location. The solar pyrolysis reactor, placed at the focal line of the parabola, reached a peak temperature of 465°C, and a total weight loss of 79% was achieved on the orange peel at an average irradiance of 12.55 kW/m^2. In this study, it was possible to pyrolyze the orange peel to a liquid (77.64 wt%) and a non-condensable gas (1.43 wt%), leaving 20.93 wt% biochar in the reactor. Furthermore, the work demonstrated the possibility for obtaining valuable chemical and pharmaceutical products, such as diisooctyl phthalate, squalene, D-limonene, (Z)-0-octadecenamide, and phenol, in addition to the production of combustible gases, such as hydrogen and carbon monoxide, generated through solar radiation augmented by conventional heating sources, including microwave and plasma.

Li et al. (2016) conducted solar pyrolysis in a laboratory-scale solar reactor to produce fuel gas from pine sawdust, peach pit, grape stalk, and grape marc. The solar energy was concentrated to a temperature of 800°C–2,000°C using solar dish. The samples were prepared to a cylinder of size 10 mm in diameter by 5 mm height, with an approximate weight of 0.3 g. For each type of biomass, the influences of final temperature, heating rate, and lignocellulose composition were analyzed. The results showed an increased gas yield and tar decomposition with temperature and heating rates, whereas the liquid yield progressed oppositely. The highest gas yield of 63.5wt% was obtained from pine sawdust at 1,200°C and 50°C/s. The remaining 37%wt were tar and char. A higher lignin content promoted char production, whereas higher cellulose and hemicellulose contents increased the gas yields. The H$_2$:CO ratio was always greater than one for both grape by-products, grape marc and grape stalk.

Joardder et al. (2014) designed a laboratory-scale, solar-assisted fast pyrolysis reactor in which part of the reaction heat came from a solar concentrator. Dried date seeds ground to sizes of 0.2–0.6 cm^3 in size were used as feedstock. It was found that 50%wt of the liquids were produced at 500°C with a running time of 120 min. It was also found that solar energy would contribute to the reduction of both CO$_2$ emissions and fuel cost by 32.4%.

A semi-static parabolic solar concentrator with a surface area of 1.37 m^2 was tested by Ramos and Perez-Marquez (2014). During the experiment, the temperature in the receiver was above 270°C. The prototype was able to produce 70 g of biochar out of 180 g of wood in five hours on a sunny day, which implies a conversion efficiency of 38%.

In a separate work, solar radiation was concentrated by a Fresnel lens to a maximum temperature of 850°C in a simulated solar radiation intensity of 1,500 W/m^2 (Zeaiter et al., 2015). This solar system, integrated with an automated solar tracking electronic system, was able to pyrolyze scrap tires at a temperature of 550°C. The experiment was carried out in the presence of H-beta, HUSY, and TiO$_2$ catalysts. Pyrolysis with the H-beta catalyst gives high oil and high gas yields of 32.8%. The TiO$_2$ and non-catalyzed pyrolysis results in a gas-like product (isopropane) with a quantity of 76.4wt% and 88.4wt%, respectively. Previous studies on tire pyrolysis revealed that oil and gas yield increase considerably with temperature, but the effect of temperature reduces in a temperature range over 500°C (Laresgoiti et al., 2004, Murillo et al., 2006).

Solar energy, concentrated by linear mirrors, was used to drive the pyrolysis of agricultural wastes, such as wheat straw (Hans et al., 2015). The system consists of sets of linear mirrors and a rectangular hollow-section steel as a reactor is placed at 5 m from the mirrors. A maximum temperature of 500°C was reached in about 90 minutes in the reactor that contains the wheat straw. In eight hours of sunshine per day, the system produces solar carbon (charcoal) with an energy density of 24–28 MJ/kg from a biomass of 16.9 MJ/kg. More discussion on the design, construction, and concentration principles of solar linear mirrors are presented in Hans et al. (2015).

Soria et al. (2017) studied beechwood pellet degradation under fast solar pyrolysis with Computational Fluid Dynamics (CFD) modeling. Simulation results were compared to experimental tests carried out using a solar facility at temperatures ranging from 600°C to 2,000°C, and heating rates of 10°C/s and 50°C/s (Zeng et al., 2015d). Results indicated that increasing the heating rate improved both uniformity of the char profile and intraparticle tar decomposition, producing more volatiles. Moreover, the higher the temperature and heating rate, the higher the gas yield, improving the intraparticle tar decomposition (Zeng et al., 2016). Solar-driven pyrolysis and gasification of algae, wheat straw, and sewage sludge were analyzed with a high-flux solar simulator (Arribas et al., 2017). The facility consisted of a 7 kW xenon short-arc lamp, flat mirror, two ellipsoidal mirrors, and a stainless steel reactor. The arc discharge was located in one of the ellipsoid loci; then emitted radiation was reflected by the flat mirror and concentrated on the second ellipsoid mirror to give maximum flux of 5,800 kW/m^2 at the focal plane. Released gases contained syngas in the range of 63 vol%–74 vol% for pyrolysis (highest for sludge) and 82 vol%–90 vol% for gasification (highest for algae).

12.4.3   SOLAR-ASSISTED GASIFICATION PROCESSES

There are a limited number of studies with various degrees of success on the integration of concentrated solar radiations with biomass gasification, summarized in Table 12.3. Pozzobon et al. (2016)

TABLE 12.3
Summary of Solar-Assisted Gasification

Reference	Heat Source	Power	Temperatures (°C)	Samples Studied	Yield Summary
(Pozzobon et al., 2016)	Artificial radiation from xenon arc lamp	1,000 suns	//	Beechwood	H$_2$ (38%vol), CO (31%vol), CH$_4$ (13%vol), and CO$_2$ (8.5%vol)
(Bai et al., 2015)	Laboratory-scale heliostat concentrator		727–1,227	Fossil fuel	Syngas and methanol; system efficiency was 56.9%
(Liu et al., 2016)	Parabolic trough		350	Biomass	Char; system efficiency was 19.2%
(Manenti et al., 2014, Ravaghi-Ardebili and Manenti, 2015)	Parabolic trough		400–410	Biomass	H$_2$ (24.1%), CO (34.2%), CO$_2$ (33.8%), and CH$_4$ (7.7%)
(Liao and Guo, 2015)	Dish concentrator	3.7–7.2 kW	500–600	Ethylene glycol, ethanol, glycerine and glucose	H$_2$ (10–26 mol/kg); gasification efficiency ranges 48.5%–105.8%
(Maag and Steinfeld, 2010, Yadav and Banerjee, 2016)	//	1,500 suns	1,227	Carbonaceous feedstock	37% (solar to chemical conversion efficiency)
(Service, 2009)	//	1 MW	1,200–1,300	Wood waste	Hydrogen at 13% conversion efficiency
(Nzihou et al., 2012)	Concentrate solar system	25–80 kW/m^2	//	Beechwood	Char (45%–20%), liquid (55%), and gas (10%–25%)
(Adinberg et al., 2004)	Electrically heated reactor		800–915	Cellulose	Syngas with 94% conversion efficiency

developed a gasification reactor system comprised of artificial sun, a xenon arc lamp capable of producing heat fluxes higher than 1,000 suns, and a new reaction chamber. This system allowed investigation of the thermal gasification behavior of thick beechwood when exposed to radiative heat. The impact of moisture content and wood fiber orientation relative to solar flux was tested and showed that increasing the sample's moisture content led to direct-drying steam gasification of the char. With 50%wt moisture content of the beechwood, the gasification products were H_2 38%vol, 31%vol CO, and 13%vol CH_4, while CO_2 was 8.5%vol. Up to 72% of the incident solar power was captured in the chemical form in this work. The wood fiber orientation was found to have no major impact on the production rates and gas composition.

Ravaghi-Ardebili et al. (2015) re-designed a low-temperature, solar-driven steam gasifier and investigated the impact of residence time of solid fuel and gas phase, as well as the amount of injected oxygen and steam on the gasification performance. Manenti and Ravaghi-Ardebili (2013) performed dynamic simulation and control of concentrating solar plants with the aim of defining a reasonably simplified layout, as well as highlighting the main issues for characterization of the process dynamics of these energy systems and their related energy storage capabilities. When operating parameters such as feedstock size, ratio of steam to biomass, type of biomass, geometry of reactors, air, and temperature effects are optimized, then heat energy collected from a solar concentrator can be stored on a working fluid and generate power for biomass gasification activities.

Bai et al. (2015) proposed a solar-driven biomass gasification system with the generation of methanol and electricity. The system consisted of three main parts, power generation subsystem (laboratory-scale heliostat concentrator), methanol production, and gasification subsystems. The endothermic reactions of the fossil fuel gasification were driven by the concentrated solar thermal energy of the heliostat in a range of 1,000 K–1,500 K, where a syngas from the biomass gasification was used to produce methanol through a synthesis reactor. Results indicated the syngas produced by the solar-driven gasification has a higher H_2:CO (1.43–1.89) molar ratio, which satisfies the requirement for methanol synthesis. Moreover, the produced syngas has better chemical energy quality than the conventional gasification technologies. The energy efficiency of the system was found to be 56.9%, which makes it a promising approach for the efficient utilization of the abundant solar and biomass resources.

A novel tri-generation system, coupled with a biomass gasification and solar thermal system, was investigated by Li et al. (2016). It was comprised of biomass gasification, a steam generation subsystem made of a parabolic trough solar collector, and an internal combustion engine subsystem. Ground biomass was preheated at 200°C, then fed into the gasifier. Steam at 350°C, generated from the solar collector, was fed into the gasifier with biomass. After removing the ash, char, and certain purification processes, the gas was fed to an internal combustion engine for electricity generation. In this study, the efficiency of this system was determined to be 19.2%, but introduction of the solar collector reduced the excess consumption of the biomass and improved the efficiency to 29%.

The feasibility of solar steam–supplied biomass gasification was demonstrated by Ravaghi-Ardebili and Manenti (2015) and Manenti et al. (2014). The study was aimed at storing concentrated solar energy generated from parabolic troughs for the purpose of steam production to accomplish biomass gasification. The parabolic trough was modelled and simulated to generate steam (approximately 400°C–410°C) and supplied to produce syngas consisting of H_2 (24.1%), CO (34.2%), CO_2 (33.8%) and CH_4 (7.7%). The syngas was further converted to methanol/dimethyl ether by means of a one-step synthesis process.

Solar energy can be stored in chemicals. Hydrogen production driven by solar chemical reaction is one of the ways to store solar energy. Liao and Guo (2015) developed a solar receiver integrated with a dish concentrator for gasification of ethylene glycol, ethanol, glycerine, and glucose in supercritical water. A series of outdoor experiments were conducted at 500°C–600°C (supercritical water state) with solar power input ranging from 3.1 kW to 7.2 kW. At 600°C, H_2, CH_4, and CO_2 generated gases of 41.2%, 15.1%, and 34.7%, respectively. The gasification efficiency was observed to increase from 48.5% to 105.8% following the radiation increase from 3.1 kW to 7.2 kW.

Maag and Steinfeld (2010) and Yadav and Banerjee (2016) investigated solar-to-chemical conversion efficiencies of carbonaceous feedstock. For an optimized reactor geometry and a desired outlet temperature of 1,500 K, the solar-to-chemical conversion efficiency was 37% for 1,500 suns solar concentration.

Pilot-scale solar biomass gasification was demonstrated at the University of Colorado Boulder and Sundrop Fuels (Service, 2009). The demonstration was conducted in tubular solar reactors, which can operate at 1 MW and 1,473 to 1,573K, and presented sunlight into hydrogen conversion efficiencies of wood waste at more than 13% efficiency.

Production of synthetic fuels through biomass gasification was studied by Nzihou et al. (2012). It was demonstrated that the efficiency of the process can be improved by supplying process heat from concentrated solar systems. Beechwood cylinders 40 mm in diameter were irradiated using concentrated solar radiation. The production of char fell from 45% to 20% when the irradiation level was increased from 25 kW/m^2 to 80 kW/m^2, while the liquid yield increased to 55% and the gas yield increased slowly from 10% to 25%.

Gasification of cellulose particles heated in a molten sodium carbonate and potassium carbonate medium at 800°C–915°C was investigated in a laboratory-scale, electrically heated reactor (Adinberg et al., 2004). Approximately 94%wt of the biomass was converted to syngas, primarily composed of H_2, CO_2, CH_4, and CO with 26 vol% hydrogen. It was reported that the same gasification process can be operated using concentrated solar energy supplied from solar collectors. The preliminary assessments of the processes performed for a commercial prototype demonstrated that gasification of biomass particles, dispersed in a molten salt phase and heated by solar energy, is a feasible and promising option for clean production of synthesis gas.

12.4.4 Solar-Assisted Distillation

The initial concentration of ethanol achieved by fermentation is approximately 7%–10% v/v (volume/volume), whereas the initial concentration required for use as fuel should be higher than 99.5% v/v (Jareanjit et al., 2014). For this reason, solar distillation can be applied to achieve the ethanol concentration required for achieving the standards. Solar distillation is a relatively matured technology used to increase ethanol concentrations to appreciable levels. Vorayos et al. (2006) performed analysis of solar ethanol distillation using a flat plate and evacuated heat pipe solar collectors to generate sufficient heat for ethanol distillation. Accordingly, 4 m^2 evacuated pipe solar collector was able to concentrate solar heat to enhance the ethanol concentration from 10% to 80% (v/v) using the solar distillation process. Similarly, Jareanjit et al. (2015) performed a solar distillation experiment to manage ethanol waste from the solar distillation process. The system consisted of three solar distillation stages operated in a batch, each contributing to reduce the amount of feed materials (cassava broth) in the system and increase the ethanol concentration from 8% to 80% (v/v).

12.5 CONCLUSION

Biomass thermochemical processing requires heat, which is typically supplied either by combusting part of the biomass or from non-renewable energy sources, such as combustion of fossil fuels or using electricity from the grid. This, in turn, decreases the efficiency of the conversion process by approximately 35% and challenges the sustainability of the biofuel production. Thus, it is important to develop an alternative, clean, and environmentally friendly source of energy for production of biofuels. This chapter reviewed the current state of research for the use of solar technologies for biomass processing and conversion to biofuels. Parabolic dish has the highest capture of solar energy, with optical efficiency reaching 94%, followed by parabolic trough. The solar parabolic dish, if integrated with the appropriate receiver/reactor systems, selective coatings, and reflective structures, can supply the required heat for thermochemical processing of biomass.

The chapter also reviews the solar-assisted pyrolysis, gasification, and distillation researches performed to date. Solar-assisted pyrolysis was applied to different types of biomass fuels to produce 1.43%–63.00% of bio-gas, 28.00%–77.64% of bio-oils, and 21%–62% of biochar. The heating rate and the final temperature were identified as the most important parameters that defined the distribution of the biofuel fractions. The solar-assisted biomass gasification process has been used to produce several high-value fuels, such as hydrogen-rich fuel gas and methane at concentrations ranging from 24% to 38% and 7% to 13%, respectively. The solar-assisted distillation process is a relatively mature technology used to increase the concentration of ethanol to achieve the required level for use as fuel. The solar-assisted biofuel extraction is an emerging technology that needs a technical breakthrough to overcome the challenges of the process. This implies developing standalone solar technologies with efficient concentration and storage capacity for extracting the biofuels. As biomass is low energy density, building small systems that can easily move to biomass-available sites can remove transporting bulk biomass and maximize the usability and distribution of the solar technologies.

After resolving these challenges in the future, the solar extraction of biofuels has the potential to produce high-grade energy products that can fully substitute fossil-derived fuels and also generate valuable chemicals. This review revealed that solar-assisted thermochemical conversion of biomass is a new area of research attracting significant interest for its potential. In particular, most of the research studies on the solar-assisted pyrolysis processes were performed in a laboratory environment using artificial sun, which needs to be validated with outdoor research using natural sun to realize the possible contribution of solar energy in the process of biofuel extractions. Efficient technologies for extracting biofuels from biomass using solar energy as process heat need to be further developed and examined.

ACKNOWLEDGEMENT

This chapter was reprinted from Weldekidan et al., Renewable and Sustainable Energy Reviews, 88, 184–192, 2018 with permission from Elsevier.

REFERENCES

Abid, M., Ratlamwala, T. A. H. and Atikol, U. 2016. Performance assessment of parabolic dish and parabolic trough solar thermal power plant using nanofluids and molten salts. *International Journal of Energy Research*, 40, 550–563.

Abu Bakar, R. B., Froome, C., Sup, B. A., Zainudin, M. F., Ali, T. Z. S., Bakar, R. A. and Ming, G. L. 2015. 2nd international conference on sustainable energy engineering and application (ICSEEA) 2014 sustainable energy for green mobility effect of rim angle to the flux distribution diameter in solar parabolic dish collector. *Energy Procedia*, 68, 45–52.

Adinberg, R., Epstein, M. and Karni, J. 2004. Solar gasification of biomass: A molten salt pyrolysis study. *Journal of Solar Energy Engineering*, 126, 850.

Agrawal, R. and Singh, N. R. 2010. Solar energy to biofuels. In: Prausnitz, J. M., Doherty, M. F. and Segalman, R. A. (Eds.) *Annual Review of Chemical and Biomolecular Engineering*, Vol 1. Palo Alto, CA: Annual Reviews.

Alonso, E. and Romero, M. 2015. Review of experimental investigation on directly irradiated particles solar reactors. *Renewable and Sustainable Energy Reviews*, 41, 53–67.

Arribas, L., Arconada, N., Gonzalez-Fernandez, C., Lohrl, C., Gonzalez-Aguilar, J., Kaltschmitt, M. and Romero, M. 2017. Solar-driven pyrolysis and gasification of low-grade carbonaceous materials. *International Journal of Hydrogen Energy*, 42, 13598–13606.

Asmelash, H., Bayray, M., Kimambo, C. Z. M., Gebray, P. and Adam, S. 2014a. Performance test of parabolic trough solar cooker for indoor cooking. *Momona Ethiopian Journal of Science (MEJS)*, 6, 39–54.

Asmelash, H., Kebedom, A., Bayray, M. and Mustofa, A. 2014b. Performance investigation of offset parabolic solar cooker for rural applications. *International Journal of Engineering Research & Technology (IJERT)*, 3, 920–923.

Authier, O., Ferrer, M., Mauviel, G., Khalfi, A.-E. and Lédé, J. 2009. Wood fast pyrolysis: Comparison of Lagrangian and Eulerian modeling approaches with experimental measurements. *Industrial and Engineering Chemistry Research*, 48, 4796–4809.

Auti, A. B., Pangavane, D. R., Singh, T. P., Sapre, M. and Warke, A. S. 2015. Study on reflector material optimization of a parabolic solar concentrator. In: Kamalakannan, C., Suresh, L. P., Dash, S. S. and Panigrahi, B. K. (Eds.) *Power Electronics and Renewable Energy Systems*. New Delhi, India: Springer.

Bai, Z., Liu, Q., Lei, J., Li, H. and Jin, H. 2015. A polygeneration system for the methanol production and the power generation with the solar–biomass thermal gasification. *Energy Conversion and Management*, 102, 190–201.

Baral, S., Kim, D., Yun, E. and Kim, K. C. 2015. Experimental and thermoeconomic analysis of small-scale solar organic Rankine cycle (SORC) system. *Entropy*, 17, 2039–2061.

Bernardo, L. R., Davidsson, H. and Karlsson, B. 2012. Retrofitting domestic hot water heaters for solar water heating systems in single-family houses in a cold climate: A theoretical analysis. *Energies*, 5, 4110.

Blanco, M. E., Gomez-Leal, E. and Gordon, J. M. 1986. Asymmetric CPC solar collectors with tubular receiver: Geometric characteristics and optimal configurations. *Solar Energy*, 37, 49–54.

Boutin, O., Ferrer, M. and Lédé, J. 1998. Radiant flash pyrolysis of cellulose—Evidence for the formation of short life time intermediate liquid species. *Journal of Analytical and Applied Pyrolysis*, 47, 13–31.

Boutin, O., Ferrer, M. and Lédé, J. 2002. Flash pyrolysis of cellulose pellets submitted to a concentrated radiation: experiments and modelling. *Chemical Engineering Science*, 57, 15–25.

Bulushev, D. A. and Ross, J. R. H. 2011. Catalysis for conversion of biomass to fuels via pyrolysis and gasification: A review. *Catalysis Today*, 171, 1–13.

Chen, W.-H., Peng, J. and Bi, X. T. 2015. A state-of-the-art review of biomass torrefaction, densification and applications. *Renewable and Sustainable Energy Reviews*, 44, 847–866.

Cronshaw, I. 2015. World energy outlook 2014 projections to 2040: Natural gas and coal trade, and the role of China. *Australian Journal of Agricultural and Resource Economics*, 59, 571–585.

Demirbas, A. 2007. Progress and recent trends in biofuels. *Progress in Energy and Combustion Science*, 33, 1–18.

Digrazia, M., Gee, R., Jorgensen, G. and ASME. 2009. Reflectech (R) mirror film attributes and durability for CSP applications. *Es2009: Proceedings of the ASME 3rd International Conference on Energy Sustainability*, Vol 2, San Francisco, CA, pp. 677–682.

Duffie, J. A. and Beckman, W. A. 2013. *Solar Engineering of Thermal Processes*, Wiley.

Elliott, D. C., Beckman, D., Bridgwater, A. V., Diebold, J. P., Gevert, S. B. and Solantausta, Y. 1991. Developments in direct thermochemical liquefaction of biomass: 1983–1990. *Energy and Fuels*, 5, 399–410.

Fend, T., Jorgensen, G. and Kuster, H. 2000. Applicability of highly reflective aluminium coil for solar concentrators. *Solar Energy*, 68, 361–370.

Feuermann, D. and Gordon, J. M. 1991. Analysis of a two-stage linear Fresnel reflector solar concentrator. *Journal of Solar Energy Engineering*, 113, 272–279.

Flores, V. and Almanza, R. 2004. Direct steam generation in parabolic trough concentrators with bimetallic receivers. *Energy*, 29, 645–651.

Gu, X. G., Taylor, R. A. and Rosengarten, G. 2014. Analysis of a new compound parabolic concentrator-based solar collector designed for methanol reforming. *Journal of Solar Energy Engineering-Transactions of the ASME*, 136.

Han, J. and Kim, H. 2008. The reduction and control technology of tar during biomass gasification/pyrolysis: An overview. *Renewable and Sustainable Energy Reviews*, 12, 397–416.

Hans, G., Marta, B., Marco, C., Marina, C., Enrico, E., Elvis, K. and Andrea, P. 2015. Solar biomass pyrolysis with the linear mirror II. *Smart Grid and Renewable Energy*, 6, 179–186.

Hertwich, E. G. and Zhang, X. 2009. Concentrating-solar biomass gasification process for a 3rd generation biofuel. *Environmental Science and Technology*, 43, 4207–4212.

Hoogwijk, M., Faaij, A., Eickhout, B., De Vries, B. and Turkenburg, W. 2005. Potential of biomass energy out to 2100, for four IPCC SRES land-use scenarios. *Biomass and Bioenergy*, 29, 225–257.

Hopkins, M. W., Antal, M. J. and Kay, J. G. 1984a. Radiant flash pyrolysis of biomass using a xenon flashtube. *Journal of Applied Polymer Science*, 29, 2163–2175.

Hopkins, M. W., Dejenga, C. and Antal, M. J. 1984b. The flash pyrolysis of cellulosic materials using concentrated visible light. *Solar Energy*, 32, 547–551.

Hotz, N., Zimmerman, R., Weinmueller, C., Lee, M.-T., Grigoropoulos, C. P., Rosengarten, G. and Poulikakos, D. 2010. Exergetic analysis and optimization of a solar-powered reformed methanol fuel cell micro-powerplant. *Journal of Power Sources*, 195, 1676–1687.

Hunjan, M. S., Mok, W. S. L. and Antal, M. J. 1989. Photolytic formation of free radicals and their effect on hydrocarbon pyrolysis chemistry in a concentrated solar environment. *Industrial and Engineering Chemistry Research*, 28, 1140–1146.

Jareanjit, J., Siangsukone, P., Wongwailikhit, K. and Tiansuwan, J. 2014. Development of a mathematical model and simulation of mass transfer of solar ethanol distillation in modified brewery tank. *Applied Thermal Engineering*, 73, 723–731.

Jareanjit, J., Siangsukone, P., Wongwailikhit, K. and Tiansuwan, J. 2015. Management of ethanol waste from the solar distillation process: Experimental and theoretical studies. *Energy Conversion and Management*, 89, 330–338.

Joardder, M. U. H., Halder, P. K., Rahim, A. and Paul, N. 2014. Solar assisted fast pyrolysis: A novel approach of renewable energy production. *Journal of Engineering*, 2014, 1–9.

Kan, T., Strezov, V. and Evans, T. J. 2016. Lignocellulosic biomass pyrolysis: A review of product properties and effects of pyrolysis parameters. *Renewable and Sustainable Energy Reviews*, 57, 1126–1140.

Kaygusuz, K. 2001. Renewable energy: Power for a sustainable future. *Energy Exploration and Exploitation*, 19, 603–626.

Kenarsari, S. D. and Zheng, Y. 2014. CO_2 gasification of coal under concentrated thermal radiation: A numerical study. *Fuel Processing Technology*, 118, 218–227.

Kennedy, C. E. 2002. *Review of Mid- to High-Temperature Solar Selective Absorber Materials*. Golden, CO: National Renewable Energy Laboratory.

Kennedy, C. E. and Terwilliger, K. 2005. Optical durability of candidate solar reflectors. *Journal of Solar Energy Engineering-Transactions of the ASME*, 127, 262–269.

Khan, A. A., De Jong, W., Jansens, P. J. and Spliethoff, H. 2009. Biomass combustion in fluidized bed boilers: Potential problems and remedies. *Fuel Processing Technology*, 90, 21–50.

Kodama, T., Gokon, N., Enomoto, S.-I., Itoh, S. and Hatamachi, T. 2010. Coal coke gasification in a windowed solar chemical reactor for beam-down optics. *Journal of Solar Energy Engineering*, 132, 041004.

Kraemer, D., Poudel, B., Feng, H.-P., Caylor, J. C., Yu, B., Yan, X., Ma, Y. et al. 2011. High-performance flat-panel solar thermoelectric generators with high thermal concentration. *Nature Materials*, 10, 532–538.

Kumar, A., Pachauri, R. K., Chauhan, Y. K. and IEEE. 2015. Analysis and performance improvement of solar PV system by solar irradiation tracking. *2015 International Conference on Energy Economics and Environment (ICEEE)*.

Laresgoiti, M. F., Caballero, B. M., De Marco, I., Torres, A., Cabrero, M. A. and Chomón, M. J. 2004. Characterization of the liquid products obtained in tyre pyrolysis. *Journal of Analytical and Applied Pyrolysis*, 71, 917–934.

Lee, M.-T., Werhahn, M., Hwang, D. J., Hotz, N., Greif, R., Poulikakos, D. and Grigoropoulos, C. P. 2010. Hydrogen production with a solar steam–methanol reformer and colloid nanocatalyst. *International Journal of Hydrogen Energy*, 35, 118–126.

Li, R., Zeng, K., Soria, J., Mazza, G., Gauthier, D., Rodriguez, R. and Flamant, G. 2016. Product distribution from solar pyrolysis of agricultural and forestry biomass residues. *Renewable Energy*, 89, 27–35.

Liao, B. and Guo, L. J. 2015. Concentrating solar thermochemical hydrogen production by biomass gasification in supercritical water. In: Wang, Z. (Ed.) *International Conference on Concentrating Solar Power and Chemical Energy Systems, Solarpaces 2014*.

Liu, P., Zhao, Y., Guo, Y., Feng, D., Wu, J., Wang, P. and Sun, S. 2016. Effects of volatile–char interactions on char during pyrolysis of rice husk at mild temperatures. *Bioresource Technology*, 219, 702–709.

Lovejoy, D., Johansson, T., Kelly, H., Reddy, A. and Williams, R. 1993. Renewable energy—sources for fuels and electricity. *Natural Resources Forum*, 17, 244–245.

Maag, G. and Steinfeld, A. 2010. Design of a 10 MW particle-flow reactor for syngas production by steam-gasification of carbonaceous feedstock using concentrated solar energy. *Energy and Fuels*, 24, 6540–6547.

Manenti, F., Leon-Garzon, A. R., Ravaghi-Ardebili, Z. and Pirola, C. 2014. Assessing thermal energy storage technologies of concentrating solar plants for the direct coupling with chemical processes. The case of solar-driven biomass gasification. *Energy*, 75, 45–52.

Manenti, F. and Ravaghi-Ardebili, Z. 2013. Dynamic simulation of concentrating solar power plant and two-tanks direct thermal energy storage. *Energy*, 55, 89–97.

Melchior, T., Perkins, C., Lichty, P., Weimer, A. W. and Steinfeld, A. 2009. Solar-driven biochar gasification in a particle-flow reactor. *Chemical Engineering and Processing: Process Intensification*, 48, 1279–1287.

Mettler, M. S., Vlachos, D. G. and Dauenhauer, P. J. 2012. Top ten fundamental challenges of biomass pyrolysis for biofuels. *Energy and Environmental Science*, 5, 7797–7809.

Metzger, J. O. and Hüttermann, A. 2009. Sustainable global energy supply based on lignocellulosic biomass from afforestation of degraded areas. *Naturwissenschaften*, 96, 279–288.

Moon, J., Kim, T. K., Vansaders, B., Choi, C., Liu, Z., Jin, S. and Chen, R. 2015. Black oxide nanoparticles as durable solar absorbing material for high-temperature concentrating solar power system. *Solar Energy Materials and Solar Cells*, 134, 417–424.

Morales, S., Miranda, R., Bustos, D., Cazares, T. and Tran, H. 2014. Solar biomass pyrolysis for the production of bio-fuels and chemical commodities. *Journal of Analytical and Applied Pyrolysis*, 109, 65–78.

Murillo, R., Aylón, E., Navarro, M. V., Callén, M. S., Aranda, A. and Mastral, A. M. 2006. The application of thermal processes to valorise waste tyre. *Fuel Processing Technology*, 87, 143–147.

Ni, G., Li, G., Boriskina, S. V., Li, H., Yang, W., Zhang, T. and Chen, G. 2016. Steam generation under one sun enabled by a floating structure with thermal concentration. *Nature Energy*, 1, 16126.

Nixon, J. D., Dey, P. K. and Davies, P. A. 2010. Which is the best solar thermal collection technology for electricity generation in north-west India? Evaluation of options using the analytical hierarchy process. *Energy*, 35, 5230–5240.

Nostell, P., Roos, A. and Karlsson, B. 1997. Optical characterisation of solar reflecting surfaces. In *Proceedings of SPIE*, San Diego, CA, pp. 163–172.

Nzihou, A., Flamant, G. and Stanmore, B. 2012. Synthetic fuels from biomass using concentrated solar energy – A review. *Energy*, 42, 121–131.

Philibert, C. 2011. *Solar Energy Prospective*. Paris, France: International Energy Agency.

Piatkowski, N. and Steinfeld, A. 2008. Solar-driven coal gasification in a thermally irradiated packed-bed reactor. *Energy and Fuels*, 22, 2043–2052.

Piatkowski, N. and Steinfeld, A. 2011. Solar gasification of carbonaceous waste feedstocks in a packed-bed reactor-Dynamic modeling and experimental validation. *AIChE Journal*, 57, 3522–3533.

Pozzobon, V., Salvador, S. and Bézian, J. J. 2016. Biomass gasification under high solar heat flux: Experiments on thermally thick samples. *Fuel*, 174, 257–266.

Rahman, M. M., B. Mostafiz, S., Paatero, J. V. and Lahdelma, R. 2014. Extension of energy crops on surplus agricultural lands: A potentially viable option in developing countries while fossil fuel reserves are diminishing. *Renewable and Sustainable Energy Reviews*, 29, 108–119.

Ramos, G. and Pérez-Márquez, D. 2014. Design of semi-static solar concentrator for charcoal production. *Energy Procedia*, 57, 2167–2175.

Rapagnà, S., Jand, N. and Foscolo, P. U. 1998. Catalytic gasification of biomass to produce hydrogen rich gas. *International Journal of Hydrogen Energy*, 23, 551–557.

Ravaghi-Ardebili, Z. and Manenti, F. 2015. Unified modeling and feasibility study of novel green pathway of biomass to methanol/dimethylether. *Applied Energy*, 145, 278–294.

Ravaghi-Ardebili, Z., Manenti, F., Corbetta, M., Pirola, C. and Ranzi, E. 2015. Biomass gasification using low-temperature solar-driven steam supply. *Renewable Energy*, 74, 671–680.

Schissel, P., Kennedy, C. and Goggin, R. 1995. Role of inorganic oxide interlayers in improving the adhesion of sputtered silver film on PMMA. *Journal of Adhesion Science and Technology*, 9, 413–424.

Service, R. F. 2009. Sunlight in your tank. *Science*, 326, 1472–1475.

Singh, N. R., Delgass, W. N., Ribeiro, F. H. and Agrawal, R. 2010. Estimation of liquid fuel yields from biomass. *Environmental Science and Technology*, 44, 5298–5305.

Soria, J., Zeng, K., Asensio, D., Gauthier, D., Flamant, G. and Mazza, G. 2017. Comprehensive CFD modelling of solar fast pyrolysis of beech wood pellets. *Fuel Processing Technology*, 158, 226–237.

Strezov, V. and Evans, T. J. 2015. *Biomass Processing Technologies*. Boca Raton, FL: CRC Press (Taylor & Francis Group).

Tabatabaie-Raissi, A. and Antal, M. J. 1986. Design and operation of a 30KWe/2KWth downward facing beam ARC image furnace. *Solar Energy*, 36, 419–429.

Tesfay, A. H., Kahsay, M. B. and Nydal, O. J. 2014. Design and development of solar thermal injera baking: Steam based direct baking. *Energy Procedia*, 57, 2946–2955.

Tsoutsos, T., Gekas, V. and Marketaki, K. 2003. Technical and economical evaluation of solar thermal power generation. *Renewable Energy*, 28, 873–886.

Ullah, K., Kumar Sharma, V., Dhingra, S., Braccio, G., Ahmad, M. and Sofia, S. 2015. Assessing the lignocellulosic biomass resources potential in developing countries: A critical review. *Renewable and Sustainable Energy Reviews*, 51, 682–698.

Vorayos, N., Kiatsiriroat, T. and Vorayos, N. 2006. Performance analysis of solar ethanol distillation. *Renewable Energy*, 31, 2543–2554.

Weldekidan, H., Strezov, V. and Town, G. 2017. Performance evaluation of absorber reactors for solar fuel production. *Chemical Engineering Transactions*, 61, 1111–1116. doi:10.3303/CET1761183.

Weldekidan, H., Strezov, V. and Town, G. 2018. Review of solar energy for biofuel extraction. *Renewable and Sustainable Energy Reviews*, 88, 184–192.

Williams, A., Jones, J. M., Ma, L. and Pourkashanian, M. 2012. Pollutants from the combustion of solid biomass fuels. *Progress in Energy and Combustion Science*, 38, 113–137.

Yadav, D. and Banerjee, R. 2016. A review of solar thermochemical processes. *Renewable and Sustainable Energy Reviews*, 54, 497–532.

Zeaiter, J., Ahmad, M. N., Rooney, D., Samneh, B. and Shammas, E. 2015. Design of an automated solar concentrator for the pyrolysis of scrap rubber. *Energy Conversion and Management*, 101, 118–125.

Zeng, K., Flamant, G., Gauthier, D. and Guillot, E. 2015a. Solar pyrolysis of wood in a lab-scale solar reactor: influence of temperature and sweep gas flow rate on products distribution. In: Wang, Z. (Ed.) *International Conference on Concentrating Solar Power and Chemical Energy Systems, Solarpaces 2014*.

Zeng, K., Gauthier, D., Li, R. and Flamant, G. 2015b. Solar pyrolysis of beech wood: Effects of pyrolysis parameters on the product distribution and gas product composition. *Energy*, 93, 1648–1657.

Zeng, K., Gauthier, D., Lu, J. D. and Flamant, G. 2015c. Parametric study and process optimization for solar pyrolysis of beech wood. *Energy Conversion and Management*, 106, 987–998.

Zeng, K., Gauthier, D., Soria, J., Mazza, G. and Flamant, G. 2017. Solar pyrolysis of carbonaceous feedstocks: A review. *Solar Energy*, 156, 73–92.

Zeng, K., Minh, D. P., Gauthier, D., Weiss-Hortala, E., Nzihou, A. and Flamant, G. 2015d. The effect of temperature and heating rate on char properties obtained from solar pyrolysis of beech wood. *Bioresource Technology*, 182, 114–119.

Zeng, K., Soria, J., Gauthier, D., Mazza, G. and Flamant, G. 2016. Modeling of beech wood pellet pyrolysis under concentrated solar radiation. *Renewable Energy*, 99, 721–729.

Zhang, Z., Liu, J., Shen, F., Yang, Y. and Liu, F. 2016. On-line measurement and kinetic studies of sodium release during biomass gasification and pyrolysis. *Fuel*, 178, 202–208.

Zhu, X.-G., Long, S. P. and Ort, D. R. 2008. What is the maximum efficiency with which photosynthesis can convert solar energy into biomass? *Current Opinion in Biotechnology*, 19, 153–159.

13 Hydrogen Production from Biomass

Tao Kan and Vladimir Strezov

CONTENTS

13.1 INTRODUCTION

As a storage and delivery medium for renewable energy sources, hydrogen will play a significant role in the future energy market. It is estimated that renewable sources will contribute approximately 69% of the total global energy demand, with the hydrogen share of 34%, by 2050 (Balat and Kirtay 2010). Until 2008, approximately 95% of hydrogen production was from reforming and gasification of fossil fuels (mainly natural gas, followed by heavy oils, naphtha, and coal), while the share from water electrolysis and biomass were only 4% and 1%, respectively (Marban and Vales-Solis 2007, Das et al. 2008). Water electrolysis for hydrogen generation has been also investigated for a long time. Biomass as a renewable and sustainable energy source can be another significant source due to its considerable content of hydrogen, although it is combined with carbon, oxygen, and other elements.

Hydrogen is the simplest element with the lowest density among all fuels, and with very high energy density. Hydrogen's lower heating value (LHV) per mass unit is as high as 119.96 MJ/kg, which is approximately 2.5, 2.8, and 5.3 times that of natural gas (47.13 MJ/kg), gasoline (43.44 MJ/kg), and coal (22.73 MJ/kg), respectively (Huang et al. 2007, Hydrogen Tools 2018). Other inherent

TABLE 13.1

Application Options of Hydrogen in the Hydrogen Economy

Sectors	Application Options
Industrial sector	Chemical processing, such as ammonium manufacture, petro-refining, and fatty acid hydrogenation
	Power generation
Power generation sector	Co-generation
	Power generation
Transportation sector	Internal combustion engines
	Fuel cells
Building sector	Residential, commercial, industrial, etc.

Source: Gnanapragasam, N. V. and Rosen, M. A. 2017. *Biofuels-UK* 8: 725–745.

properties, such as wide resource availability, very long-term viability for future applications, eco-friendly combustion product of water, and acceptance by a wide range of consumers (Barbir 2009, Mazloomi and Gomes 2012), have also endowed hydrogen with the potential to be a promising energy carrier. Hydrogen has applications in a variety of industries. Table 13.1 describes the application options of hydrogen in the hydrogen economy.

The technologies and processes for biomass-based hydrogen production can be categorized into two main groups, thermochemical processes (conventional gasification, hydrothermal gasification, pyrolysis, steam reforming of pyrolytic oils, combination of different processes, fuel cells, and other emerging technologies) and biologic processes (fermentation, photosynthesis, and biological water–gas shift [WGS] process) (Kalinci et al. 2009). Hydrogen may be directly or indirectly produced from biomass by one of the above processes. In later cases, further processing of the biomass-based intermediate products is required to produce hydrogen. Figure 13.1 depicts the main routes of hydrogen production from biomass (Milne et al. 2002).

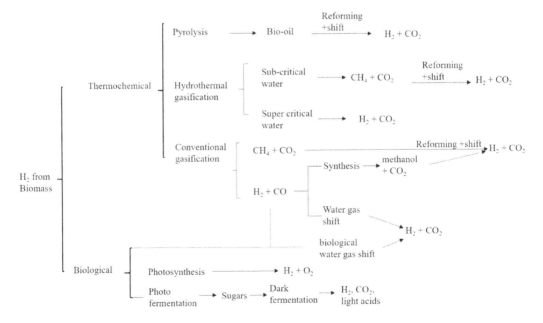

FIGURE 13.1   Main routes of hydrogen production from biomass.

It can be seen from Figure 13.1 that secondary treatment of raw product gas from primary processing is required, especially for the thermochemical processes. The final target product is H_2 together with CO_2, the ultimate oxide of carbon, as the removable by-product. Thermochemical routes are generally more efficient than biological, with a higher hydrogen yield and production rate. In this chapter, all the above-mentioned biomass-based hydrogen production technologies are reviewed, with emphases on the thermochemical routes. The comprehensive comparison between biomass-based hydrogen production and other hydrogen production processes has also been made in aspects of production cost, energy efficiency, and environmental performance.

13.2 CONVENTIONAL GASIFICATION

Gasification has a long history since the middle of 1800s, starting with coal gasification for home heating purposes. During World War II, Germany utilized the coal producer gas to synthesize transportation liquid fuels via Fischer–Tropsch (n CO + [$2n$ + 1] H_2 = C_nH_{2n+2} + n H_2O) and Bergius (n C + [n + 1] H_2 = C_nH_{2n+2}) reactions. As a spin-off technology, biomass gasification has been developed for more than 50 years.

Biomass gasification is usually carried out at high temperatures of typically above 800°C, generating a hydrogen-rich gas mixture called *producer gas* (also known as syngas). The gas has a typical volumetric composition of hydrogen (H_2, 12%–20%), carbon monoxide (CO, 17%–22%), methane (CH_4, 2%–3%), carbon dioxide (CO_2, 9%–15%), and water vapor (H_2O), as well as unwanted impurities of nitrogen (N_2), tar vapor and dust on a N_2-free basis (McKendry 2002). Biomass gasification is a partial oxidization process that efficiently converts biomass in oxidative atmospheres. The gasifying agents are generally air, oxygen, CO_2, steam, or their mixture.

Biomass gasification is a complex process involving a series of chemical reactions and physical changes in a gasifier. The gasification process can be typically divided into four different zones.

1. Drying zone, where the drying of biomass occurs. Both the surficial and inherent moisture evaporates at the zone temperature of 100°C–200°C, decreasing the moisture content to a low level of less than 5 wt% (Puig-Arnavat et al. 2010).
2. Pyrolysis zone, where decomposition of biomass takes place at around 200°C–700°C. Major biomass constitutes, such as cellulose, hemicellulose, and lignin, degrade to medium and then small molecules (H_2 + CH_4 + C_2H_6 + CO + CO_2 + H_2O + NH_3 + N_2 + H_2S, etc.), accompanied by formation of solid char. Tars and bio-oils are also formed through the polymerization and aggregation of some medium molecules.
3. Partial combustion zone, where highly exothermic oxidation reactions of char, small gaseous molecules, tar, and oil vapors with the gasifying agents dominate. Elevated temperatures of 1,200°C–1,500°C or even higher are formed in the zone. CO, CO_2, H_2O, N_2, and SO_2 are the major products in this zone.
4. Gasification zone, where char and methane are gasified by CO_2 and/or H_2O to form the components of producer gas, including H_2, CO, and CH_4. The dominant reactions are given in Table 13.2 (Kan and Strezov 2015).

Beside the upstream biomass gasification, the whole process also consists of downstream processing of producer gas, and end-use of the final gas product. The raw producer gas usually contains undesired impurities, especially heavy tars, which may cause equipment blocking and fouling if untreated. The treated producer gas can be used for power generation in internal combustion engines, heat supply in boilers, transportation fuels via chemical synthesis, and production of chemicals, such as H_2 and fertilizers.

TABLE 13.2
Dominant Reaction in Gasification Zone

Reaction Name	Reactions	Equation No.
Boudouard	$C(s) + CO_2 = 2CO$	13.1
Water gas	$C(s) + H_2O = H_2 + CO$	13.2
Water-gas shift	$CO + H_2O = H_2 + CO_2$	13.3
Steam reforming	$CH_4 + H_2O = 3H_2 + CO$	13.4
	$CH_4 + 2H_2O = 4H_2 + CO_2$	13.5
Hydrogasification	$C(s) + 2H_2 = CH_4$	13.6
Methane formation	$CO + 3H_2 = CH_4 + H_2O$	13.7
Other reactions	$NH_3 = 1/2\,N_2 + 3/2\,H_2$	13.8
	$SO_2 + 3H_2 = H_2S + 2H_2O$	13.9

13.2.1 INDICATORS FOR GAS QUALITY

There are indicators for evaluating the quality of gasification producer gas/final gas product. The quality of the raw producer gas is determined by several parameters, including the gas composition and some derivative indicators (e.g., H_2 and CO contents, as well as the H_2:CO ratio), impurity types, and contents (e.g., tar, solid carbon particles, entrained ash, and metals), and energy content, expressed as LHV. The end-use applications of the producer gas have determined the desirable gas composition and the subsequent measurements for tuning the gas to meet requirements.

For H_2 production, various gasification parameters can be adjusted to achieve the maximum H_2 yield and proportion in producer gas. The H_2 content can be further elevated by adding in-situ or ex-situ catalysts or other additives, such as CaO, which will be discussed in other sections of this chapter.

13.2.2 EFFECTS OF PARAMETERS

13.2.2.1 Biomass Material

Thermochemical methods can accept a wide variety of biomass feedstocks, unlike biochemical processes, which require starch- or sugar-rich biomass, excluding lignocellulosic materials (Basu 2010). For the most common lignocellulosic biomass materials, the three basic constituents of cellulose, hemicellulose, and lignin differ in contents. Yang et al. (2007) investigated the contribution of constituents to the composition of producer gas. It was found that H_2 generated from lignin was approximately four times that from hemicellulose, with the smallest H_2 contribution being from cellulose. Similar results were also obtained by Tian et al. (2017). Gasification of cellulose, hemicellulose, and lignin was performed in an updraft fixed-bed reactor. Over the temperature range of 1,000°C–1,200°C, lignin generated a higher hydrogen yield of 1.0–1.1 Nm³/kg, compared to the values of hemicellulose and cellulose (0.25–0.35 and 0.2–0.3 Nm³/kg, respectively). Smaller biomass particle size improves H_2 and CO production due to the enhanced heat and mass transfer, resulting in improvement in water gas reaction, Boudouard reaction, and carbon gasification reaction (Parthasarathy and Narayanan 2014).

Ash content of the biomass also plays a significant role. Generally, the alkali content of biomass exerts negative effects on the biomass gasification. Alkali salts will limit the gasification temperature due to the ash's softening and agglomeration (Corella et al. 2008). The nitrogen-containing compounds in biomass determine the NH_3 and NO_x contents of the producer gas. NH_3 can also be formed through Equation 13.8 (Table 13.2).

13.2.2.2 Gasifying Agent

Pure oxygen as a gasifying agent is favorable, and high reaction temperatures of 1,000°C–1,400°C can be achieved, resulting in a high-quality producer gas of mainly H_2 and CO, with an LHV of 12–20 MJ/m^3 (McKendry 2002). However, cost and safety may become barriers for pure oxygen's application (Saxena et al. 2008).

When cheap air is used as the gasifying agent, the producer gas may contain up to 60% N_2, which compromises the LHV (dropping to 4–6 MJ/Nm^3) and limits the further application of producer gas (Kalinci et al. 2009). In the market, the availability of mature technology and acceptable cost for N_2 removal will be important issues to resolve.

Compared to oxygen or air gasification, which retains no char, steam can act as a much weaker oxidizing agent for biomass gasification. Steam gasification is an endothermic process, generating a mixture gas of H_2, CO, CO_2, CH_4, light hydrocarbons, and tar, as well as a considerable yield of char. The LHV of the producer gas falls in the medium range of around 12–13 MJ/Nm^3 (Kalinci et al. 2009). In view of the distinct product distribution, this process is sometimes termed *steam pyrolysis of biomass* (Putun et al. 2008, Kan and Li 2015, Duman and Yanik 2017).

Steam gasification is one of the most efficient technologies for hydrogen production from biomass. Steam gasification of biomass can be described by the following reaction:

$$\text{Biomass} + \text{steam} \rightarrow H_2 + CO_2 + CO + \text{light gaseous hydrocarbons} \left(\text{e.g.,} \ CH_4\right) + \text{char} + \text{tar}$$

(Equation 13.10)

Among all the thermochemical technologies, steam gasification can produce the highest stoichiometric hydrogen yield. The producer gas from steam gasification contains approximately 40% of H_2, 25% of CO_2, and 25% of CO. However, the use of steam also causes several problems, such as corrosion to metal-gasifying reactors, poisoning of catalysts, and a decrease in thermal efficiency due to the energy consumption for heating water to high-temperature steam.

13.2.2.3 Temperature

Gasification temperature plays a key role for the energy efficiency and producer gas distribution. Generally, a higher temperature favors production of H_2 and CO, since biomass gasification is an endothermal process. Gasification is also beneficial for elimination of the tar content.

At lower temperatures, the producer gas contains H_2, CO, and CO_2 and also non-negligible carbon particles and CH_4. The presence of fine carbon particles will deactivate the catalyst bed if installed downstream due to the carbon deposition onto the catalyst grains. Results from the previous studies showed that the increase in temperature facilitates the reforming of carbon and CH_4 to H_2 and CO (Equations 13.2 and 13.4), which benefits the H_2 production (Mahishi 2006). The maximum H_2 yield was achieved at around 757°C. Further increase in temperature reduced the H_2 yield due to the exothermic WGS reaction (Equation 13.3 of Table 13.2) moving to the left, at the cost of H_2 consumption.

13.2.2.4 Pressure

Gasifiers could be operated in an atmospheric, pressurized, or low-pressure mode. Table 13.3 compares the advantages and disadvantages of pressurized conditions to the conventional atmospheric operation.

Pressurized gasification possibly enables the direct utilization of gas product in certain downstream, high-pressure processes with compression, such as methanol synthesis and Fischer–Tropsch synthesis. Other benefits may involve energy efficiency and smaller plant space. In addition, the high pressure affects the gasification chemistry in many aspects, such as rates of various reactions, producer gas composition, and tar removal. For example, it was observed that the methane production could be enhanced by elevating the pressure (Kitzler et al. 2011). However, the H_2 production

TABLE 13.3

Biomass Gasification under Pressurized Conditions in Comparison to the Conventional Atmospheric Operation

Advantages	Disadvantages
Increase in reaction rate constants of char-steam gasification, water gas shift reaction	—
Enhanced tar elimination	—
Energy efficiency if combined with high-temperature gas cleaning	More-complex high-temperature gas cleaning process
Improved heat transfer in gasifiers	Increased difficulty in operation, such as biomass feeding issues
Smaller space of plants	Higher capital operation and maintenance costs
Reduction in heat loss due to smaller surface area	Lack in experience

Sources: Chen, G. X. et al. 1992. *Ind Eng Chem Res* 31: 2764–2768; Goransson, K. et al. 2011. *Renew Sust Energ Rev* 15: 482–492.

may be suppressed at higher pressures, as the chemical equilibrium of the reactions described in Equations 13.2 and 13.4–13.9 (see Table 13.2) are shifted to the side of gas volume reduction. In another study, the tar yield was reduced from 4.3 wt% to 3.3 wt% dry tree chips when increasing the reaction pressure from 8.0 bar to 21.4 bar (Knight 2000). The contained phenol compounds (yield of 0.5 wt% dry biomass) were almost removed under these conditions.

On the other hand, low pressure was believed to not have obvious effects on the gasification. Previous research results indicated that, when decreasing the pressure to below 1 atm or even to 0.1 atm, negligible contribution was observed to the H_2 yield (Mahishi 2006).

13.2.2.5 Steam:Biomass Ratio

Steam:biomass ratio (SBR) is defined by the molar ratio of steam fed into the gasifier to biomass. At low SBR values, some amounts of undesirable carbon particles and CH_4 tend to form in the producer gas. When increasing the steam supply, sufficient steam is used to gasify the biomass, thus reducing the formation of carbon particles and methane while increasing the H_2 formation by facilitating WGS reaction and reforming of methane and char. However, excessive steam supply may compromise the gasification temperature, resulting in the increase in tar production.

13.2.2.6 Addition of Catalysts or Sorbents

In-bed catalysts have been extensively used to convert the tar into useful gas products, including H_2 and CO, therefore increasing the biomass conversion efficiency. The most relevant reactions are listed below:

$$\text{Steam reforming of tar: } C_nH_m(\text{tar}) + H_2O \rightarrow CO + H_2 \quad \text{(Equation 13.11)}$$

$$\text{Water gas shift reaction: } CO + H_2O \rightleftharpoons H_2 + CO_2 \quad \text{(Equation 13.3)}$$

$$\text{Dry reforming of tar: } C_nH_m(\text{tar}) + CO_2 \rightarrow CO + H_2 \quad \text{(Equation 13.12)}$$

$$\text{Cracking of tar: } C_nH_m(\text{tar}) \rightarrow C + H_2 \quad \text{(Equation 13.13)}$$

Although the dominant majority of previous studies of catalysts focused on reduction of tar rather than increasing hydrogen yield, the catalysts generally enhanced the hydrogen production

(Parthasarathy and Narayanan 2014). Many types of sorbents were also added to remove the unwanted impurities, such as alkali metals, sulfur, and CO_2.

The attempted catalysts/sorbents include both synthetic ones and minerals. Alkali metal–based catalysts are able to increase the hydrogen yield effectively when they are directly added into biomass feedstocks. Nickel-based catalysts can facilitate the WGS reaction and increase the hydrogen generation. Clay minerals and limestone/dolomite have been utilized as alkali and sulfur sorbents, respectively (Goransson et al. 2011). The removal of CO_2 will drive the WGS reaction (Equation 13.3 in Table 13.2) rightwards to enhance the hydrogen production. Hydrogen yield, total gas yield, and carbon conversion efficiency increased by approximately 50%, 60%, and 80%, respectively, at gasification temperature of 600°C by adding calcium oxide (Mahishi and Goswami 2007). In the process, calcium oxide acted as CO_2 sorbent and catalyst for tar hydrocarbon reforming.

13.3 HYDROTHERMAL GASIFICATION

Water exhibits special properties after it reaches the critical point (T = 374°C and pressure = 22 MPa). It becomes one phase between water and gas, even when changing the temperature and pressure. Hydrothermal gasification generally occurs at high temperatures of 400°C–800°C and pressures of 4.3–50.0 MPa (typically >20 MPa) (Osada et al. 2006, Peterson et al. 2008, Pavlovic et al. 2013). It exhibits several advantages, including the following:

- It accepts a wide range of biomass materials and is especially suitable for wet materials, which avoids energy-consuming drying pre-treatment.
- It can obtain high selectivity of desirable products by changing the temperature and pressure.
- The system can achieve high energy efficiency.

Hydrothermal gasification mainly produces gases consisting of H_2, CH_4, CO, and CO_2 with certain amounts of tars, chars, and bio-oils, as well as other impurities. Water temperature and pressure strongly influence the hydrothermal gasification process and gas products. Accordingly, hydrothermal gasification can be divided into three groups (Correa and Kruse 2018):

1. Low-temperature gasification at T = 215°C–265°C. The major products are H_2 and CO_2, which are much lower temperatures than conventional thermochemical processes. However, the biomass hydrolysis rate is low. This can be improved by adding noble metal catalysts (such as Pt/Al_2O_3).
2. Near-critical gasification at T = 300°C–500°C in presence of catalysts. In this case, CH_4 and CO_2 are the target gaseous products, rather than H_2 and CO_2. The catalysts are able to gasify the reactive intermediates (mainly phenols and furfurals) from the biomass hydrolysis and dehydration reactions and catalyze the CO hydrogenation to form methane (Peterson et al. 2008).
3. Supercritical water gasification at T > 374°C (usually 500°C–800°C) without catalysts or in presence of non-metal catalysts, such as activated carbon (Peterson et al. 2008). The resulting gaseous products are mainly H_2 and CO_2. The high-temperature condition largely enhances the free radical reactions and WGS reaction, resulting in more H_2 formation. The catalyst addition also facilitates the WGS reaction for hydrogen production. However, the effect of pressure on gas composition is limited. It was found that lower biomass concentrations and high temperatures are favorable to a higher hydrogen yield. In a study, glucose was processed at conditions of T = 767°C and pressure = 25 MPa, and the H_2 yield was 87.5%–93.3%, with 10.5–11.2 mol H_2/mol glucose (Susanti et al. 2012).

In addition to temperature and pressure, the performance of hydrothermal gasification for hydrogen production is also affected by other parameters, such as biomass composition and structure. It was concluded that cellulose contributed more to H_2 production than hemicellulose, and the presence of lignin decreased the H_2 yield due to the hydrogenation reaction of lignin, which consumed H_2 (Yoshida and Matsumura 2001). The effect of biomass composition on hydrogen yield is somewhat opposite to the conventional gasification of biomass as above-mentioned, which might be partly due to the presence of water at hydrothermal conditions and needs further investigation. The existence of inorganic salts, especially potassium species, facilitates the catalysis of WGS reaction to enhance hydrogen production (Sinag et al. 2003). Other chemical additives, such as butanol and boric acid, in the feedstock were also observed to increase the hydrogen formation by either driving the WGS reaction or facilitating the depolymerization of lignin (Ana et al. 2012, Yoshikawa et al. 2013).

Catalysts have been extensively used to enhance the hydrogen production by decomposing larger compounds to smaller ones, supressing the polymerization reactions, and driving the WGS reaction. Supercritical water gasification of sugarcane bagasse was performed, and the H_2 yield was promoted from 0.5 mmol/g to 3.2 mmol/g after adding Ru/C catalyst (Osada et al. 2012). In another study, the H_2 yield was found to be enhanced by different types of catalysts in the order of Pd > Ru > Pt > Rh > Ni (Correa and Kruse 2018).

13.4 PYROLYSIS FOLLOWED BY STEAM REFORMING

13.4.1 BIOMASS PYROLYSIS

Pyrolysis, as one of the most promising thermochemical technologies for energy extraction from biomass, is a processing technology to thermally decompose biomass in an oxygen-free inert atmosphere aiming to produce pyrolytic gas, pyrolysis oil (bio-oil), and biochar (Mulligan et al. 2015). The proportion of pyrolytic gas can be 20% or higher, depending on a variety of factors such as biomass type, reactor type, operating conditions (e.g., temperature and biomass residence time), and added catalyst and type. Pyrolytic gas is generally composed of CO_2, CO, H_2, light alkanes, and alkenes (C_xH_y, where x ≤4), and small amounts of other organics, such as heteroatom-containing compounds with small carbon numbers (Kan et al. 2014). Hydrogen-rich gas can be achieved by biomass pyrolysis, followed by steam reforming of bio-oil.

Fast or flash pyrolysis is preferable for maximizing the bio-oil yield and for carbon extraction from the biomass feedstock to form bio-oil products. The bio-oil yield can be affected by several parameters. It can be as high as 75%–80% under optimized pyrolysis conditions (Bridgwater and Peacocke 2000). The influencing parameters may include but are not limited to the following:

1. Biomass material. Cellulose and hemicellulose contents are mainly responsible for the bio-oil production with lignin for char generation. Solvent-extractable non-structural extractives in biomass also contribute to higher bio-oil yields.
2. Biomass pre-treatment. Physical pre-treatments, such as grinding or milling for smaller particle preparation, and thermal pre-treatment, such as drying, torrefaction, steam explosion, ultrasound, and microwave irradiation, have been widely used to improve bio-oil yields (Kan et al. 2016). In addition, chemical pre-treatment (e.g., inorganic salts and ionic liquids) and biological pre-treatment (e.g., fungal, microbial consortium, and enzymes) also find their applications.
3. Reaction atmosphere. The influence of N_2, CO_2, CO, CH_4, and H_2 atmospheres on the pyrolysis product yields were investigated in the past. It was observed that the highest bio-oil yield of 58.7% was reached in the CH_4 atmosphere, while CO gave the lowest value of 49.6% (Zhang et al. 2011).
4. Temperature is the most influential parameter. Bio-oil yield generally reaches its peak value at around 400°C–550°C and then declines when further increasing the temperature.

When the temperature increases higher than 700°C, the carbon content of bio-oils will be elevated correspondingly with temperature due to the enhanced formation of aromatic hydrocarbons (PAHs).

5. Heating rate. Slow pyrolysis is favorable to char formation rather than the bio-oil production. In a study using sawdust feedstock, it was observed that the increase in heating rate from 500°C/min to 700°C/min resulted in an increase in bio-oil yield by 8% (Salehi et al. 2009).

13.4.2 STEAM REFORMING OF BIO-OIL

The organic oxygenated compounds in bio-oils can be simply expressed as a chemical formula of $C_nH_mO_k$, as C, H, and O are the major elements. Steam reforming for hydrogen production can be described by the following reaction, followed by the WGS reaction:

$$C_nH_mO_k + H_2O \rightarrow H_2 + CO \qquad \text{(Equation 13.14)}$$

Thus, the overall stoichiometric steam reforming reaction can be described by

$$C_nH_mO_k + (2n-k) H_2O \rightarrow (2n + m/2 - k) H_2 + nCO_2 \qquad \text{(Equation 13.15)}$$

As can be seen, the stoichiometric maximum hydrogen yield is 2n + m/2 − k moles per n moles of carbon element in the bio-oil feed. The hydrogen yield can be defined as the ratio of practical hydrogen amount obtained to the stoichiometric maximum hydrogen. Other reactions may also take place during the reforming process, for example:

$$C_nH_mO_k \rightarrow C_xH_yO_z + \text{gases} + \text{coke} \qquad \text{(Equation 13.16)}$$

$$CO \rightarrow CO_2 + \text{solid carbon} \qquad \text{(Equation 13.17)}$$

During steam reforming of bio-oils, catalysts are generally required. Noble metal–based catalysts, such as Pt-based or Rh-based catalysts, could usually give higher hydrogen yields of 50%–70% than the commonly used Ni-based catalysts (<50%) (Rioche et al. 2005, van Rossum et al. 2007, Domine et al. 2008, Valle et al. 2018). Besides catalyst type, the H_2 production performance is also affected by other parameters. Higher processing temperature is more beneficial for the conversion of bio-oils on catalysts. However, high temperatures require more energy to heat the system and steam, and thus decrease the viability of the technology. Further increase in temperature above 500°C–600°C tends to lower the CO conversion due to the exothermic nature of the WGS reaction (Kan et al. 2010). Thus, a reaction temperature of 700°C–800°C was mostly used in previous studies. A higher catalyst/bio-oil ratio is favorable to the conversion of bio-oils, which is limited by the cost of catalysts. Previous researchers used a steam:carbon ratio ranging around 0.5–11.0. To achieve a satisfying hydrogen yield, the steam:carbon ratio of above 6 is recommended.

The product gas could contain 50%–70% of H_2 and a 15%–25% CO_2 content. The maximum H_2 yield ranged within 45%–80%, corresponding to 20–60 mol H_2/kg bio-oil. For example, in a previous study, two-stage processes, including thermal heating of bio-oil on silica sands and steam reforming of resulted intermedium organic vapors on NiCuZnAl catalysts, were employed to minimize the carbon deposit on reforming catalysts (Kan et al. 2010). The integrative reaction system achieved the maximum hydrogen yield of 81.4% and carbon conversion of 87.6% at 800°C. The final product gas mainly contained hydrogen (around 70%) and CO_2 (around 25%), with small amounts of CO and CH_4. In a most recent study, H_2-rich (66%) product gas with a very high hydrogen yield of 87% (on a basis of crude bio-oil) was obtained on Ni/La$_2$O$_3$-Al$_2$O$_3$ catalysts at conditions of 700°C, catalyst:bio-oil ratio of 0.38, and steam:carbon ratio of 6 (Valle et al. 2018).

The most challenging issue for steam reforming of crude bio-oils is the severe and fast catalyst deactivation due to the coke deposit on catalysts. Previous studies used the light volatile fraction of bio-oils and the relevant model compounds (e.g., phenols, acetic acid, ethanol, acetone, and glucose) other than crude bio-oils for hydrogen production (Wang et al. 2007a, 2007b, Garcia-Garcia et al. 2015, Xie et al. 2015, Ma et al. 2017, Sahebdelfar 2017, Xue et al. 2017, Yao et al. 2018).

Steam reforming of bio-oils is a developed industrial process, has no need for oxygen input, has the lowest reaction temperature, and provides the best H_2:CO ratio, with the disadvantage of higher emissions and lower energy efficiency (Nabgan et al. 2017).

13.5 BIOLOGICAL TECHNOLOGIES

Up to date, three main biological technologies for biohydrogen production from biomass have been developed, including fermentation, photosynthesis, and biological WGS process. All of the three conversion methods involve use of enzymes.

13.5.1 FERMENTATION

Hydrogen production from biomass fermentation generally involves two stages (Datar et al. 2007, Claassen et al. 2010, Eker and Sarp 2017).

1. Biomass rich in carbohydrate is firstly subjected to acid or enzymatic hydrolysis (or other pre-treatments) to obtain hydrolysate liquid (or other fermentables) with high sugar concentrations. Non-fermentables are also produced as by-products.
2. The hydrolysate liquid is then converted to H_2, CO_2, and volatile organic acids by anaerobic fermentation (also known as *dark fermentation*).

Further stages may also involve the following (Claassen et al. 2010):

- The organic acids can be further processed by photoheterotrophic fermentation (also known as *photo fermentation*) to enhance the generation of H_2 and CO_2.
- The produced H_2-rich gas is upgraded to meet the requirements for downstream end-uses.
- Non-fermentables can be utilized to produce valuable energy or chemicals.

Fermentation of biomass for hydrogen production has attracted extensive research interest. In previous studies, the H_2 yield varied in a wide range, from a few to more than 300 mL H_2/g material (Kumar et al. 2017). Acid-hydrolyzed reed canary grass (RCG) was fermented to generate biohydrogen, with the highest yield being 4.5 mL H_2/g RCG (Lakaniemi et al. 2011). In another study, aging corn pre-treated by solid microbe additives was used as material, and the maximum H_2 yield of 346 mL of H_2/g ash-free corn and H_2 production rate of 11.8 mL/h were obtained, respectively.

13.5.2 PHOTOSYNTHESIS

Photosynthesis refers to hydrogen production from phototropic organisms assisted by solar energy. The process can be briefly described by the following reaction:

$$H_2O + CO_2 \rightarrow \text{Sugars} + H_2 + O_2 \qquad \text{(Equation 13.18)}$$

The organisms absorb solar energy and produce protons and electrons, and then electrons take part in the formation of ferredoxin (Saxena et al. 2009). The protons and electrons subsequently combine with each other and generate molecular hydrogen by the hydrogen-generating enzyme hydrogenase (Allakhverdiev et al. 2010). The organisms with this function may include green algae, cyanobacteria, green bacteria, purple bacteria, and others.

Due to the primary product gas species of hydrogen and oxygen, safety is one of the major concerns. The simultaneous production of oxygen limits the activity of hydrogenase and lowers the hydrogen production, which substantially leads to the process' weak durability and low hydrogen-production efficiency (Allakhverdiev et al. 2010). Bench-scale hydrogen production from timber waste through photosynthesis was realized with the assistance of platinum nano-particles (Amao et al. 2016). Total hydrogen of 23 µmol was obtained during five hours of irradiation, corresponding to a hydrogen production rate of 0.5 mL/hr.

Many efforts have been devoted to improving hydrogen production. Since most hydrogenases are hypersensitive to oxygen, the suppression of oxygen evolution is a natural option. For example, in a previous study, the photochemical reduction and inactivation of oxygen generation was achieved in photosystems (Kruse et al. 2005). Modular devices without the oxygen evolution function, rather than intact cells, were developed to eliminate the impacts of oxygen evolution (Wenk et al. 2002). Light-to-H_2 photon conversion efficiency could be increased by reducing the antenna size to suppress undesirable fluorescence and heat losses. Proper cultivation and maintenance of organisms are also essential to reinforce their resistance to environmental stress conditions, benefiting hydrogen production. In addition, genetic engineering (screening of more-efficient mutants) has been introduced to improve hydrogen production efficiency (Allakhverdiev et al. 2010).

13.5.3 BIOLOGICAL WATER GAS SHIFT PROCESS

During biological WGS (BWGS) process, certain bacteria are employed to drive the WGS reaction for hydrogen production (Kalinci et al. 2009), as described below.

$$\text{Biomass-derived } CO + H_2O \ = \ H_2 + CO_2 \qquad \text{(Equation 13.3)}$$

Some bacteria (e.g., *Rubrivivax gelatinosus*) can perform WGS reactions at atmospheric conditions. Nearly 100% of CO can be utilized by the bacteria to produce adenosine triphosphate (ATP), coupling the CO oxidation with the reduction of protons from water to H_2 (Saxena et al. 2009). Compared to the biohydrogen production rate of 400–1,000 mL H_2 L^{-1} h^{-1} for dark fermentation process, a much higher value of 2,100 mL H_2 L^{-1} h^{-1} by the BWGS route has been reported.

13.6 SECONDARY PROCESSING OF PRODUCT GAS

Secondary measures are generally required to further process the producer gas to generate the ultimate hydrogen, especially for the large-scale hydrogen production. Gnanapragasam and Rosen (2017) depicted an entire typical flow of pure hydrogen production, from solid fuels feeding and gasification to hydrogen storage and transportation, as shown in Figure 13.2.

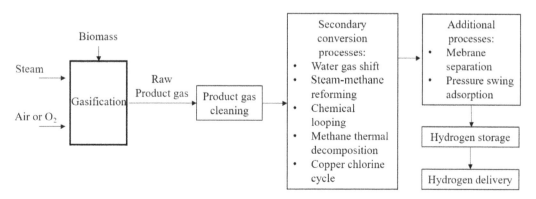

FIGURE 13.2 A typical flow of pure hydrogen production from thermochemical conversion of solid fuels with gasification taken as example.

During the solid fuel (biomass) injection, steam and air (or oxygen) are simultaneously fed into the gasifier as reactant agents and carrier gases. The secondary processing may include several steps, as follows:

1. Gas cleaning. The ash from the gasifier goes to the channel of waste management, while the raw producer gas carrying undesirable impurities (e.g., tars, solid carbon particles, H_2S, and HCl) enter the syngas cleaning processes.

   A large variety of gas cleaning technologies have been developed, which may include ash removal, syngas quench/cooling, and scrubbing (Chiesa et al. 2005). After cleaning, the syngas is mainly composed of CO, H_2, CH_4, H_2O, CO_2, N_2, and O_2.

2. Chemical conversion. The hot, clean syngas is then subjected to one or more of the following chemical conversion processes to enhance the hydrogen production and/or further removal of impurities.
   * WGS process

$$CO + H_2O \rightarrow H_2 + CO_2 \qquad \text{(Equation 13.3)}$$

   * Steam-methane reforming

$$CH_4 + H_2O \rightarrow 3H_2 + CO \qquad \text{(Equation 13.4)}$$

$$CH_4 + 2H_2O \rightarrow 4H_2 + CO_2 \qquad \text{(Equation 13.5)}$$

   * Chemical looping process
   Metallic iron or nickel can react with steam to generate hydrogen and act as the medium during the chemical looping process.

$$Fe + H_2O \rightarrow FeO + H_2 \qquad \text{(Equation 13.19)}$$

   * Thermal decomposition of methane

$$CH_4 \rightarrow C + 2H_2$$

   (reverse reaction of Equation 13.6)
   * Copper–chlorine cycle

$$Cu + HCl \rightarrow CuCl + \tfrac{1}{2} H_2 \qquad \text{(Equation 13.20)}$$

3. After chemical conversion, the effluent gas then goes through gas coolers and H_2S removal.
4. Separation and purification of H_2. Additional processes are essential to obtain purified hydrogen. Membrane separation technologies are being investigated as technologies to purify hydrogen. Currently, membrane separation for hydrogen purification is mainly at the stage of research, with some technical barriers on its way to commercialization, such as insufficient durability caused by contamination (Hannula 2009). The product gas is firstly purged through membranes to remove CO_2 and other minor impurity gases, then compressed and subjected to a pressure swing adsorption (PSA) process to finally acquire high-purity hydrogen. Depending on the end-use options, the hydrogen product may be directly used on-site (e.g., fuel cells) or sent to storage and delivery.

13.7   MARKET STATUS AND DEVELOPMENT CHALLENGES

Biological-based technologies for hydrogen production are still in development. Ren et al. (2006) carried out pilot-scale biohydrogen production in a continuous dark anaerobic fermentation reactor with a duration of more than 200 days. The substrate of molasses was fed into the system at organic loading rates (OLR) of around 3–86 kg COD/m^3 reactor per day, where COD represents chemical oxygen demand. The hydrogen content of product gas ranged between 40 vol% and 52 vol%. The hydrogen yield increased with the OLR in the range of 3–68 kg COD/m^3 reactor per day, then decreased when further increasing the OLR, with the maximum hydrogen production rate of 5.6 m^3 H$_2$/m^3 reactor per day. The Indian Institute of Technology Kharagpur operated a pilot-scale immobilized whole-cell bioreactor with the presence of ethanol in the substrate to generate biohydrogen (Das et al. 2008). A hydrogen production rate of around 80 mmol H$_2$/L^{-1} per hour was obtained.

In addition to the H$_2$ generation from biomass, many other technologies have also been developed. During the technological, economic, and environmental analyses of H$_2$ production, energy efficiency, production cost, global warming potential, and social cost of carbon have been extensively assessed. H$_2$ production cost involves many uncertainties, and it can be influenced by a variety of factors, such as the following:

- Market requirement of H$_2$
- Maturity and advancement level of the used technology
- Availability and prices of biomass, gasifying agents, catalysts, and other materials
- Plant-scale related capital investment, labor, electricity, water, maintenance fees, etc. (Huisman et al. 2011, Yao et al. 2017)

Global warming potential has been used in lifecycle assessment (LCA) of H$_2$ production, which is an indicator for assessing environmental impact of CO$_2$ emissions (Ozbilen et al. 2013, Bhandari et al. 2014). Social cost of carbon refers to the "marginal external cost of a unit of CO$_2$ emissions," and CO$_2$ is well known as a cause of environmental damage (Dincer and Acar 2015). In Table 13.4, 18 technologies for H$_2$ production are summarized and compared, although other H$_2$ production methods may also exist (Dincer and Acar 2015). The relevant energy efficiency, production cost, and social cost of carbon are listed in Table 13.4.

The technologies with the highest energy efficiency are fossil fuel (mainly natural gas) reforming (83%), plasma arc decomposition (70%), biomass gasification (65%), and coal gasification (63%). Biofuel reforming and other biological alternatives (e.g., dark fermentation, biophotolysis, and photofermentation) are currently not showing the advantage in energy efficiency. Photochemical-based methods, such as photocatalysis, photoelectrochemistry, artificial photosynthesis, and photoelectrolysis, are the least efficient, with energy efficiency values below 10%.

When considering the production cost, the major commercial dominant H$_2$ production methods include fossil fuel reforming (0.8 USD/kg H$_2$) and coal gasification (1.0 USD/kg H$_2$). Plasma arc decomposition also showed high competitiveness in the H$_2$ production cost (0.9 USD/kg H$_2$). Thermochemical technologies of biomass-based H$_2$ production (i.e., biomass gasification and biofuel reforming) can generate H$_2$ at the cost of 2.8 and 2.1 USD/kg H$_2$, respectively. The biomass-based biological alternatives have similar medium H$_2$ production costs of around 2–3 USD/kg H$_2$.

To achieve sustainable development goals in modern society, environmental performance of industrial processes will need to be taken into consideration. Although the most established technologies of fossil fuel reforming, coal gasification, and plasma arc decomposition are characterized with the highest energy efficiencies and lowest production costs, their environmental performances are the lowest due to the largest contributions to the global warming potential (9/12/11 kg CO$_2$/kg H$_2$) and social cost of carbon (1.43/1.9/1.76 USD/kg H$_2$). Photochemical and biological hydrogen production methods exhibit the minimum environmental impact, with global warming potential of 0.5 kg CO$_2$/kg H$_2$ and social cost of carbon 0.08 USD/kg H$_2$, respectively. In terms of biomass gasification and biofuel reforming, medium values of global warming potential (4.5–5.0 kg CO$_2$/kg H$_2$) and social cost of carbon (0.7–0.8 USD/kg H$_2$) are identified.

H$_2$ production from biomass may become competitive in the future market, if the market and technology conditions become ready. Large-scale system integration may provide a solution for

TABLE 13.4

Evaluation of H_2 Production from Different Technologies Based on Energy Efficiency, Production Cost, Global Warming Potential, and Social Cost of Carbon

		Evaluation			
Number	Technologies	Energy Efficiency (%)	Production Cost (USD/kg H_2)	Global Warming Potential (kg CO_2/kg H_2)	Social Cost of Carbon (USD/kg H_2)
1	Coal gasification	63	1.0	12	1.9
2	Fossil fuel reforming	83	0.8	9	1.43
3	Thermolysis	50	4.0	3	0.47
4	Hydrothermal routes	42	2.0	1	0.16
5	Biomass gasification	65	2.8	5	0.8
6	Biofuel reforming	38	2.1	4.5	0.7
7	Plasma arc decomposition	70	0.9	11	1.76
8	Hybrid thermochemical cycles	53	2.7	0.7	0.1
9	PV electrolysis	12	5.7	3	0.48
10	Photocatalysis	2	4.9	0.5	0.08
11	Photoelectrochemistry	7	10.3	0.5	0.08
12	Dark fermentation	12	2.6	0.5	0.08
13	Photofermentation	15	2.5	0.5	0.08
14	Photoelectrolysis	7	3.0	2	0.32
15	Biophotolysis	13	2.9	3	0.48
16	Artificial photosynthesis	8	2.6	0.5	0.08
17	High-temperature electrolysis	28	4.5	2.5	0.4
18	Electrolysis	53	2.8	8	1.28

Note: The figures in the table were adopted from Figures 3–6 in Dincer, I. and Acar, C. 2015. *Int J Hydrogen Energ* 40: 11094–11111.

the more-economical H_2 production. However, some challenges are to be overcome, including the following:

1. Costly transportation and pre-treatment of biomass are considered as one of the main challenges (Hosseini et al. 2015). The transportation issue can be alleviated by building the hydrogen production plant near the biomass sources.
2. Increasing the efficiency of processes parameters for all technologies.
3. For pyrolysis followed by steam reforming of bio-oils route, coke-resistant and more-efficient catalysts are always one of the main research priorities. Better reaction system configuration, integrating biomass pyrolysis and steam reforming of bio-oils, as well as the facility scale-up with acceptable energy efficiency, are also significant challenges (Nabgan et al. 2017).
4. Hydrothermal technologies generally remain at the research and development stage, with some demonstration facilities around the world. Special reactor materials are required. Salt precipitation is a severe problem due to the decreased solubility of salt content in biomass. Additionally, it may clog reactors and deactivate catalysts.
5. For biological routes, to make further progress in biohydrogen production, efforts are required in areas of application of cheaper materials, process combination, and optimization to address scaling up of the production system, and more-efficient and adaptable microorganisms (Das et al. 2008).

6. New downstream technologies of selective hydrogen separation from the mixture gas product are required. Catalysts are generally required to process the raw product gas from the thermochemical treatment of biomass, which are more robust, with strong resistance to deactivation and high product selectivity.

13.8 CONCLUSIONS

The hydrogen economy is expected to play an important role in the future global market. Renewable and sustainable energy sources are being widely investigated as actions against and alternatives to the gradual depletion of fossil fuels. In this chapter, the renewable biomass-based hydrogen production technologies were comprehensively described and discussed. Thermochemical routes of hydrogen production from biomass mainly include conventional gasification, hydrothermal gasification, and biomass pyrolysis followed by steam reforming of pyrolytic oils. Generally, the hydrogen production performance is determined by a number of factors, such as biomass material type and property, reaction atmosphere, temperature, pressure (especially in cases of hydrothermal gasification), steam:biomass ratio (for conventional gasification), steam:bio-oil ratio (for biomass pyrolysis followed by steam reforming of pyrolytic oils), and the type and amount of catalysts or absorbents. Steam gasification of biomass is considered as the most efficient approach for hydrogen production, and low-temperature hydrothermal gasification is attractive, but requires more research inputs. Raw product gas cleaning, secondary chemical conversion, and the subsequent hydrogen separation and purification are also essential for obtaining high-purity hydrogen product suitable for downstream applications. Biological technologies for biohydrogen production mainly involve dark and photo fermentation, photosynthesis, and biological WGS processes. These technologies are characterized by low hydrogen production efficiency and high production cost, but better environmental performance, compared to thermochemical routes. Demonstration operations of biohydrogen production are still limited, with scaling-up as one of the major barriers.

REFERENCES

Allakhverdiev, S. I., Thavasi, V., Kreslavski, V. D., Zharmukhamedov, S. K., Klimov, V. V., Ramakrishna, S., Los, D. A., Mimuro, M., Nishihara, H. and Carpentier, R. (2010). "Photosynthetic hydrogen production." *Journal of Photochemistry and Photobiology C: Photochemistry Reviews* 11(2): 101–113.

Amao, Y., Sakai, Y. and Takahara, S. (2016). "Solar hydrogen production from cellulose biomass with enzymatic and artificial photosynthesis system." *Research on Chemical Intermediates* 42(11): 7753–7759.

Ana, T., Luis, S. and Jalel, L. (2012). "Organosolv lignin depolymerization with different base catalysts." *Journal of Chemical Technology & Biotechnology* 87(11): 1593–1599.

Balat, H. and Kirtay, E. (2010). "Hydrogen from biomass - Present scenario and future prospects." *International Journal of Hydrogen Energy* 35(14): 7416–7426.

Barbir, F. (2009). "Transition to renewable energy systems with hydrogen as an energy carrier." *Energy* 34(3): 308–312.

Basu, P. (2010). Chapter 1 - Introduction. In *Biomass Gasification and Pyrolysis*. Boston, MA: Academic Press, pp. 1–25.

Bhandari, R., Trudewind, C. A. and Zapp, P. (2014). "Life cycle assessment of hydrogen production via electrolysis – A review." *Journal of Cleaner Production* 85: 151–163.

Bridgwater, A. V. and Peacocke, G. V. C. (2000). "Fast pyrolysis processes for biomass." *Renewable and Sustainable Energy Reviews* 4(1): 1–73.

Chen, G. X., Sjostrom, K. and Bjornbom, E. (1992). "Pyrolysis gasification of wood in a pressurized fluidized-bed reactor." *Industrial & Engineering Chemistry Research* 31(12): 2764–2768.

Chiesa, P., Consonni, S., Kreutz, T. and Williams, R. (2005). "Co-production of hydrogen, electricity and CO_2 from coal with commercially ready technology. PartA: Performance and emissions." *International Journal of Hydrogen Energy* 30(7): 747–767.

Claassen, P. A. M., de Vrije, T., Urbaniec, K. and Grabarczyk, R. (2010). "Development of a fermentation-based process for biomass conversion to hydrogen gas." *Zuckerindustrie* 135(4): 218–221.

Corella, J., Toledo, J. M. and Molina, G. (2008). "Biomass gasification with pure steam in fluidised bed: 12 variables that affect the effectiveness of the biomass gasifier." *International Journal of Oil Gas and Coal Technology* 1(1–2): 194–207.

Correa, C. R. and Kruse, A. (2018). "Supercritical water gasification of biomass for hydrogen production review." *Journal of Supercritical Fluids* 133: 573–590.

Das, D., Khanna, N. and Veziroglu, T. N. (2008). "Recent developments in biological hydrogen production processes." *Chemical Industry & Chemical Engineering Quarterly* 14(2): 57–67.

Datar, R., Huang, J., Maness, P. C., Mohagheghi, A., Czemik, S. and Chornet, E. (2007). "Hydrogen production from the fermentation of corn stover biomass pretreated with a steam-explosion process." *International Journal of Hydrogen Energy* 32(8): 932–939.

Dincer, I. and Acar, C. (2015). "Review and evaluation of hydrogen production methods for better sustainability." *International Journal of Hydrogen Energy* 40(34): 11094–11111.

Domine, M. E., Iojoiu, E. E., Davidian, T., Guilhaume, N. and Mirodatos, C. (2008). "Hydrogen production from biomass-derived oil over monolithic Pt- and Rh-based catalysts using steam reforming and sequential cracking processes." *Catalysis Today* 133: 565–573.

Duman, G. and Yanik, J. (2017). "Two-step steam pyrolysis of biomass for hydrogen production." *International Journal of Hydrogen Energy* 42(27): 17000–17008.

Eker, S. and Sarp, M. (2017). "Hydrogen gas production from waste paper by dark fermentation: Effects of initial substrate and biomass concentrations." *International Journal of Hydrogen Energy* 42(4): 2562–2568.

Garcia-Garcia, I., Bizkarra, E. A. K., de Ilarduya, J. M., Requies, J. and Carnbra, J. F. (2015). "Hydrogen production by steam reforming of m-cresol, a bio-oil model compound, using catalysts supported on conventional and unconventional supports." *International Journal of Hydrogen Energy* 40(42): 14445–14455.

Gnanapragasam, N. V. and Rosen, M. A. (2017). "A review of hydrogen production using coal, biomass and other solid fuels." *Biofuels-UK* 8(6): 725–745.

Goransson, K., Soderlind, U., He, J. and Zhang, W. N. (2011). "Review of syngas production via biomass DFBGs." *Renewable & Sustainable Energy Reviews* 15(1): 482–492.

Hannula, I. (2009). "Hydrogen production via thermal gasification of biomass in near-to-medium term." Retrieved 10 May 2018, from http://www.vtt.fi.

Hosseini, S. E., Wahid, M. A., Jamil, M. M., Azli, A. A. M. and Misbah, M. F. (2015). "A review on biomass-based hydrogen production for renewable energy supply." *International Journal of Energy Research* 39(12): 1597–1615.

Huang, Z. H., Wang, J. H., Liu, B., Zeng, K., Yu, J. R. and Jiang, D. M. (2007). "Combustion characteristics of a direct-injection engine fueled with natural gas-hydrogen blends under different ignition timings." *Fuel* 86(3): 381–387.

Huisman, G. H., Van Rens, G. L. M. A., De Lathouder, H. and Cornelissen, R. L. (2011). "Cost estimation of biomass-to-fuel plants producing methanol, dimethylether or hydrogen." *Biomass & Bioenergy* 35: S155–S166.

Hydrogen Tools. (2018). "Lower and higher heating values of fuels." Retrieved 17 April 2018, from https://www.h2tools.org/hyarc/calculator-tools/lower-and-higher-heating-values-fuels.

Kalinci, Y., Hepbasli, A. and Dincer, I. (2009). "Biomass-based hydrogen production: A review and analysis." *International Journal of Hydrogen Energy* 34(21): 8799–8817.

Kan, T., Grierson, S., de Nys, R. and Strezov, V. (2014). "Comparative assessment of the thermochemical conversion of freshwater and marine micro- and macroalgae." *Energy & Fuels* 28(1): 104–114.

Kan, T. and Li, Q. (2015). "Influence of temperature on properties of products from steam pyrolysis of rice husk in a fluidized bed." *Energy Sources Part a-Recovery Utilization and Environmental Effects* 37(17): 1883–1889.

Kan, T. and Strezov, V. (2015). Gasification of biomass. In *Biomass Processing Technologies*, V. Strezov and T. J. Evans (Eds.). Boca Raton, FL: CRC Press/Taylor & Francis Group, pp. 81–122.

Kan, T., Strezov, V. and Evans, T. J. (2016). "Lignocellulosic biomass pyrolysis: A review of product properties and effects of pyrolysis parameters." *Renewable & Sustainable Energy Reviews* 57: 1126–1140.

Kan, T., Xiong, J. X., Li, X. L., Ye, T. Q., Yuan, L. X., Torimoto, Y., Yamamoto, M. and Li, Q. X. (2010). "High efficient production of hydrogen from crude bio-oil via an integrative process between gasification and current-enhanced catalytic steam reforming." *International Journal of Hydrogen Energy* 35(2): 518–532.

Kitzler, H., Pfeifer, C. and Hofbauer, H. (2011). "Pressurized gasification of woody biomass-Variation of parameter." *Fuel Processing Technology* 92(5): 908–914.

Knight, R. A. (2000). "Experience with raw gas analysis from pressurized gasification of biomass." *Biomass & Bioenergy* 18(1): 67–77.

Kruse, O., Rupprecht, J., Mussgnug, J. H., Dismukes, G. C. and Hankamer, B. (2005). "Photosynthesis: A blueprint for solar energy capture and biohydrogen production technologies." *Photochemical & Photobiological Sciences* 4(12): 957–970.

Kumar, G., Sivagurunathan, P., Sen, B., Mudhoo, A., Davila-Vazquez, G., Wang, G. Y. and Kim, S. H. (2017). "Research and development perspectives of lignocellulose-based biohydrogen production." *International Biodeterioration & Biodegradation* 119: 225–238.

Lakaniemi, A. M., Koskinen, P. E. P., Nevatalo, L. M., Kaksonen, A. H. and Puhakka, J. A. (2011). "Biogenic hydrogen and methane production from reed canary grass." *Biomass & Bioenergy* 35(2): 773–780.

Ma, Z., Xiao, R. and Zhang, H. Y. (2017). "Catalytic steam reforming of bio-oil model compounds for hydrogen-rich gas production using bio-char as catalyst." *International Journal of Hydrogen Energy* 42(6): 3579–3585.

Mahishi, M. (2006). Theoretical and experimental investigation of hydrogen production by gasification of biomass. PhD, University of Florida.

Mahishi, M. R. and Goswami, D. Y. (2007). "An experimental study of hydrogen production by gasification of biomass in the presence of a CO_2 sorbent." *International Journal of Hydrogen Energy* 32(14): 2803–2808.

Marban, G. and Vales-Solis, T. (2007). "Towards the hydrogen economy?" *International Journal of Hydrogen Energy* 32(12): 1625–1637.

Mazloomi, K. and Gomes, C. (2012). "Hydrogen as an energy carrier: Prospects and challenges." *Renewable & Sustainable Energy Reviews* 16(5): 3024–3033.

McKendry, P. (2002). "Energy production from biomass (part 2): Conversion technologies." *Bioresource Technology* 83(1): 47–54.

Milne, T., Elam, C. and Evans, R. (2002). Hydrogen from biomass: State of the art and research challenges. Report for IEA, IEA/H2/ TR-02/001. Golden, CO: National Renewable Energy Laboratory.

Mulligan, C. J., Strezov, L. and Strezov, V. (2015). Pyrolysis of biomass. In *Biomass Processing Technologies*, V. Strezov and T. J. Evans (Eds.). Boca Raton, FL: CRC Press/Taylor & Francis Group, pp. 123–154.

Nabgan, W., Abdullah, T. A. T., Mat, R., Nabgan, B., Gambo, Y., Ibrahim, M., Ahmad, A., Jalil, A. A., Triwahyono, S. and Saeh, I. (2017). "Renewable hydrogen production from bio-oil derivative via catalytic steam reforming: An overview." *Renewable & Sustainable Energy Reviews* 79: 347–357.

Osada, M., Sato, T., Watanabe, M., Shirai, M. and Arai, K. (2006). "Catalytic gasification of wood biomass in subcritical and supercritical water." *Combustion Science and Technology* 178(1–3): 537–552.

Osada, M., Yamaguchi, A., Hiyoshi, N., Sato, O. and Shirai, M. (2012). "Gasification of sugarcane bagasse over supported ruthenium catalysts in supercritical water." *Energy & Fuels* 26(6): 3179–3186.

Ozbilen, A., Dincer, I. and Rosen, M. A. (2013). "Comparative environmental impact and efficiency assessment of selected hydrogen production methods." *Environmental Impact Assessment Review* 42: 1–9.

Parthasarathy, P. and Narayanan, K. S. (2014). "Hydrogen production from steam gasification of biomass: Influence of process parameters on hydrogen yield - A review." *Renewable Energy* 66: 570–579.

Pavlovic, I., Knez, Z. and Skerget, M. (2013). "Hydrothermal reactions of agricultural and food processing wastes in sub- and supercritical water: A review of fundamentals, mechanisms, and state of research." *Journal of Agricultural and Food Chemistry* 61(34): 8003–8025.

Peterson, A. A., Vogel, F., Lachance, R. P., Froling, M., Antal, M. J. and Tester, J. W. (2008). "Thermochemical biofuel production in hydrothermal media: A review of sub- and supercritical water technologies." *Energy & Environmental Science* 1(1): 32–65.

Puig-Arnavat, M., Bruno, J. C. and Coronas, A. (2010). "Review and analysis of biomass gasification models." *Renewable & Sustainable Energy Reviews* 14(9): 2841–2851.

Putun, E., Ates, F. and Putun, A. E. (2008). "Catalytic pyrolysis of biomass in inert and steam atmospheres." *Fuel* 87(6): 815–824.

Ren, N., Li, J., Li, B., Wang, Y. and Liu, S. (2006). "Biohydrogen production from molasses by anaerobic fermentation with a pilot-scale bioreactor system." *International Journal of Hydrogen Energy* 31(15): 2147–2157.

Rioche, C., Kulkarni, S., Meunier, F. C., Breen, J. P. and Burch, R. (2005). "Steam reforming of model compounds and fast pyrolysis bio-oil on supported noble metal catalysts." *Applied Catalysis B-Environmental* 61(1–2): 130–139.

Sahebdelfar, S. (2017). "Steam reforming of propionic acid: Thermodynamic analysis of a model compound for hydrogen production from bio-oil." *International Journal of Hydrogen Energy* 42(26): 16386–16395.

Salehi, E., Abedi, J. and Harding, T. (2009). "Bio-oil from sawdust: Pyrolysis of sawdust in a fixed-bed system." *Energy & Fuels* 23(7): 3767–3772.

Saxena, R. C., Adhikari, D. K. and Goyal, H. B. (2009). "Biomass-based energy fuel through biochemical routes: A review." *Renewable & Sustainable Energy Reviews* 13(1): 167–178.

Saxena, R. C., Seal, D., Kumar, S. and Goyal, H. B. (2008). "Thermo-chemical routes for hydrogen rich gas from biomass: A review." *Renewable & Sustainable Energy Reviews* 12(7): 1909–1927.

Sinag, A., Kruse, A. and Schwarzkopf, V. (2003). "Formation and degradation pathways of intermediate products formed during the hydropyrolysis of glucose as a model substance for wet biomass in a tubular reactor." *Engineering in Life Sciences* 3(12): 469–473.

Susanti, R. F., Dianningrum, L. W., Yum, T., Kim, Y., Lee, B. G. and Kim, J. (2012). "High-yield hydrogen production from glucose by supercritical water gasification without added catalyst." *International Journal of Hydrogen Energy* 37(16): 11677–11690.

Tian, T., Li, Q., He, R., Tan, Z. and Zhang, Y. (2017). "Effects of biochemical composition on hydrogen production by biomass gasification." *International Journal of Hydrogen Energy* 42(31): 19723–19732.

Valle, B., Aramburu, B., Benito, P. L., Bilbao, J. and Gayubo, A. G. (2018). "Biomass to hydrogen-rich gas via steam reforming of raw bio-oil over Ni/La$_2$O$_3$-alpha Al$_2$O$_3$ catalyst: Effect of space-time and steam-to-carbon ratio." *Fuel* 216: 445–455.

van Rossum, G., Kersten, S. R. A. and van Swaaij, W. P. M. (2007). "Catalytic and noncatalytic gasification of pyrolysis oil." *Industrial & Engineering Chemistry Research* 46(12): 3959–3967.

Wang, Z. X., Dong, T., Yuan, L. X., Kan, T., Zhu, X. F., Torimoto, Y., Sadakata, M. and Li, Q. X. (2007a). "Characteristics of bio-oil-syngas and its utilization in Fischer–Tropsch synthesis." *Energy & Fuels* 21(4): 2421–2432.

Wang, Z. X., Pan, Y., Dong, T., Zhu, X. F., Kan, T., Yuan, L. X., Torimoto, Y., Sadakata, M. and Li, Q. X. (2007b). "Production of hydrogen from catalytic steam reforming of bio-oil using C12A7-O--based catalysts." *Applied Catalysis A-General* 320: 24–34.

Wenk, S. O., Qian, D. J., Wakayama, T., Nakamura, C., Zorin, N., Rogner, M. and Miyake, J. (2002). "Biomolecular device for photoinduced hydrogen production." *International Journal of Hydrogen Energy* 27(11–12): 1489–1493.

Xie, H. Q., Yu, Q. B., Yao, X., Duan, W. J., Zuo, Z. L. and Qin, Q. (2015). "Hydrogen production via steam reforming of bio-oil model compounds over supported nickel catalysts." *Journal of Energy Chemistry* 24(3): 299–308.

Xue, Y. P., Yan, C. F., Zhao, X. Y., Huang, S. L. and Guo, C. Q. (2017). "Ni/La$_2$O$_3$-ZrO$_2$ catalyst for hydrogen production from steam reforming of acetic acid as a model compound of bio-oil." *Korean Journal of Chemical Engineering* 34(2): 305–313.

Yang, H., Yan, R., Chen, H., Lee, D. H. and Zheng, C. (2007). "Characteristics of hemicellulose, cellulose and lignin pyrolysis." *Fuel* 86(12): 1781–1788.

Yao, J. G., Kraussler, M., Benedikt, F. and Hofbauer, H. (2017). "Techno-economic assessment of hydrogen production based on dual fluidized bed biomass steam gasification, biogas steam reforming, and alkaline water electrolysis processes." *Energy Conversion and Management* 145: 278–292.

Yao, X., Yu, Q. B., Xie, H. Q., Duan, W. J., Han, Z. R., Liu, S. H. and Qin, Q. (2018). "The production of hydrogen through steam reforming of bio-oil model compounds recovering waste heat from blast furnace slag." *Journal of Thermal Analysis and Calorimetry* 131(3): 2951–2962.

Yoshida, T. and Matsumura, Y. (2001). "Gasification of cellulose, xylan, and lignin mixtures in supercritical water." *Industrial & Engineering Chemistry Research* 40(23): 5469–5474.

Yoshikawa, T., Yagi, T., Shinohara, S., Fukunaga, T., Nakasaka, Y., Tago, T. and Masuda, T. (2013). "Production of phenols from lignin via depolymerization and catalytic cracking." *Fuel Processing Technology* 108: 69–75.

Zhang, H., Xiao, R., Wang, D., He, G., Shao, S., Zhang, J. and Zhong, Z. (2011). "Biomass fast pyrolysis in a fluidized bed reactor under N$_2$, CO$_2$, CO, CH$_4$ and H$_2$ atmospheres." *Bioresource Technology* 102(5): 4258–4264.

14 Biomass-Fueled Direct Carbon Fuel Cells

Tao Kan, Vladimir Strezov, Graham Town, and Peter Nelson

CONTENTS

14.1 INTRODUCTION

With the gradual depletion of fossil fuels, increasing global energy demand, and stricter environmental regulations, the exploitation of sustainable energy-generation technologies other than conventional ones is attracting more research interest, leading to technology innovations. Fuel cells are considered as one of the energy-conversion technologies with significant potential due to their large energy efficiencies and superior environmental performance.

As a power-generation technology with long development, conventional fuel cells generally use gaseous fuels, such as methane and hydrogen, to produce power (Kacprzak et al. 2014). Fuel cells have current or potential applications in areas of large- or small-scale power supply for commercial or residential purpose, transportation vehicles, and portable power sources.

Direct carbon fuel cells (DCFCs) have been developed based on the same electrochemical principals as conventional fuel cells. DCFCs directly convert chemical energy of solid carbonaceous materials (e.g., coal, carbon black, graphite, biomass, biomass chars, and municipal solid wastes) into electrical energy. This one-step conversion has no requirement for a former step of transforming fuel feedstocks to gases.

Being similar to conventional fuel cells, DCFCs also generally consist of a cathode compartment, anode compartment, electrolyte, current collectors, and external loads. The DCFC structure is depicted in Figure 14.1.

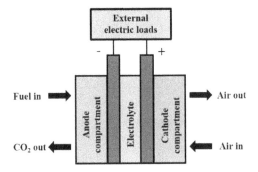

FIGURE 14.1 Schematic diagram of the DCFC structure. (Modified from Cao, D.X. et al., *J. Power Sources*, 167, 250–257, 2007; Kacprzak, A. et al., *J. Power Sources*, 255, 179–186, 2014.)

In a DCFC, the oxygen-bearing species (commonly O_2 in the air) is used as the oxygen source that combines with the carbon atom to produce CO_2. Cathodic reaction (typically Reaction 14.1) consumes electrons, and the reduction of the oxygen-bearing species to electron-carrying anion (e.g., O^{2-} and OH^-) occurs:

$$O_2 + 4e^- \rightarrow 2O^{2-}$$ (Reaction 14.1)

Anodic reaction of fuel oxidation (typically Reaction 14.2) produces electrons that then flow through the external load to do electrical work.

$$C + 2O^{2-} \rightarrow CO_2 + 4e^-$$ (Reaction 14.2)

The simplified overall cell reaction can be expressed in Reaction 14.3 for all DCFC types. The Boudouard reaction (Reaction 14.4) wherein the carbon material is gasified by CO_2 to CO also takes place as a side reaction, which is favoured at temperatures greater than 750°C.

$$C + O_2 \rightarrow CO_2 \quad E^\circ = 1.02\,V$$ (Reaction 14.3)

$$C + CO_2 \leftrightarrow 2CO$$ (Reaction 14.4)

DCFC technology has several advantages over conventional power-generation technologies. For conventional power generation (e.g., gas turbines), the chemical energy of fuels is transformed to kinetic energy, which is then used to generate power. The efficiency of the process is limited by the Carnot cycle. However, DCFCs convert chemical energy to electrical energy in one step, without the intermediate formation of kinetic energy, which eliminates the limitation of the Carnot cycle. The theoretical thermal efficiency (φ^{th}) of Reaction 14.3 can even exceed 100% in cases of negative entropy change for the fuel reaction, according to Equation 14.1 (Kacprzak et al. 2016).

$$\varphi^{th} = 1 - T\frac{\Delta S}{\Delta H}$$ (Equation 14.1)

where T is temperature in K, ΔS is entropy change, and ΔH is enthalpy change.

The actual energy efficiency of DCFCs can be in the range of 80%–95% (Muthuvel et al. 2009). After considering the internal and external losses, Giddey et al. (2012) estimated that the practical overall electric efficiency of a DCFC system is approximately 70%, which is almost twice the conventional routes for power generation (Hemmes et al. 2013).

The DCFC technology powered by biomass materials has recently attracted much research interest (Ahn et al. 2013, Dudek et al. 2013, Lima et al. 2013a, Munnings et al. 2014, Kacprzak et al. 2016). Biomass resources are often abundant, cheap, and widely dispersed. The utilization of biomass to generate energy and fuels can benefit the environment due to its wide acceptance as a carbon-neutral source of energy.

14.2 DIRECT CARBON FUEL CELL TECHNOLOGIES

There are three main types of fuel cells for DCFC application, including direct carbon molten hydroxide fuel cell (DC-MHFC), direct carbon molten carbonate fuel cell (DC-MCFC) and direct carbon solid oxide fuel cell (DC-SOFC). They are catalogued and named on the basis of the electrolyte material. Recently, hybrid direct carbon fuel cells (H-DCFCs) have been emerging as a new DCFC type and attracting great research attention.

14.2.1 DIRECT CARBON MOLTEN HYDROXIDE FUEL CELL

The DC-MHFC was developed in the late 1800s and patented by William W. Jacques (Cao et al. 2007). DC-MHFCs are typically operated at about 500°C–650°C with molten hydroxide (most commonly NaOH) as the electrolyte. The DC-MHFC can provide several advantages. The moderate operation temperature enables the application of relatively inexpensive cell materials, such as stainless steel (Steele and Heinzel 2001). Besides, the molten hydroxide also provides high ionic conductivity, which is essential for the high performance of electrochemical carbon reactions and cell operation. In addition, the activity of electrochemical reactions is high, which ensures the prompt consumption of carbon fuels to generate satisfactory current intensity.

The anode and cathode reactions can be expressed by Reactions 14.5 and 14.6, respectively.

$$C + 4OH^- \rightarrow CO_2 + 2H_2O + 4e^- \qquad \text{(Reaction 14.5)}$$

$$C + 2H_2O + 4e^- \rightarrow 4HO^- \qquad \text{(Reaction 14.6)}$$

However, the side reactions of carbonate formation are undesirable, which include Reactions 14.7 and 14.8 (Goret and Tremillon 1967, Cao et al. 2007):

$$CO_2 + 2OH^- = CO_3^{2-} + H_2O \qquad \text{(Reaction 14.7)}$$

$$C + 6OH^- \rightarrow CO_3^{2-} + 3H_2O + 4e^- \qquad \text{(Reaction 14.8)}$$

14.2.2 DIRECT CARBON MOLTEN CARBONATE FUEL CELL

DC-MCFCs have similar cell arrangement to DC-MHFCs. However, compared with DC-MHFCs, DC-MCFCs provide lower ion conductivity and carbon oxidation rate (Zecevic et al. 2005). DC-MCFCs operate at temperatures of around 600°C–800°C, using molten carbonate salts (Li_2CO_3, Na_2CO_3, and K_2CO_3) as the electrolyte. Due to the highly corrosive electrolyte at high temperatures, the cell component materials should have excellent corrosion resistance. Ni-based alloys, such as Inconel and Hastelloy, are among the most commonly used materials (Rady et al. 2012). The anode and cathode materials are commonly Ni (or Ni-Cr alloy) and a NiO-Li mixture, respectively (Giddey et al. 2012). In addition, the sulfur impurities contained in carbonaceous fuels are supposed to be removed before the loading of fuels in cells.

The anode and cathode electrochemical reactions can be described by Reactions 14.9 and 14.10, respectively (Li et al. 2009).

$$C + 2CO_3^{2-} \rightarrow 3CO_2 + 4e^- \text{ (anode reaction)} \qquad \text{(Reaction 14.9)}$$

$$C + 2CO_2 + 4e^- \rightarrow 2CO_3^{2-} \text{ (cathode reaction)} \qquad \text{(Reaction 14.10)}$$

The side chemical reaction (see Reaction 14.4) may take place when the operation temperature is above 750°C or the cell is in a standby mode. It may also occur when the fuel particles do not contact the anode well or there is no contact between them (Cherepy et al. 2005, Hao et al. 2014). In this case, the carbon fuel is consumed to form CO without the contribution of electrochemical reactions with the electrolyte. Another CO formation route was proposed by Li et al. (2009), as depicted in Reaction 14.11.

$$C + CO_3^{2-} \rightarrow 3CO_2 + CO + 2e^- \qquad \text{(Reaction 14.11)}$$

According to Li et al. (2009), this reaction was believed to occur at 800°C with activated carbon (AC), HNO_3-treated AC, or HCl-treated AC as carbon fuels. In addition, the performance of DC-MCFCs is largely affected by the surface oxygen functional groups.

14.2.3 DIRECT CARBON SOLID OXIDE FUEL CELL

DC-SOFCs are operated in a temperature range of 500°C–1,000°C (typically 800°–1,000°C) to acquire satisfying ionic conductivity of the solid oxide electrolyte. The solid electrolyte is generally porous ceramic material. YSZ (Y_2O_3–ZrO_2) is commonly used as the electrolyte in SOFCs. The high temperature ensures the desirable ionic conductivity; however, this brings some issues, such as cell instability and corrosion by impurities. Great research efforts have been devoted to the decrease in the operation temperature while maintaining the ionic conductivity at an acceptable level (Zhan et al. 2011). Another concern is related to the thickness of the electrolyte to balance between the current leakage (if too thin) and considerable resistance (if too thick) to achieve a compromise. The anode material is commonly made of solid electrolyte and one metal (e.g., Ni/YSZ), while the cathode is composed of electrolyte and lanthanum strontium manganite (LSM) (Carrette et al. 2001, Stambouli and Traversa 2002).

Based on the fuel-anode contact mode, the DC-SOFCs can be subdivided into three main types: (1) direct contact, (2) carbon gasification, and (3) carbon-deposited (Rady et al. 2012, Deleebeeck and Hansen 2014). Beside these DC-SOFCs fueled directly by solid fuels, some liquid anode media may also be employed to enhance the reaction between the carbon fuel and O^{2-}. The media can be inert molten silver or reactive molten metals (e.g., Sn and Sb) at 900°C–1,000°C (Deleebeeck and Hansen 2014). Molten carbonates are another option of media, and related DC-SOFCs are commonly named by Hybrid DCFC and have been given special attention (see Section 14.2.4).

14.2.3.1 Type I: Direct Contact

For type I, the fuel makes direct physical contact with the anode. Solid fuel particles may be pressed firmly onto the anode surface. Another typical option is the fluidized bed, where the direct contact between fuel particles and the anode can also be achieved. Besides, the fluidized bed provides additional flexibility for fuel recharge. Reactions 14.12 and 14.13 occur during the direct contact fuel cell.

$$C + O^{2-} \rightarrow CO + 2e^- \qquad \text{(Reaction 14.12)}$$

$$CO + O^{2-} \rightarrow CO_2 + 2e^- \qquad \text{(Reaction 14.13)}$$

The reaction mechanism, however, requires further investigation on the extent of the direct electrochemical oxidation of fuel, as described by Reaction 14.12 (Rady et al. 2012).

14.2.3.2 Type II: Carbon Gasification

In the type II DC-SOFCs, the fuel has no direct physical contact with the anode. It is well accepted that in the anode chamber, the fuel is gasified by CO_2 to CO via the Boudourd reaction ($C + CO_2 \leftrightarrow 2CO$, Reaction 14.4). In the meantime, the generated CO is oxidized by O^{2-} to CO_2 and releases electrons ($CO + O^{2-} \rightarrow CO_2 + 2e^-$, Reaction 14.13) at the triple-phase boundary (TPB), which is an active interface of carbon fuel/anode/electrolyte (Gür 2013). The gasification may take place in-situ in the anode chamber. Fuel particles can also be gasified ex-situ, which means Reaction 14.4 is arranged in an external compartment (Giddey et al. 2012).

14.2.3.3 Type III: Carbon-Deposited

In the type III DC-SOFCs, the hydrocarbon gas supplied or generated from pyrolysis of carbonaceous materials is degraded thermally, which leaves a carbon deposit on the anode. The carbon element then reacts with O^{2-} to produce CO and/or CO_2 at the TPB to trigger the cell reaction cycle consisting of $C + O^{2-} \rightarrow CO + 2e^-$ (Reaction 14.12) and/or $C + 2O^{2-} \rightarrow CO_2 + 4e^-$ (Reaction 14.2).

14.2.4 Hybrid DCFC

Hybrid DCFC (H-DCFC) is a type of DC-SOFC with both solid oxide and molten carbonate used as the electrolyte, and solid fuel particles distributed in the molten carbonate. It combines the electrolyte of MCFC and the arrangement of SOFC. The fuel particles can take part in electrochemical reactions within the molten carbonate due to its ionic conductivity, which is a large extension to the limited reaction area of TPB in DC-SOFCs (Jain et al. 2009, Deleebeeck and Hansen 2014). In this way, the reaction kinetics can be considerably improved when compared to the DC-SOFCs.

Both O^{2-} and CO_3^{2-} are believed to take part in the electrochemical reactions. The anodic reactions are as follows (Elleuch et al. 2013b):

$$C + 2O^{2-} \rightarrow CO_2 + 4e^- \qquad \text{(Reaction 14.2)}$$

$$C + 2CO_3^{2-} \rightarrow 3CO_2 + 4e^- \qquad \text{(Reaction 14.9)}$$

At the cathode TPBs, oxygen molecules accept electrons to form oxygen anions, as below:

$$O_2 + 4e^- \rightarrow 2O^{2-} \qquad \text{(Reaction 14.1)}$$

Other reactions with CO as both a product and another fuel in H-DCFCs were also proposed by researchers (Yu et al. 2013):

$$C + CO_2 \leftrightarrow 2CO \qquad \text{(Reaction 14.4)}$$

$$CO + O^{2-} \rightarrow CO_2 + 2e^- \qquad \text{(Reaction 14.13)}$$

$$CO + CO_3^{2-} \rightarrow 2CO_2 + 2e^- \qquad \text{(Reaction 14.14)}$$

If CO_2 is also introduced into the cathode chamber, the following reaction may also occur (Elleuch et al. 2013b).

$$O_2 + CO_2 + 4e^- \rightarrow 2CO_3^{2-} \qquad \text{(Reaction 14.15)}$$

14.3   BIOMASS-FUELED DIRECT CARBON FUEL CELLS

14.3.1   Biomass and Biomass Chars as Direct Carbon Fuel Cell Fuels

The use of biomass and biomass-derived chars as fuel materials in DCFCs may greatly exploit the potential of distributed power generation, which is deemed as one of the future power generation options. As a widely used renewable and sustainable energy resource, the available biomass is estimated to be up to about 10^{11} toe (tonnes of oil equivalent) annually (de Wit and Faaij 2010, Williams et al. 2012). It contributes approximately 10% of the global primary energy generation (Strezov 2014). Several thermochemical technologies have been applied to generate energy (e.g., power) and chemicals from biomass, such as combustion, gasification, pyrolysis, and hydrothermal pretreatment (Gassner et al. 2011a, 2011b, Mettler et al. 2012). Currently, the efficiency of power generation from biomass is typically 25%–40% (Elleuch et al. 2013b, Kacprzak et al. 2016).

Biomass is generally not suitable for serving as a DCFC fuel due to its chemical and physical nature (Kacprzak et al. 2016). The low carbon density and the inherent water in biomass severely degrade its oxidation reactivity in DCFCs. There are only limited examples of DCFCs that directly used untreated biomass as fuel. Jang et al. (2015) applied dried coffee grounds waste in a DC-SOFC with Ni-YSZ, 8YSZ, and LSM ($La_{0.8}Sr_{0.2}MnO_3$) as the anode, electrolyte, and cathode, respectively. At the cell operation T = 900°C, a maximum power density of 88 mW/cm² was obtained, with the corresponding cell voltage at 0.65 V and current density at 170 mA/cm². However, the authors did not present the data of carbonized coffee grounds as the fuel in the above fuel cell. Lima et al. (2013b) designed a pellet fuel cell by pressing the dry powder with samarium-doped ceria (SDC)–(Li/Na)₂CO₃ composite as electrolyte, a mixture of lignin (or lignin + active carbon) and electrolyte at mass ratio of 1:1 as anode, and Cu-Ni-Zn-Li catalyst plus electrolyte composite (mass ratio of 4:1) as cathode. Results showed that at cell operation T = 560°C, maximum current densities of 57 and 43 mA/cm² were achieved by using lignosulfonate and kraft lignin as fuels, respectively. Adding 17% AC into the lignin fuels could increase the above current densities to 101 and 85 mA/cm², respectively. In another study by Dudek and Socha (2014), a DC-SOFC of Ni-10 GDC as anode), 8YSZ as electrolyte), and LSM as cathode at 850°C fed with raw beechwood and acacia wood was tested, and the maximum power densities were 100 and 90 mW/cm², respectively. However, the carbonized wood fuel was not investigated. Cantero-Tubilla et al. (2016) compared the performance of raw switchgrass and torrefied switchgrass in a DC-SOFC operated at 800°C, and found that the torrefaction pre-treatment increased the maximum power density from 8.0 to 120.5 mW/cm².

Selective pre-treatments of biomass is usually required to achieve better DCFC performance. With respect to biomass processing technologies for DCFCs, gasification and pyrolysis are the most commonly adopted. Gasification of biomass converts biomass particles to combustible gases, such as hydrogen and carbon monoxide, at high temperatures (usually >800°C) in oxidative atmosphere (typically oxygen and steam) (Kan and Strezov 2014). The product gas is then processed and subsequently used as the actual fuel in anode compartments of fuel cells, especially SOFCs.

The conversion of biomass into biochars with negligible moisture content and satisfying carbon density can be realized through a variety of thermochemical carbonization processes, such as torrefaction, pyrolysis, and hydrothermal processing. Torrefaction can be regarded as a treatment route to achieve deep drying and initial decomposition of biomass. As another efficient biomass processing route, pyrolysis can remove the excessive oxygen atoms and produce biomass chars that can fuel the DCFC anode. Generally, pyrolysis of biomass refers to "the thermal decomposition of the biomass organic matrix in non-oxidising atmospheres resulting in liquid bio-oil, solid biochar, and non-condensable gas products" at temperatures of around 350°C–550°C or higher (Kan et al. 2016, 2017).

14.3.2 Fuel Characterization and Cell Performance

A large variety of methods and standards have been developed to characterize the physical and chemical properties of biomass and biomass char fuels fed in DCFCs. Table 14.1 lists the most used analysis methods. These analyses may help to determine the correlation between the solid fuel properties and the DCFC performance, which can guide the screening of biomass materials as well as the biomass pre-treatment conditions. The effects of the fuel properties on the DCFC performance will be discussed in the following Section 14.3.3.

TABLE 14.1
Analysis Methods Employed for the Biomass/Biomass Char Fuels

Properties	Analysis Methods	References
Proximate analysis (moisture, ash, volatile matter, and fixed carbon contents)	ASTM D1762-84, 2007; ASTM D3172-07; etc.	Munnings et al. (2014)
Ultimate analysis (elemental analysis of CHN, and O by difference)	ASTM standard D3176; etc.	Dudek et al. (2013), Cantero-Tubilla et al. (2016)
Higher heating value (HHV)	European standard UNI EN 14918:2010	Kacprzak et al. (2016)
Ash content and composition	X-ray fluorescence (XRF), inductively coupled plasma atomic emission spectroscopy (ICP-AES)/inductively coupled plasma optical emission spectroscopy (ICP-OES)	Elleuch et al. (2013b), Munnings et al. (2014), Cantero-Tubilla et al. (2016)
Elemental surface distribution	Scanning electron microscopy with energy dispersive X-ray spectroscopy (SEM/EDX)	Dudek et al. (2013)
Quantitative surface composition (e.g., surface oxygen/carbon ratio), chemical state of surface elements (e.g., valence state of element)	X-ray photoelectron spectroscopy (XPS)	Ahn et al. (2013), Dudek et al. (2013), Dudek and Socha (2014), Hao et al. (2014)
Functional groups	FT-IR	Lima et al. (2013b)
Aromaticity	^{13}C NMR	Nabae et al. (2008)
Mass/heat change during heating in different atmospheres (N_2, O_2, air, etc.)	Thermogravimetric analysis with differential scanning calorimetry (TGA/DSC)	Dudek et al. (2013), Munnings et al. (2014), Yu et al. (2014)
Crystalline phases, and their qualitative and semiquantitative data	X-ray diffraction (XRD)	Dudek et al. (2013), Elleuch et al. (2013b), Munnings et al. (2014), Yu et al. (2014)
Surface area, porous structure	Brunauer–Emmett–Teller (BET) method; mercury porosimetry analysis	Dudek et al. (2013), Elleuch et al. (2013b), Cantero-Tubilla et al. (2016)
Surface morphology, particle size distribution	Scanning electron microscopy (SEM)	Yu et al. (2014), Cantero-Tubilla et al. (2016)
Permeability coefficient (k)	Calculated according to Darcy's law	Elleuch et al. (2013b)
Biochar oxidation characteristics	Temperature-programmed oxidation (TPO) by a thermogravimetric analyzer with air as the reaction gas	Elleuch et al. (2013b)
Properties of carbon surface oxygen complex, such as CO_x evolution from oxygen-containing surface groups in biochar	Temperature-programmed desorption (TPD) in a furnace connected to a gas chromatograph	Elleuch et al. (2013b)
Electrical conductivity, which greatly depends on the fuel structure, especially the graphitic crystallinity.	Frequency response analyzer	Rady et al. (2012)

Similar to the other fuel cells, DCFC performance is commonly characterized using an impedance analyzer (i.e., for electrochemical impedance spectroscopy [EIS]), and to determine the cell
resistance. The static i-V curve can be obtained by measuring the voltage with stepwise increasing
current density in the cell, and the cell power density (P) subsequently determined. Sometimes, the
anode off-gas is also analyzed using gas chromatography to monitor the reaction process (Yu et al.
2014). The practical efficiency is calculated by the measured electric work divided by the total
chemical energy of fuel consumed (Kacprzak et al. 2016).

The i-V curve may be divided into three main regions:

1. Activation polarization region. The initial fast decrease in voltage starting from the open
   circuit voltage (OCV) in the i-V curve mainly arises from the activation polarization to
   overcome the activation energy barrier before the electrochemical reactions can take place
   (Ahn et al. 2013).
2. Ohmic region. Then the i-V curve moves to the ohmic region. Ohmic resistance is calculated by the slope of the "linear central region" of the i-V curve (Adeniyi and Ewan
   2012). It includes all the resistances from cell components (i.e., the electrode and electrolyte materials, as well as current collectors), other mechanical connections in the current
   circuit, and the fuel particles. Generally, at the same operating conditions, the fuels with
   higher surface area, pore volume, and/or more functional groups enable their i-V curves to
   reach the ohmic region earlier. (Cao et al. 2010, Ahn et al. 2013).
3. Eventually, the voltage declines quickly at the high current density zone, which is due to
   mass transport or concentration losses (Adeniyi and Ewan 2012).

14.3.3 PROCESS PARAMETERS

Power generation in DCFCs is a complex process that depends on a large number of parameters, such
as the biomass source, biomass pre-treatment conditions, DCFC type and configuration, and fuel
cell operation condition (e.g., temperature, stirring speed, reaction atmosphere, and gas flow rate).
Table 14.2 summarizes the performance data of biomass-fueled DCFCs, including biomass/char
source, DCFC configuration, DCFC operation conditions, OCV, maximum current density, and
power density and resistances.

As shown in Table 14.2, a wide range of biomass/biomass char sources have been investigated
as potential DCFC fuels, such as poultry excrement, coconut, switchgrass, wood (including hardwood, poplar wood, beechwood, apple tree chips, willow, and pine), corn stover, corn cob, sunflower
husks, almond shell, and some commercial bio-chars with unknown original source. Some raw biomass fuels were directly introduced into the anode compartment. For example, Dudek et al. (2013)
fed beechwood chips into a DC-SOFC at 800°C and claimed that the chips were pyrolyzed inside
the DCFC compartment. In most cases, the biomass materials were carbonized to bio-chars prior to
being fed into the DCFCs. In terms of the DCFC type, DC-SOFC, DC-MCFC, and H-DCFC were
adopted, with DC-SOFC being the most widely used.

The OCV can be affected by the fuel type. For example, the biochar from carbonization (500°C,
30 min.) of pine gave a distinctly lower OCV (0.8666 V) than other biomass resources (1.0047,
1.0090, and 1.0526 V for apple tree chips, willow, and sunflower husks, respectively) at the same
operating conditions (Kacprzak et al. 2014). This is because the pine was resin-rich compared to
other resources and might still contain "tar-like impurities" after carbonization. For the investigated
experiments, the maximum current densities varied in a wide range between 35 and 480 mA/cm^2,
depending on the biomass source, biomass pre-treatment conditions, DCFC types and structure,
and DCFC operating conditions. Similarly, the maximum power density ranged between 8 mW/cm^2
and 340 mW/cm^2. The highest current and power densities were obtained by Yu et al. (2014), using
corn cob char as fuel in an H-DCFC. More research work on biomass-fueled H-DCFCs is needed to

TABLE 14.2

Summary of Biomass Char Source and Direct Carbon Fuel Cell Performance

Biomass/Char Source	Direct Carbon Fuel Cell (DCFC) Configuration	DCFC Operation Conditions	Open Circle Voltage (OCV)	Max Current Density (mA/cm²)	Max Power Density (mW/cm²)	Resistance (R) at OCV, Ω·cm²			References
						Ohmic R	Polarization Resistance (Rp)	Total Cell Resistance	
Biochar from slow pyrolysis of poultry excrement and bedding materials	Button cell of SOFC: Same anode and cathode material: lanthanum strontium cobalt ferrite (LSCF); electrolyte: thin disc of 8-YSZ	800°C, N$_2$	1.07	39	9				Kulkarni et al. (2012), Munnings et al. (2014)
Commercial biochar: Vulcan XC72-R			1.00	100	50				
Coconut char			0.87	37	12	3.1 (at 0.5 V)	22.4 (at 0.5V)	25.5 (at 0.5V)	
Coconut char		800°C, CO	1.1	185	68	2.4 (at 0.5 V)	2.0 (at 0.5V)	4.4 (at 0.5V)	
Coconut char		850°C, N$_2$	0.9	85	24	3 (at OCV)			
Raw switchgrass	SOFC: Planar electrolyte: 8-YSZ; Cathode: multilayer LSCF; Anode: double gadolinium–doped ceria layers	800°C, Ar	1.1		8 (at 0.7 V)				Cantero-Tubilla et al. (2016)
Raw switchgrass		650°C, Ar	1.03						
Switchgrass torrefied at 250°C		800°C, Ar	1.15	380	120.5 (at 0.599 V)	2.37	0.28	2.59	

(Continued)

TABLE 14.2 (*Continued*)

Summary of Biomass Char Source and Direct Carbon Fuel Cell Performance

Biomass/Char Source	Direct Carbon Fuel Cell (DCFC) Configuration	DCFC Operation Conditions	Open Circle Voltage (OCV)	Max Current Density (mA/cm²)	Max Power Density (mW/cm²)	Resistance (R) at OCV, Ω·cm²			References
						Ohmic R	Polarization Resistance (Rp)	Total Cell Resistance	
Switchgrass torrefied at 250°C		650°C, Ar	1.08	125	30				
Switchgrass pyrolyzed at 900°C in N$_2$		650°C, Ar	1.01		23 (at 0.495 V)				
Switchgrass pyrolyzed at 900°C in N$_2$		800°C, Ar	1.13			2.87	0.81	3.68	
Hardwood torrefied at 250°C		800°C, Ar			112 (at 0.7 V)	2.39	0.3	2.69	
Corn stover torrefied at 250°C		800°C, Ar	1.1	370	112	2.57	0.29	2.86	
Corn stover torrefied at 250°C		650°C, Ar	1.05	70	15			15	
Commercial charcoal	SOFC: solid electrolyte (8YSZ) and LSM (La$_{0.8}$Sr$_{0.2}$MnO$_3$)-GDC (Gd$_2$O$_3$ in CeO$_2$) composite cathode but varied in material of anode (Ni-YSZ, Ni-GDC, (Ni$_{0.9}$-Fe$_{0.1}$)-GDC and La$_{0.8}$Ca$_{0.1}$Sr$_{0.1}$CrO$_3$ (LCCr))	660°C	0.95	125	27				Dudek et al. (2013)
Beechwood chips pyrolyzed in-situ		750°C	1.0	235	50				
		800°C	1.02	255	60				
		800°C	1.0	/	35				

(*Continued*)

TABLE 14.2 (Continued)
Summary of Biomass Char Source and Direct Carbon Fuel Cell Performance

Biomass/Char Source	Direct Carbon Fuel Cell (DCFC) Configuration	DCFC Operation Conditions	Open Circle Voltage (OCV)	Max Current Density (mA/cm²)	Max Power Density (mW/cm²)	Resistance (R) at OCV, Ω·cm²			References
						Ohmic R	Polarization Resistance (Rp)	Total Cell Resistance	
Biochar from carbonization (500°C, 30 min.) of apple tree chips	MHFC: Electrolyte: binary eutectic mixture of NaOH and LiOH; The anode was also made from Nickel® 201, while the cathode was Ni-based Inconel® alloy 600	450°C, air: 0.5 L/m	1.0047	55	23			~10.9	Kacprzak et al. (2014, 2016)
Biochar from willow			1.009	50	22			~12.2	
Biochar from sunflower husks			1.0526	60	18			~13.3	
Biochar from pine			0.8666	35	19			~6.7 (Id < 30 mA/cm²), ~70 (Id > 30 mA/cm²)	
Commercial biochar			1.0749	75.2	35.1			~7.1	
Almond shell biochar	H-DCFC: Electrolyte: ceria-carbonate composite; Anode: biochar and carbonate (Li_2CO_3/Na_2CO_3 eutectic mixture.) in a mass ratio of 1:9; Cathode: 30 wt% Electrolyte and 70 wt% lithiated NiO ($LiNiO_2$) powders; Cathode gas: a mixture of O_2 and CO_2	600°C	1.0	100	25			10.4	Elleuch et al. (2013b)
		650°C	1.03	250	55			4.7	
		700°C	1.07	450	105			2.5	
		750°C	1.05	480	120			~2.3	
Commercial Activated carbon		750°C	1.05	240	100			/	

(Continued)

TABLE 14.2 (Continued)
Summary of Biomass Char Source and Direct Carbon Fuel Cell Performance

Biomass/Char Source	Direct Carbon Fuel Cell (DCFC) Configuration	DCFC Operation Conditions	Open Circle Voltage (OCV)	Max Current Density (mA/cm²)	Max Power Density (mW/cm²)	Resistance (R) at OCV, Ω·cm²			References
						Ohmic R	Polarization Resistance (Rp)	Total Cell Resistance	
Corn cob char (corn cob at 700°C in N₂ for 2 hr.)	H-DCFC (three-layer cell pellet): Electrolyte: composite (binary carbonate eutectic of Li₂CO₃ and Na₂CO₃ +; Anode: Ce₀.₈Sm₀.₂O₁.₉ (SDC) and NiO; Cathode: lithiated NiO and composite electrolyte	750°C	1.05	480	340	0.17 Ω			Yu et al. (2014)
Switchgrass straws (pyrolyzed at 800°C)	MCFC: Electrolyte: zirconia cloth saturated in a mixture of Li₂CO₃ + K₂CO₃	800°C	0.87	74.00	21.60	9.99			Adeniyi and Ewan (2012)
Poplar wood chips (pyrolyzed at 800°C)		800°C	1.08	81.53	23.91	12.69			
Wood char	MCFC: Electrolyte powder (62 mol% Li₂CO₃ and 38 mol% K₂CO₃); working electrode (WE) was made of a silver layer; silver counter electrode	700°C with 300 rpm of stirring	0.9	119	40.8				Ahn et al. (2013)

MCFC: molten carbonate fuel cell; SOFC: solid oxide fuel cell.

verify the advantage of using H-DCFCs. The total cell resistance was roughly comprised of ohmic resistance and polarization resistance, ranging from 2.3 to 25.5 $\Omega \cdot cm^2$.

A great deal of effort has been made to link the biochar properties to their performance in DCFCs. Table 14.3 lists the effects of various physical and chemical properties of fuels on the DCFC performance. The effects of cell operating conditions and cell components were also

TABLE 14.3
Parameters Affecting Performance of Biomass-Fueled Direct Carbon Fuel Cells

Category	Parameter	Influence/Trend	References
Fuel property	Surface area	Higher surface area leads to higher carbon discharge rate.	Giddey et al. (2012), Rady et al. (2012)
	Crystallization	Generally, carbon with more surface defects, less crystallization, more crystal disorder, and more dislocations is more reactive.	Stevens et al. (1998)
	Surface functional groups, especially oxygen and nitrogen-containing ones	Surface functional groups can increase the reactivity of fuel particles.	Kacprzak et al. (2014)
	Mineral contents and other impurities	The impurities have promotive or inhibitive effects on the performance of direct carbon fuel cells (DCFCs). Ca, Mg, Fe, Na, Cr, Cu, Ni, Co may catalyze the electrochemical and Boudouard reactions, while Al and Si tend to lower the DCFC performance. Sulfur element is proven to poison some anode materials, especially the Ni-based ones.	Jain et al. (2009), Li et al. (2010), Munnings et al. (2014)
	Alkais and alkaline metals	Alkais and alkaline metals react with silica to form sticky slag, which is corrosive to the cell components.	Cantero-Tubilla et al. (2016)
	Electrical conductivity	Better electrical conductivity can reduce the ohmic polarization and promote the carbon electrochemical reactions.	Cherepy et al. (2005)
Operating conditions	CO_2 partial pressure	For molten carbonate fuel cells (MCFCs), the cell voltage is calculated.	Cao et al. (2007)
	External contact pressure	The increase in contact pressure can effectively enhance the contact among fuel particles and the contact between fuel and the anode, thus reducing the resistance and increasing the open circuit voltage (OCV).	Desclaux et al. (2013)
	Introduction of CO to the anode chamber	CO increases OCV and peak power density.	Munnings et al. (2014)
	Operation temperature	The increase in operation T decreases in the viscosity of electrolyte and accelerates the mass transfer and diffusion of ions in electrolyte, which decreases the cell ohmic resistance.	Adeniyi and Ewan (2012), Elleuch et al. (2013b)
	Stirring in the anode	Stirring (in the cases of liquid electrolyte) enhances the mass transfer and the three-phase boundary area.	Li et al. (2008)
Cell components	Electrode materials	Preferably, the anode materials should feature a porous structure that can selectively catalyze the fuel oxidation to CO_2, guarantee quick O^{2-} diffusion, and maintain mechanical and thermodynamic stability during cell operation.	Dudek et al. (2013), Elleuch et al. (2013a)
	Electrolyte	High conductivity and ion transport ability are the basic requirements.	Giddey et al. (2012), Rady et al. (2012)

discussed. The oxidation rate of carbon particles is highly dependent on the physical and chemical properties. For example, Kacprzak et al. (2014) compared five types of biochars in a DC-MHFC and found that the power density values with different biochars decreased in order of commercial biochar > charred apple chips > charred willow > charred sunflower husks ~ charred pine. The oxygen content of these fuels followed the same order. The authors explained that the higher content of surface oxygen–containing functional groups in different fuels resulted in higher power density. However, the correlation between pore size distribution/pore volume/surface area and power density could not be established for different biochars.

14.4   STATUS AND FUTURE DEVELOPMENT

Currently, biomass-fueled DCFCs are still at an early stage of research and development. There are a number of technologic and economic challenges to be addressed before this technology can move to the next stage of development:

- The electrochemical and chemical reaction mechanisms are still not clear and require further confirmation through research, which also contributes to the complexity of the biomass-fueled DCFC process.
- Biomass materials cover a wide variety of substances, which on one hand provides the benefit of feedstock availability and selectivity; on the other hand, it increases the difficulty in the feedstock screening. The previous studies generally investigated the DCFC performance of a very limited number of biomass materials. Few studies have indicated the most suitable biomass materials for power generation in DCFCs.
- Pre-treatment of raw biomass is costly. Systematic investigation into the effects of pre-treatment methods and conditions is required to balance the biomass pre-treatment cost and the preferred DCFC performance.
- Effect of ash on the DCFC performance requires further clarification. Minerals may exert different influences on the DCFC process, which can be promotive or suppressive.
- Generally the power density with biomass or biomass char fuels is low (Giddey et al. 2012), which is a common issue for most DCFC systems. It is possibly improved by choosing the suitable biomass type after pre-treatment. It is also highly dependent on the progress of fuel cell technologies.
- Scale-up of DCFCs may give rise to "a high IR (voltage) drop and a large unit size" (Cao et al. 2007). In the same time, a simple design of the DCFC system is preferred to avoid the potential technologic issues.
- Other common fuel cell issues, such as cell components' degradation due to high temperature corrosion, cell system lifetime, reliability, and operation cost (Carrette et al. 2001, Elleuch et al. 2013a, Sharaf and Orhan 2014). For example, some fundamental aspects of MCFC technology need to be addressed, such as the wetting properties of different carbon fuels with molten carbonates and kinetics of oxygen reduction at different cell materials (Selman 2006). Besides, the cell components' corrosion mechanism and abatement methods are also one of the research focuses. Developing alternative fuel cell materials has become the central task for accelerating the progress of fuel cells (e.g., the exploitation of new anode materials that are resistant to the chemical degradation) (Steele and Heinzel 2001, Ruiz-Morales et al. 2010).

14.5   CONCLUSIONS

In this chapter, the major aspects of biomass-fueled DCFCs were comprehensively reviewed. According to the nature of electrolytes, DCFCs can be divided into four main types, including molten carbonate, molten hydroxide, solid oxide, and hybrid. The solid fuel characterization technologies, such as X-ray diffraction, thermogravimetric analysis, and scanning electron microscopy, are helpful to explain the fuels' performance in DCFCs, as well as the selection of suitable biomass fuels and processing conditions. The DCFC performance indicators include OCV, maximum current density, maximum power density, and internal impedance. The performance of DCFCs with various biomass sources, biomass processing conditions, fuel cell configurations (cell type, arrangement and materials), and cell operating conditions were then summarized. The research on the DCFC types of solid oxide and hybrid is gaining more attention, due to their higher flexibility in fuel acceptance and cell configuration, compared to the more traditional DCFC types of molten carbonate and molten hydroxide. Currently, biomass-fueled DCFC technology is generally at an early stage of research and development. Further research and development are required on the reaction mechanism, process optimization (such as biomass material screening, the effect of biomass processing conditions on DCFC performance, and the role of metals in biomass), and further development of DCFC technologies.

REFERENCES

Adeniyi, O. D. and Ewan, B. C. R. (2012). "Electrochemical conversion of switchgrass and poplar in molten carbonate direct carbon fuel cell." *International Journal of Ambient Energy* 33(4): 204–208.

Ahn, S. Y., Eom, S. Y., Rhie, Y. H., Sung, Y. M., Moon, C. E., Choi, G. M. and Kim, D. J. (2013). "Utilization of wood biomass char in a direct carbon fuel cell (DCFC) system." *Applied Energy* 105: 207–216.

Cantero-Tubilla, B., Sabolsky, K., Sabolsky, E. M. and Zondlo, J. W. (2016). "Investigation of pretreated switchgrass, corn stover, and hardwood fuels in direct carbon fuel cells." *International Journal of Electrochemical Science* 11(1): 303–321.

Cao, D. X., Sun, Y. and Wang, G. L. (2007). "Direct carbon fuel cell: Fundamentals and recent developments." *Journal of Power Sources* 167(2): 250–257.

Cao, D. X., Wang, G. L., Wang, C. Q., Wang, J. and Lu, T. H. (2010). "Enhancement of electrooxidation activity of activated carbon for direct carbon fuel cell." *International Journal of Hydrogen Energy* 35(4): 1778–1782.

Carrette, L., Friedrich, K. A. and Stimming, U. (2001). "Fuel cells–Fundamentals and applications." *Fuel Cells* 1(1): 5–39.

Cherepy, N. J., Krueger, R., Fiet, K. J., Jankowski, A. F. and Cooper, J. F. (2005). "Direct conversion of carbon fuels in a molten carbonate fuel cell." *Journal of The Electrochemical Society* 152(1): A80–A87.

de Wit, M. and Faaij, A. (2010). "European biomass resource potential and costs." *Biomass and Bioenergy* 34(2): 188–202.

Deleebeeck, L. and Hansen, K. K. (2014). "Hybrid direct carbon fuel cells and their reaction mechanisms–A review." *Journal of Solid State Electrochemistry* 18(4): 861–882.

Desclaux, P., Schirmerl, H. C., Woiton, M., Stern, E. and Rzepka, M. (2013). "Influence of the carbon/anode interaction on direct carbon conversion in a SOFC." *International Journal of Electrochemical Science* 8: 9125–9132.

Dudek, M. and Socha, R. (2014). "Direct electrochemical conversion of the chemical energy of raw waste wood to electrical energy in tubular direct carbon solid oxide fuel cells." *International Journal of Electrochemical Science* 9(12): 7414–7430.

Dudek, M., Tomczyk, P., Socha, R., Skrzypkiewicz, M. and Jewulski, J. (2013). "Biomass fuels for direct carbon fuel cell with solid oxide electrolyte." *International Journal of Electrochemical Science* 8(3): 3229–3253.

Elleuch, A., Boussetta, A., Halouani, K. and Li, Y. D. (2013a). "Experimental investigation of direct carbon fuel cell fueled by almond shell biochar: Part II. Improvement of cell stability and performance by a three-layer planar configuration." *International Journal of Hydrogen Energy* 38(36): 16605–16614.

Elleuch, A., Boussetta, A., Yu, J. S., Halouani, K. and Li, Y. D. (2013b). "Experimental investigation of direct carbon fuel cell fueled by almond shell biochar: Part I. Physicochemical characterization of the biochar fuel and cell performance examination." *International Journal of Hydrogen Energy* 38(36): 16590–16604.

Gassner, M., Vogel, F., Heyen, G. and Marechal, F. (2011a). "Optimal process design for the polygeneration of SNG, power and heat by hydrothermal gasification of waste biomass: Process optimisation for selected substrates." *Energy & Environmental Science* 4(5): 1742–1758.

Gassner, M., Vogel, F., Heyen, G. and Marechal, F. (2011b). "Optimal process design for the polygeneration of SNG, power and heat by hydrothermal gasification of waste biomass: Thermo-economic process modelling and integration." *Energy & Environmental Science* 4(5): 1726–1741.

Giddey, S., Badwal, S. P. S., Kulkarni, A. and Munnings, C. (2012). "A comprehensive review of direct carbon fuel cell technology." *Progress in Energy and Combustion Science* 38(3): 360–399.

Goret, J. and Tremillon, B. (1967). "Propriétés chimiques et électrochimiques en solution dans les hydroxydes alcalins fondus—IV. Comportement électrochimique de quelques métaux utilisés comme électrodes indicatrices." *Electrochimica Acta* 12(8): 1065–1083.

Gür, T. M. (2013). "Critical review of carbon conversion in 'carbon fuel cells'." *Chemical Reviews* 113(8): 6179–6206.

Hao, W., He, X. and Mi, Y. (2014). "Achieving high performance in intermediate temperature direct carbon fuel cells with renewable carbon as a fuel source." *Applied Energy* 135: 174–181.

Hemmes, K., Cooper, J. F. and Selman, J. R. (2013). "Recent insights concerning DCFC development: 1998–2012." *International Journal of Hydrogen Energy* 38(20): 8503–8513.

Jain, S. L., Lakeman, J. B., Pointon, K. D., Marshall, R. and Irvine, J. T. S. (2009). "Electrochemical performance of a hybrid direct carbon fuel cell powered by pyrolysed MDF." *Energy & Environmental Science* 2(6): 687–693.

Jang, H., Ocon, J. D., Lee, S., Lee, J. K. and Lee, J. (2015). "Direct power generation from waste coffee grounds in a biomass fuel cell." *Journal of Power Sources* 296: 433–439.

Kacprzak, A., Kobylecki, R., Wlodarczyk, R. and Bis, Z. (2014). "The effect of fuel type on the performance of a direct carbon fuel cell with molten alkaline electrolyte." *Journal of Power Sources* 255: 179–186.

Kacprzak, A., Kobylecki, R., Wlodarczyk, R. and Bis, Z. (2016). "Efficiency of non-optimized direct carbon fuel cell with molten alkaline electrolyte fueled by carbonized biomass." *Journal of Power Sources* 321: 233–240.

Kan, T. and Strezov, V. (2014). *Biomass Processing Technologies*. V. Strezov and T. Evans (Eds.). Boca Raton, FL: CRC Press, pp. 81–122.

Kan, T., Strezov, V. and Evans, T. (2017). "Fuel production from pyrolysis of natural and synthetic rubbers." *Fuel* 191: 403–410.

Kan, T., Strezov, V. and Evans, T. J. (2016). "Lignocellulosic biomass pyrolysis: A review of product properties and effects of pyrolysis parameters." *Renewable and Sustainable Energy Reviews* 57: 1126–1140.

Kulkarni, A., Ciacchi, F. T., Giddey, S., Munnings, C., Badwal, S. P. S., Kimpton, J. A. and Fini, D. (2012). "Mixed ionic electronic conducting perovskite anode for direct carbon fuel cells." *International Journal of Hydrogen Energy* 37(24): 19092–19102.

Li, X., Zhu, Z. H., Chen, J. L., De Marco, R., Dicks, A., Bradley, J. and Lu, G. Q. (2009). "Surface modification of carbon fuels for direct carbon fuel cells." *Journal of Power Sources* 186(1): 1–9.

Li, X., Zhu, Z. H., De Marco, R., Bradley, J. and Dicks, A. (2010). "Evaluation of raw coals as fuels for direct carbon fuel cells." *Journal of Power Sources* 195(13): 4051–4058.

Li, X., Zhu, Z. H., De Marco, R., Dicks, A., Bradley, J., Liu, S. M. and Lu, G. Q. (2008). "Factors that determine the performance of carbon fuels in the direct carbon fuel cell." *Industrial & Engineering Chemistry Research* 47(23): 9670–9677.

Lima, R. B., Li, J. B. and Brouwe, J. (2013a). "Modeling and studies in direct carbon-biomass fuel cell for power generation." *Abstracts of Papers of the American Chemical Society* 245.

Lima, R. B., Raza, R., Qin, H., Li, J., Lindstrom, M. E. and Zhu, B. (2013b). "Direct lignin fuel cell for power generation." *RSC Advances* 3(15): 5083–5089.

Mettler, M. S., Vlachos, D. G. and Dauenhauer, P. J. (2012). "Top ten fundamental challenges of biomass pyrolysis for biofuels." *Energy & Environmental Science* 5(7): 7797–7809.

Munnings, C., Kulkarni, A., Giddey, S. and Badwal, S. P. S. (2014). "Biomass to power conversion in a direct carbon fuel cell." *International Journal of Hydrogen Energy* 39(23): 12377–12385.

Muthuvel, M., Jin, X. and Botte, G. G. (2009). Fuel cells–Exploratory fuel cells | Direct carbon fuel cells. In *Encyclopedia of Electrochemical Power Sources*. G. Jurgen (Ed.). Amsterdam, the Netherlands: Elsevier, pp. 158–171.

Nabae, Y., Pointon, K. D. and Irvine, J. T. S. (2008). "Electrochemical oxidation of solid carbon in hybrid DCFC with solid oxide and molten carbonate binary electrolyte." *Energy & Environmental Science* 1(1): 148–155.

Rady, A. C., Giddey, S., Badwal, S. P. S., Ladewig, B. P. and Bhattacharya, S. (2012). "Review of fuels for direct carbon fuel cells." *Energy & Fuels* 26(3): 1471–1488.

Ruiz-Morales, J. C., Marrero-Lopez, D., Galvez-Sanchez, M., Canales-Vazquez, J., Savaniu, C. and Savvin, S. N. (2010). "Engineering of materials for solid oxide fuel cells and other energy and environmental applications." *Energy & Environmental Science* 3(11): 1670–1681.

Selman, J. R. (2006). "Molten-salt fuel cells–Technical and economic challenges." *Journal of Power Sources* 160(2): 852–857.

Sharaf, O. Z. and Orhan, M. F. (2014). "An overview of fuel cell technology: Fundamentals and applications." *Renewable & Sustainable Energy Reviews* 32: 810–853.

Stambouli, A. B. and Traversa, E. (2002). "Solid oxide fuel cells (SOFCs): A review of an environmentally clean and efficient source of energy." *Renewable and Sustainable Energy Reviews* 6(5): 433–455.

Steele, B. C. H. and Heinzel, A. (2001). "Materials for fuel-cell technologies." *Nature* 414(6861): 345–352.

Stevens, F., Kolodny, L. A. and Beebe, T. P. (1998). "Kinetics of graphite oxidation: Monolayer and multilayer etch pits in HOPG studied by STM." *The Journal of Physical Chemistry B* 102(52): 10799–10804.

Strezov, V. (2014). *Biomass Processing Technologies*. V. Strezov and T. Evans (Eds.). Boca Raton, FL: CRC Press, pp. 1–32.

Williams, A., Jones, J. M., Ma, L. and Pourkashanian, M. (2012). "Pollutants from the combustion of solid biomass fuels." *Progress in Energy and Combustion Science* 38(2): 113–137.

Yu, J. S., Yu, B. L. and Li, Y. D. (2013). "Electrochemical oxidation of catalytic grown carbon fiber in a direct carbon fuel cell using $Ce_{0.8}Sm_{0.2}O_{1.9}$-carbonate electrolyte." *International Journal of Hydrogen Energy* 38(36): 16615–16622.

Yu, J. S., Zhao, Y. C. and Li, Y. D. (2014). "Utilization of corn cob biochar in a direct carbon fuel cell." *Journal of Power Sources* 270: 312–317.

Zecevic, S., Patton, E. M. and Parhami, P. (2005). "Direct electrochemical power generation from carbon in fuel cells with molten hydroxide electrolyte." *Chemical Engineering Communications* 192(12): 1655–1670.

Zhan, Z. L., Bierschenk, D. M., Cronin, J. S. and Barnett, S. A. (2011). "A reduced temperature solid oxide fuel cell with nanostructured anodes." *Energy & Environmental Science* 4(10): 3951–3954.

15 Integrating Renewable Energy and Biomass into Built Environment

Xiaofeng Li, Vladimir Strezov, and Hossain M. Anawar

CONTENTS

15.1 INTRODUCTION

In the broader context of increasing demand for energy associated with expanding population and urbanization, as well as the imperative for more sustainable practices, this chapter focuses on utilization of renewable energies within the built environment. During the processes of construction, occupation, and decommissioning, buildings and their surroundings have extensive environmental impacts in the short-to-long term. Operation and activities in the buildings have been confirmed as one of the principal causes of the most environmental damaging impacts. This is primarily due to the carbon footprint of fossil fuels used to produce the electricity used for operation of the buildings, which has dominated energy consumption and remained unchanged for decades. There are now many incentives for buildings to upgrade or transform into structures with the ability to implement innovative renewable energy technologies, to minimize negative impacts on the environment.

The level of sustainability a building achieves is strongly coupled with the degree that renewable energy is exploited and utilized. Historically, evidence indicates that the discovery of a new energy source has led to the sprouting of numerous relevant green technologies. However, these new technologies have to overcome competition and critical testing before they are finally accepted. The renewable energy industry in the civil engineering sector is mainly concerned with energy conversion efficiency, energy production costs, and adaptability to buildings. In this context, it is necessary to first consider the potential that these alternative energy source candidates might represent, along with the corresponding utilization techniques.

Practical evaluation of whether a renewable energy source and its applications are reliable alternatives for buildings needs to take a number of factors into account. For example, it is necessary to adjust the measures to local climates, to combine with the conditions of a particular building, including its geographical location, geometric profile, operational strategy, and occupant density and schedule. In this chapter, an educational sustainable building with large volume, complex surface area, numerous occupant mobilities, and multiple functionalities is selected as a case study to evaluate the feasibility and adaptability of the candidate renewable energy technologies that are either very close to full commercialization or have only recently been commercialized. Specifically, the large surface area of the case building offers opportunity for integration of these multiple outside and inside renewable energy sources.

15.2   GREEN BUILDING SOLUTIONS AND SUSTAINABILITY RATINGS

Green building measures are widely accepted as a sustainable, integrated, environmentally friendly approach, balancing environmental, economic, and social considerations with considerate implementation of innovative, renewable technologies. However, the level of sustainability a building achieves is hard to benchmark.

A range of building rating tools has been developed over the past few years in order to complement the knowledge about the level of sustainability. More importantly, these rating systems are still in the process of improvement (Sev 2011) with regard to the upcoming environmental concerns. The most representative and widely used building rating methods are the Building Research Establishment Environmental Assessment Method (BREEAM), Leadership in Energy and Environmental Design (LEED), and Green Star (Lee and Burnett 2008; Reed et al. 2009; Roderick et al. 2009) methods; features of each are summarized in Table 15.1. BREEAM is the oldest and was developed by the UK Building Research Establishment (BRE) as a measure of best practice in environmental building design and management (Sev 2011). LEED was developed by the US Green Building Council (USGBC) and has been the most widely accepted rating tool in the United States (Lee and Burnett 2008). Green Star is the third building sustainability rating system, which is equivalent to BREEAM and LEED. It was launched by the Green Building Council of Australia (GBCA) (Love et al. 2012).

There are many similarities between these three schemes. First of all, they take a common approach as they assess a building against multiple criteria in different categories. BREEAM evaluates a building by 10 criteria: Land Use and Ecology, Water, Energy, Health and Wellbeing, Transport, Materials, Waste, Pollution, Management, and Innovation (BREEAM 2013). LEED evaluates seven aspects: Sustainable Sites, Water Efficiency, Energy and Atmosphere, Materials and Resources, Indoor Environmental Quality, Innovation in Design, and Regional Priority (LEED 2013). Green Star focuses on nine categories, namely, Management, Indoor Environment Quality, Energy, Transport, Water, Materials, Land Use and Ecology, Emissions, and Innovation (GBCA 2008). Secondly, all these schemes are based on credit collecting systems that determine the level of sustainability and hence the building's rating classification (Reed et al. 2009). BREEAM adopts a rating scale from Pass to Excellent; LEED uses a scale of Bronze, Silver, Gold, and Platinum to indicate a ranking from low to high; while Green Star adopts a star rating from 1 to 6 accordingly. Thirdly, all three rating tools have been developed into a series aimed at rating a wide range of building types, both newly constructed and existing. Due to the functionality difference in varied

TABLE 15.1

Comparison of the Building Rating Methods

Feature	Rating Method Name		
	BREEAM	**LEED**	**Green Star**
Launch date	1990	1998	2003
Ratings	Pass/Good/Very Good/Excellent/Outstanding	Certified/Silver/Gold/Platinum	One–Six Star
Weightings	Applied to each criterion (consensus based on scientific/open consultation)	All credits equally weighted, although the number of credits related to each criterion is a de facto weighting	Applied to each criteria category (industry survey based)
Information gathering	Design/management team or assessor	Design/management team or accredited professional	Design team
Third-party valuation	BRE	N/A	GBCA
Update process	Annual	As required	Annual
Assessment collation fee	$4,000–$20,000	$75,000	$4,002–$8,004
Cost of credit appeals	Free	$500	$800
Building covered			
New	+	+	+
Interiors	–	+	+
Core & shell	–	+	+
Existing	+	+	+
Renovated	+	+	+
Mixed-use	+	+	–
Category weightings			
Management	15%	8%	12%
Energy	25%	25%	24%
Transport	N/A	N/A	12%
Health	15%	N/A	12%
Well-being	N/A	12%	N/A
Water	5%	5%	14%
Materials	10%	18%	12%
Land use	15%	N/A	8%
Ecology	N/A	5%	N/A
Pollution	15%	11%	6%
Sustainable site	N/A	16%	N/A

BRE: UK Building Research Establishment; BREEAM: Building Research Establishment Environmental Assessment Method; GBCA: Green Building Council of Australia; LEED: Leadership in Energy and Environmental Design.

building types, a comparison between the similar types of buildings with the same standard can greatly minimize the potential bias (Haapio and Viitaniemi 2008; Perez-Lombard et al. 2009).

Table 15.1 compares and summarizes the detailed similarities and unique features of these three rating methods. It demonstrates that the building rating methods have differences in their methodologies, evaluation scopes, weighting in credits, and certification processes. Some criteria have different relative importance and different assigned weightings, depending on the geographical locations and local conditions. For instance, the Water category in the Green Star rating tool possesses more weight compared to the BREEAM and LEED rating tools. This is due to the water economy measures being of higher importance in Australia, compared to the United Kingdom or United States. For BREEAM, the rating method puts more attention on Land Use and Pollution, which is in accordance with the United Kingdom's objective conditions, such as relatively smaller national territory and heavier air pollution, compared to the United States and Australia. It is also noted that buildings designed to meet the highest LEED (Platinum) or Green Star (Six Stars) ranking are likely to achieve a BREEAM result of Very Good or Good (Saunders 2008). This is probably because the building code or building regulation standards vary from country to country. The building code standards in the United States and Australia are lower than those in the United Kingdom (Reed et al. 2009), indicating a need for metrics and performance standards for assessing buildings to be standardized globally.

The Energy category in all three rating systems accounts for nearly 25% of the total weightings, with Energy savings in the operating stage considered as the most substantial benefits of the green buildings. A detailed review of 60 LEED-rated buildings confirmed that green buildings are on average 25%–30% more energy efficient compared to conventional buildings (Tsang and Jim 2011). Energy savings in green buildings are achieved primarily by reducing electricity purchases from the power grid and, secondly, by reducing the peak energy demand. The reduced energy consumption from power grids contributes to simultaneous reduction in greenhouse gas emissions. Li et al. (2009) confirmed the energy conservation potential by studying one of the most common green technologies, the photovoltaic (PV) panels in office buildings. They found that when semi-transparent PV panels are used together with dimming controls, a peak cooling load reduction of 450 kW was achieved. This was equivalent to annual building electricity savings of 120 MWh, which contributed a reduction of 12% of the annual building electricity expenditure.

Utilization of integrated renewable energy offers considerable potential to improve a building's sustainability performance. The application of on-site renewable energy production provides the possibility to reduce dependence on fossil fuel and mitigates emissions.

15.3 RENEWABLE ENERGY SOLUTIONS

15.3.1 Outdoor Energy Sources for Buildings

Renewable energy resources and their utilization in buildings are key components of sustainable development (Dincer 2000). In this context, the renewable energy resources and technologies viable for sustainable building integration are divided here into those that can generate energy from outdoor sources and indoor energy sources. The two most favored outdoor renewable resources for electricity generation are PV and wind power.

15.3.1.1 Solar-Electrical and Building Integrated Photovoltaic

Solar energy is the most abundant and inexhaustible of all the renewable energy resources (Dincer 2011). The amount of solar energy incident on the overall surface area far exceeds the total energy demand across the world. Aside from the access and abundance of the solar resource, many other positive features make solar energy an ideal source of alternative energy. First, solar energy has the potential viability to provide electricity during peak demand times, due to the correlation between solar radiation and daytime peak electricity demand. Second, solar energy technologies have high compatibility and can be easily operated within other hybrid systems (Chow 2010). Third, solar technologies can offer ancillary services distribution utilities, such as grid support (Urbanetz et al. 2011).

PV is a technology used for generating electrical power by absorbing the energy of sun and converting the energy into direct-current electricity using semiconductor materials. The electrical power converted by PV systems can be directly applied to power up most electronic appliances. Moreover, PV cells have no mechanically moving parts, hence the energy-generating system is very easy to install and requires little maintenance. The overall cost-benefits analysis of applying PV on buildings was studied via a series of lifecycle assessments, and the results were encouraging (de Wild-Scholten 2013). The environmental footprint of PV electricity was found to be insignificant compared to the carbon footprint of electricity from fossil fuel–based electricity, and the energy payback time is shorter than the expected lifetime of 30 years.

The traditional way of installing PV panels on buildings is mounting PV modules to a separate metal support structure on the roof, which is known as *building-adopted photovoltaic* (BAPV). In contrast to the BAPV approach, *building-integrated photovoltaic* (BIPV) is defined as an architecturally integrated building element: The electricity-producing modules are both a functional unit of the building and also part of the exterior building envelope, as these modules replace conventional building materials (Prasad and Snow 2005). The fields of BIPV application are the roof areas of the building, as well as building facades, such as vertical walls, skylights, windows, and external shading devices, depending on the particular features of the PV materials.

Apart from the traditional PV cells, several types of thin-film (TF) modules with varied PV materials, such as amorphous (a-Si), micromorphic (μm-Si), copper indium gallium selenide (CIGS), organic photovoltaics (OPV), and dye-sensitized cells (DSC) are accepted by the market due to their attractive features. For instance, by adding an encapsulating polymer of any color or interferential coating, the TF modules can be tailor-made into any size and shape that satisfies the specific architectural requirements (Chopra et al. 2004). The coating and printing techniques, which are also known as the *roll-to-roll printing technology*, have been employed to manufacturing the TF PV products. The manufacturing costs are expected to be much lower than the crystalline silicon solar cells in the future (Krebs et al. 2010). Additionally, a prototype transparent PV cell with an ultrahigh visible transmission has been produced in the laboratory (Lunt and Bulovic 2011; Lunt et al. 2011). By absorbing only infrared and ultraviolet light, letting visible light pass through the cells, the cell is able to reach a transmission of more than 55%±2%, which is sufficiently transparent for incorporation on architectural glass (Lunt and Bulovic 2011).

BIPV systems significantly extend the solar collecting area from roof only to the whole building skin, converting the largest vertical areas, such as exterior walls and windows, into electricity generators. They provide a vast vertical area directly exposed to the bright morning and early evening sunlight, hence the total electricity yield will be significantly increased (Peng et al. 2011).

15.3.1.2 Building Suitable Small-Scale Wind Turbines

Wind power is known as a form of kinetic energy and can reach much higher power density than solar irradiance. In addition to its environmentally friendly properties, such as cleanliness, safety, and availability in the long term, wind energy is widely accepted as a more-established and mature renewable energy alternative. However, it should be recognized that, as wind speeds are variable and can be still for prolonged periods, wind power is an intermittent energy resource, which determines its primary benefit as reduction in energy consumption, rather than addressing maximum energy demand aspects.

According to the wind energy harvesting mechanism, the power generated by a wind turbine is proportional to the cube of the wind velocity, and also directly proportional to the swept area of turbine blades. These cubic and square relationships determine that the best viability will be for large turbines in locations with high wind speeds. Therefore, the adaptability of mid- to large-scale wind turbines in urban environments faces severe challenges due to the turbulence wind profile, noise pollution, limited size, and space. On the other hand, many efforts have been made in harnessing wind energy production and optimizing the small-scale wind turbine performance in the past few years. Until now, two main classified types of wind turbines, horizontal-axis wind turbines (HAWT) and vertical-axis wind turbines (VAWT), shown in Figure 15.1, have been developed for the turbulent urban wind profile (Eriksson et al. 2008).

HAWTs dominate the majority of the wind industry in urban environments (Ishugah et al. 2014), due to their extreme high efficiency in constant wind conditions. Because their axis of rotation of the blades is in a horizontal position, which needs to be pointed into the wind direction, the turbines are very sensitive to wind direction changes. Moreover, the maximum size of HAWTs is strictly limited to 1.5–5.0 m in diameter, considered to have safety implications on birds and aircraft, as well as aesthetic and maintenance aspects (Dutton et al. 2005). On the contrary, their counterparts VAWTs do not necessarily need to point to the wind direction. Since the main rotor shaft is arranged

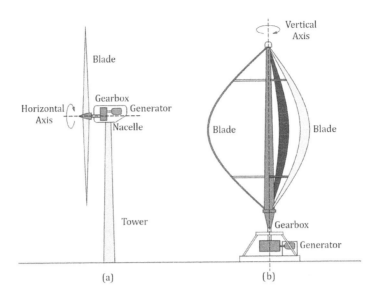

FIGURE 15.1   Small micro-wind turbines: (a) HAWT; (b) VAWT.

vertically, the vertical wind turbines can handle much higher turbulence and varied wind speeds, thus making them suitable in building mountings to overcome the turbulent nature of the wind in the urban environment. According to the size of building-mounted wind turbines, the energy generating potential is reported to be from 100 W to 30 kW in the wind speed range from 10 m/sec. to 20 m/sec. (Dutton et al. 2005).

15.3.2   Indoor Renewable Energy Alternatives

A number of indoor renewable energies, such as waste heat, flowing water in internal drainage systems, electromagnetic waves, and vibration, may also become important energy sources for buildings.

15.3.2.1   Water Flow Energy Harvesting

Water flow is a useful source of mechanical energy that is essentially constant over extended periods of time (Gilbert and Balouchi 2008). Water flow energy sources in the outdoor environment are widely used on the macro scale for electricity generation, as in hydroelectric plants. Similarly, the streams of water found in the indoor environments can also offer a great potential for energy harvesting. For instance, the grey and black water in the water pipes of a building drain system, as well as the rain water collected by the rain harvesting system installed on the building roof, have also been considered for smaller-scale harvesting applications.

The design of small-scale hydro-turbines suitable for indoor energy harvesting, Pelton (Figure 15.2a) and propeller (Figure 15.2b) turbines, is inspired by the widely used large-scale hydro-turbines (Azevedo and Santos 2012). The Pelton turbine is driven by a jet of water discharged from a nozzle and is connected to a shaft. The water stream from an upper tank or high-gravitational-potential area sluiced out through the nozzle, pushing the shaft and hence the energy is harvested in the form of electricity. It has been confirmed that the required power of 160 mW from a sensor node was obtained from a water stream of flow rate 0.14 ml/s with an efficiency of 11.3%. The energy harvest system using a propeller turbine has a relatively simpler configuration. The turbine directly installed in the middle of a water pipeline could be pushed rotating by the down-flowing water. Based on the experimental results, the turbine produced 900 mW of average power in a home irrigation system. In Hoffmann et al. (2013), a radial-flux energy harvester incorporating a three-phase generation principle is proposed for converting energy from water flow in domestic

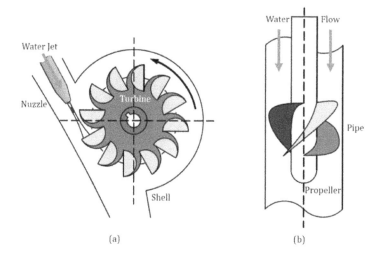

FIGURE 15.2 Hydro-turbines: (a) Pelton system; (b) propeller system.

water pipelines. The energy harvester is able to generate from 15 mW at a flow rate of 5 l/min. to 720 mW at 20 l/min. Therefore, it is possible to generate enough energy to supply micro sensors continuously using a very low water stream.

15.3.2.2 Kinetic Energy Harvesting

The vibration-based energy harvesting, termed *piezoelectric energy generation*, has received considerable attention due to its ability to capture the surrounding ambient energy and directly convert it into usable electrical energy (Erturk and Inman 2011). The piezoelectric materials are capable of generating power from the nanowatt to the watt range, depending on the piezoelectric materials and system designs.

Various types of piezoelectric harvesting systems have been developed using different modes of operation. The cantilever-type vibration energy harvester is the most commonly used design due to its simple structure and the feature that produces a large deformation under vibration (Kim et al. 2011). The majority of cantilever beam harvesters are used in micro-electro-mechanical system (MEMS) applications (Tang et al. 2011), aiming for vibration resources with high frequency and small mechanical stress. In order to extend receiving range of vibration frequencies and to improve power conversion efficiency, a two-stage energy harvester design was suggested for the very low-frequency vibration environment in the 0.2–0.5 Hz range (Peng et al. 2011), as shown in Figure 15.3a. This design consists of two main components, a mechanical energy transfer unit linked with a vibration platform and secondary vibrating units composed of additional piezoelectric elements and vibrating beams fixed on one side. Ideally, when the initial impact effects the platform, the mass attached on the mechanical energy transfer unit starts to vibrate in low frequency. The low-vibration energy is then transferred to a much higher-natural-frequency vibration in the piezoelectric elements as the mass passes over and excites the piezoelectric beams. By connecting the energy harvester with a remote sensor, the harvest can be seen as an energy source to replace the conventional battery, therefore eliminating the need for long-term maintenance.

Another effective approach is the stack-type piezoelectric harvester, which is produced by stacking multiple layers of piezoelectric materials, as shown in Figure 15.3b, resulting in a high mechanical stiffness in the stack configuration. Compared to the cantilever configuration, the stack type possesses a large capacitance and a higher capability of energy harvesting. It is suitable for a high-force environment, such as a heavy manufacturing facility or in areas of large operating machinery (Steven and Henry 2007). With these features, the stack-type piezoelectric harvester

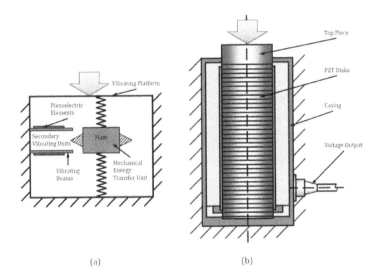

FIGURE 15.3 Two designs of the piezoelectric system: (a) a two-stage energy-harvesting approach based on the cantilever system (b) a stack-type piezoelectric harvester where PZT is Pb[Zr(x)Ti(1-x)]O3). (Reprinted from *Energ. Convers. Manage.*, 85, Li, X. and Strezov, V., Modelling piezoelectric energy harvesting potential in an educational building, 435–442, Copyright 2014, with permission from Elsevier.); (b) a stack-type piezoelectric harvester. (From Piezo-University, Basic designs of piezoelectric positioning drives/systems. Accessed February 24, 2014. http://www.physikinstrumente.com/en/products/primages.php?sortnr=400800.00&picview=2, 2013.)

breaks the limitation from micro-scale energy harvesting in MEMS to macro-scale energy generation. Innowattech, Israel, has tested this type of energy harvester on the highway to collect kinetic energy from passing cars, as well as railways (Innowattech 2010). In ideal conditions, the system is able to generate up to 200 kWh in every kilometer, which is sufficient to satisfy the energy demands for more than 800 families.

Several piezoelectric power–generating products have been released to the market. In Japan, piezoelectric floors have been in trial since the beginning of 2007 in two train stations, the Tokyo and Shibuya stations. The electricity generated from the foot traffic is used to support the electricity requirement to run the automatic ticket gates and electronic display systems (Cafiso et al. 2013). In London, a famous nightclub exploited the piezoelectric technology in its dance floor. Parts of the lighting and sound systems in the club can be powered by the energy-harvesting tiles (Arjun et al. 2011). This technology was also demonstrated at the London 2012 Olympic Games. Twelve piezoelectric tiles were installed on a temporary walkway connecting the West Ham Station to the Greenway walking route at the Olympic Park. It was estimated that these tiles would receive more than 12 million impressions from footfalls, generating 72 million Joules of energy, which equals 21 kWh of electricity. This amount of electricity was reportedly enough to illuminate the walkway for eight hours at full power during the night and 16 daylight hours at half power (Authority 2012). A similar test was conducted at the University of Beira Interior. A prototype harvester was deployed in the pavement at the main entrance of the Engineering Faculty building. It was confirmed that 525 J, or 0.15 Wh, of electric energy was harvested from 675 human steps by the piezoelectric system during a peak hour, from 9:00 a.m. to 10:00 a.m. (Duarte et al. 2013).

15.3.3 ENERGY PRODUCTION FROM WASTES

Activities within buildings contribute to considerable generation of organic waste. These wastes can be considered as a biomass and used for conversion into useful forms of energy via certain types of waste-to-energy technologies. The main technologies for biomass processing can be categorized three ways: thermochemical, biochemical, and mechanical extraction. There are three main processes of thermochemical conversion: combustion, gasification, and pyrolysis; while under the biochemical conversion category, two technologies are available, known as *digestion* and *fermentation*. In the following section, thermochemical pyrolysis and biochemical anaerobic digestion (AD) are introduced as two promising renewable energy–generating technologies that have potential to be integrated into the building sector.

15.3.3.1 Pyrolysis—A Thermo-Chemical Conversion

Pyrolysis is a thermochemical method that converts biomass to more useful fuel products (Li et al. 2014). Under pyrolytic conditions, biomass is heated to maximum temperatures of 450°C–500°C in the absence of oxygen, in order to produce a hydrocarbon-rich gas mixture (biogas), an oil-like liquid (bio-oil) and a carbon-rich solid residue (biochar) (Strezov et al. 2008).

Biochar is a light black residue of the biomass material with favorable structures of very high surface area that can retain soil-water and nutrients for plant growth. Recently, with the growing recognition of the importance of constructing human shelters that better conserve energy and water through appropriate insulation and architectural designs, biochar is used as a substrate material for green roofs due to its positive influences on both plants and roof heat insulation (Lin and Lin 2011). Another pyrolytic product, the bio-oil is a dark brown liquid fraction of biomass generated during the pyrolysis process. It is expected to play a dominant role as a substitute for crude oil because of its CO_2-neutral and low-sulphur content properties. These features promise bio-oil has a great potential to improve fuel oil security and reduce the greenhouse gas emissions from fossil oil use. There are several options for bio-oil applications as fuel materials, which include combusting in boilers, diesel engines, gas turbines, and Stirling engines, or being upgraded to higher-energy-density fuel—for example, gasoline for automobiles (Demirbas 2011). Biogas produced by pyrolytic process is a mixture of volatile gases that primarily consist of CO_2, CO, methane, and higher hydrocarbon compounds. Due to the combustible components in the biogas, it can be used as a fuel. Raveendran and Ganesh (1996) compared biogas with fossil fuels and found that the heating value of biomass pyrolysis gases are much lower than those of natural gas but are comparable with those of blast furnace gas and producer gas.

15.3.3.2 Anaerobic Digestion—A Bio-Chemical Conversion

AD is a chain of interconnected biological reactions in which organic matter is transformed into a gaseous mixture of methane, carbon dioxide, and small quantities of other gases, such as hydrogen sulphides, in an oxygen-free environment (Yilmaz and Selim 2013). During the process, the biomass is converted by bacteria in an oxygen-free environment, producing a gas with an energy content of approximately 20%–40% of the lower heating value of the feedstock (McKendry 2002).

The methane component in the biogas produced from the AD process can be combusted in internal combustion engines or micro-turbines for electricity production. De Meester et al. (2012) compared electricity produced from domestic organic waste and energy crop digestion with reference to electricity. They highlighted an effect of vaporization brought about by the AD technology, as it is able to convert almost all sources of biomass, including different types of organic wastes, slurry, and manure, to a high-calorific-value biogas. Komatsu et al. (2011) proposed a mesophilic-thermophilic hybrid flow scheme that further enhanced the electricity production from municipal sludge, resulting in electricity production at a cost of 0.05 USD/kWh, lower than the market price of 0.09 USD/kWh. In relation to individual building heating, Esen and Yuksel (2013) designed a hybrid system that integrated an AD reactor into a greenhouse. During a winter period, the system maintained a constant self-sustained temperature of 27°C within the reactor, while the greenhouse temperature was able to be maintained at approximately 23°C. Recently, an on-site prototype anaerobic digester for high-rise buildings was proposed (Ratanatamskul et al. 2014), in which it was demonstrated that the AD of sewage sludge and food waste from canteens has great potential to be used as an alternative renewable energy source.

15.4   CASE STUDY

In the above sections, a series of renewable energy sources are investigated individually regarding their energy-generating potential and the feasibility of building integration to improve buildings' sustainability performance. The extent to which the alternative renewable energy sources improve the building's energy profile was determined by employing the renewable energy-harvesting technologies together into a case study building to test their full potential in a holistic approach to meet the building's energy demands and mitigate greenhouse gas emissions.

15.4.1   CASE BUILDING DESCRIPTION AND ENERGY CONSUMPTION PROFILE

The case building shown in Figure 15.4 is an educational building located at Macquarie University Campus, Sydney, New South Wales, Australia. This educational building is a new library, represented as a flagship of the university and located at the center of the campus, where it is positioned at the south side of seven existing multi-story buildings. The library building is aiming for a Five-Star Green Building Rating from the Green Building Council of Australia (GBCA).

The building exterior uses high-performance glazing on all external façade wall areas, while inside of the building, it possesses a footprint of 6,770 m² and is composed of five stories, which account for a total gross floor area of 16,000 m². The ground level is the largest floor area in the building. Near the main entrance, a cafeteria is located in the building, providing food and drinks for the library users. Lobby meeting areas, including concourse spaces, are located after the main gates and the central cross area. They provide access to the exhibition spaces and the main collection section with open shelves. The lower ground level, ground level, level 1, level 2 and level 3 provide quiet

(a)

(b)

FIGURE 15.4   Architectural appearance of the (a) library building with (b) the adjacent buildings.

study areas and dedicated postgraduate research spaces for 3,000 undergraduate and postgraduate students. With these characteristics, the library becomes a new central hub, providing places for studying and interactive communication for not only students, but also university working staff. It makes the area the busiest spot on the campus.

The energy-consumption profile of the building consists of varied aspects, due to the size of the building as well as its multi-functional properties. The composition of the energy consumption profile includes heating, ventilating, and air conditioning (HVAC); lighting; domestic hot water supply (DHW); and lifts. The energy consumed by the HVAC system accounts for the majority of the building's total energy demand, including the energy used for indoor-climate maintenance, and the energy consumed by auxiliary systems that support the operation of the HVAC system. The energy supplied to the building, including electricity and natural gas, was obtained from the power grids. According to the energy-consumption profile, solar energy and piezoelectric energy are expected to offset the building's electricity consumption, while the biomass energy aims to produce the heating energy as a substitute for the natural gas.

15.4.2 Application of Renewable Energy Sources

15.4.2.1 Solar-Energy and Building-Integrated Photovoltaics

Solar energy is initially selected as a promising renewable energy source according to the local climate conditions, due to the abundant solar radiation throughout the year, and its highly coordinated activity schedule with building occupants' daily routine.

The new library building has a vast flat roof area of more than 1,600 m², suitable for installation of conventional PV arrays, which are also known as *multi-crystalline* (polycrystalline) *silicon solar cells*. The multi-crystalline silicon modules normally appear as opaque with a solid color ranging from blue to black. The energy-conversion efficiency of this type of module is typically around 12%–20%, according to previous studies (Green et al. 2012).

The analysis of solar energy harvesting potential for the case study building was performed via computer simulation. The results, shown in Figure 15.5, reveal a considerable amount of solar radiation falling on the building roof across the entire year period. The majority of the exposed area receives more than 1,600 kWh/m²/year of radiation, while parts of the roof reach 1,000 kWh/m²/year due to the shading effect from the adjacent building components.

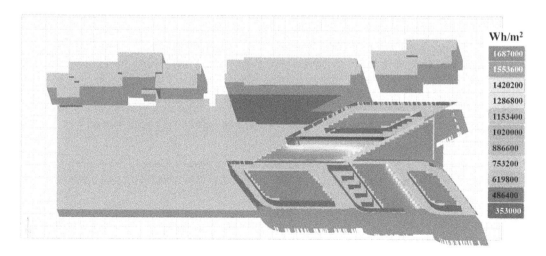

FIGURE 15.5 Annual cumulative solar radiation on the roof of the building.

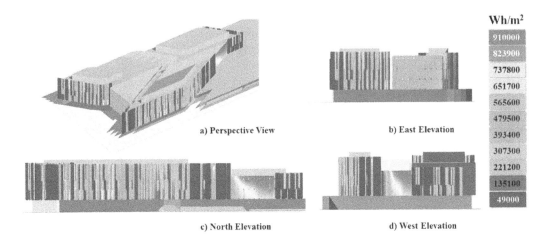

a) Perspective View

b) East Elevation

c) North Elevation

d) West Elevation

FIGURE 15.6 Annual cumulative solar radiation on the envelope of the building.

Apart from the conventional PV cells, a prototype transparent PV cell (TPC) with an ultrahigh visible transmission has been produced in the laboratory (Lunt and Bulovic 2011; Lunt et al. 2011). By absorbing only infrared and ultraviolet light, letting visible light pass through the cells, the cell is able to reach a transmission of >55% ± 2% and an efficiency of 3%–10%, which is sufficiently transparent for incorporation on architectural glass (Lunt and Bulovic 2011). With these favorable properties, the envelope of the building, including the windows, can be used as a solar electricity generator. Figure 15.6 illustrates the power generation potential using transparent PV panels on the east, north, and west building surfaces. The colorful stripes on the building envelope represent the varied cumulative amount of the solar radiation, from the lowest at 50 kWh/m²/year in the darkest area, to the highest 900 kWh/m²/year on the northwest glass curtain wall. The dark-blue areas are the glazing parts behind the vertical shading blades, which are rarely exposed to sunshine. Due to the different angles of the blade structure fixed on the external walls or windows, the shading effects to the building envelope vary. Although the shading slabs nearly cover all the building's surface area, the effective radiation area (more than 80 kWh/m²/year) still reaches a total of 1,500 m². The energy-generating potential of the TPCs is shown in Table 15.2.

15.4.2.2 Kinetic Energy Harvesting Using Piezoelectric Tiles

Piezoelectric tiles are a type of vibration-based energy harvesters with ability to capture the vibration in the surrounding environment and then directly convert the kinetic energy into usable electrical energy. In this case study, Pavegen Tile (Pavegen Systems Ltd., UK), which is one of the commercialized products newly coming to the market, was selected as the kinetic energy harvester. This

TABLE 15.2

Annual Energy-Generating Potential of the New Library Building Using Varied Photovoltaic Cells

Building Surfaces	PV Cell Types	Module Efficiency	Cell Installed (m²)	Accumulative Solar Radiation (kWh/year)	Electricity Generated (kWh/year)
Roof	polycrystalline	15%	1,500	2,300,000	350,000
Facades	TPC	8%	2,000	600,000	50,000
Total	—	—	3,500	2,900,000	400,000

PV: photovoltaic; TPC: transparent PV cell.

type of tiles is claimed to be able to generate up to 7 W of electricity per footstep, with a dimension of 600 mm × 450 mm × 82 mm. The reason for selecting the piezoelectric tiles as an indoor energy harvester for this case study building is, considering the tiles' working mechanism, that the electricity-harvesting effect is greatly dependent on the mobility and density of pedestrians. Hence, a public building with high occupant mobility, particularly the new library of Macquarie University, is an ideal place to employ this type of energy harvesters.

As described previously, this library building is a central hub for both students and university working staff. Students come to the building for book exchange and study, while the cafeteria inside the building is an attractive spot for staff. Based on the indoor mobility statistics, there are more than 6,000 people recorded entering and leaving the building every day, which provides the energy harvesters a considerable amount of kinetic energy through footsteps.

The building occupants or library users can be broadly categorized into three functional groups. The first group consists of book borrowers, who generally aim for borrowing books from the main collection areas. They spend a short time in the library and are with high mobility per time unit. The second group consists of "fixed students" who occupy the learning spaces. Although they spend longer periods of time in the library, their mobility per unit time is low. The last category is the librarian professional staff, as well as other employees in the cafeteria, with variable mobility. The mobility pattern of these three groups of users can be predicted according to users' aim and behavior. Considering the occupants' mobility behaviour, three spots were selected as the tile-deploying areas due to the highest pedestrian mobility density. They were the main entrance (central cross) area, cafeteria, and check-in/-out spot, as illustrated in Figure 15.7.

Based on the mobility statistics and the results of high traffic area location, an optimized pavement design was proposed for the tiles deployment strategy. The areas with high energy-harvesting potential are highlighted in Figure 15.8. These areas include the central cross between the main entrance and the gates towards the hall, the two doors in the cafeteria, and the pathway linking the hall and the check-in spot.

The pedestrian mobility statistics and the piezoelectric power-generating potential are illustrated in Table 15.3. It should be noted that the energy-harvesting efficiency of the piezoelectric tiles is still in rapid development. One of the promising upgrades is integrating secondary vibrating units to the tile, which will possibly enhance the harvesting efficiency up to nine times higher than the current efficiency.

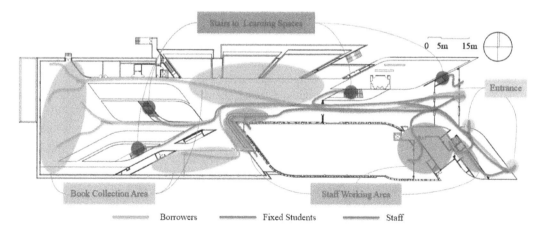

FIGURE 15.7 Main function areas of the ground floor of the new library and examples of paths followed by the library users and staff. (Reprinted from *Energ. Convers. Manage.*, 85, Li, X. and Strezov, V., Modelling piezoelectric energy harvesting potential in an educational building, 435–442, Copyright 2014, with permission from Elsevier.)

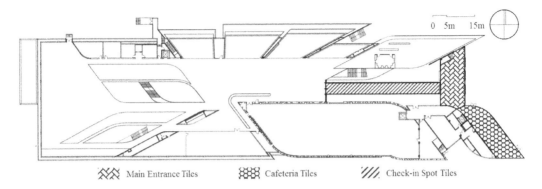

⚹⚹⚹ Main Entrance Tiles   ⚿⚿⚿ Cafeteria Tiles   ⁄⁄⁄ Check-in Spot Tiles

FIGURE 15.8 Ground-level floor plan of the new library with highlighted piezoelectric tile–deployed areas. (Reprinted from *Energ. Convers. Manage.*, 85, Li, X. and Strezov, V., Modelling piezoelectric energy harvesting potential in an educational building, 435–442, Copyright 2014, with permission from Elsevier.)

TABLE 15.3

Electricity-Generating Potential of the New Library Building Using Piezoelectric Tiles with 3.1% Total Floor Area Covered

Paving Area	Total Tiles	Pedestrian Flow (person/day)	Daily Electricity Generation (kWh/day)	Annual Electricity Generation (kWh/year)
Main entrance	614	14,035	1.09	400
Cafeteria	610	1,129	0.11	50
Check-in spot	596	11,024	1.91	700
Total	1,820	26,188	3.11	1,150

15.4.2.3 Bioenergy Produced from Thermal Conversion of Biomass

Coffee is a global beverage prepared from roasted coffee beans, with approximately 500 billion cups consumed every year (Yesil and Yilmaz 2013). At Macquarie University alone, more than 900,000 cups of coffee were consumed annually, with substantial quantities of organic waste generated as coffee grounds from this drink (Bean 2013). Due to the lack of significant market value, spent coffee grounds are currently disposed as general waste. In addition to its large amounts of organic compounds, such as fatty acids, lignin, cellulose, and hemicellulose, the disposal of coffee grounds in a landfill potentially raises environmental concerns (Pujol et al. 2013).

On the other hand, the high organic compounds make spent coffee grounds highly attractive as biomass for obtaining biofuel and valuable products (Kwon et al. 2013; Bedmutha et al. 2011). In addition to the fact that the use of coffee wastes can avoid competition with food crops compared to conventional lipid feedstock (Vardon et al. 2013), the spent coffee grounds generated in the case study building cafeteria were evaluated as a biomass type to produce heating of the building.

The most common method of biomass thermal conversion is through direct combustion. The spent coffee grounds can be subjected to a biomass boiler to generate heat energy for heating of the building during winter period. Another method, known as AD, is to digest the coffee grounds in an oxygen-free environment, producing a gas mixture containing methane and carbon dioxide. The third method is to convert the coffee grounds into varied biofuel products, which are shown in Figure 15.9, such as biochar, biogas, and bio-oil, through pyrolysis. The biochar and biogas can be combusted in a biomass boiler to generate heat, although these pyrolytic products can also

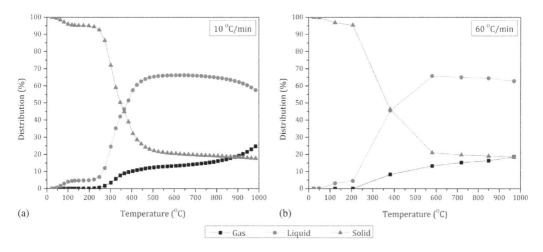

FIGURE 15.9 Distribution of gas, liquid, and solid products from pyrolysis of spent coffee grounds at heating rates of (a) 10°C/min. and (b) 60°C/min. (Reprinted from *J Anal. Appl. Pyrolysis.*, 110, Li, X. et al., Energy recovery potential analysis of spent coffee grounds pyrolysis products, 79–87, Copyright 2014, with permission from Elsevier.)

be used in turbines for electricity generation. The bio-oil product has great market potential for high-quality fuel production due to its physico-chemical properties. As a result, the bio-oil product is not considered for heat recovery in this case study.

To ensure the biomass supply has sufficient daily energy conversion demand, the spent coffee grounds was collected from the cafeteria and weighed daily for seven days. The average daily coffee grounds gathered in the case building were calculated as 21 kg. The moisture content of the coffee grounds was evaluated as 60%, under the "as received" conditions. Based on the waste statistical analysis and the pyrolytic product distribution results (Li et al. 2014), a total amount of 690 kg biochar and 375 kg biogas can be predicted annually via thermal-cracking process from the coffee waste collected in the case study building. In terms of the AD process, a biogas yield rate of 0.54 m³/kg dry spent coffee grounds was reported in previous studies (Lane 1983; Battista et al. 2016). Therefore, the annual coffee waste in the building is able to produce 1,600 m³ biogas per year, including 1,100 m³ methane.

The annual heating energy-recovery potential obtained by combustion, pyrolysis, and AD processes was calculated and is shown in Table 15.4. The results reveal that the annual heat energy that

TABLE 15.4

Annual Heating Energy Recovery Potential of Combustion and Pyrolysis with Varied Energy-Conversion Efficiencies

Biomass Conversion Technique	Conversion Products	Thermal Efficiency	Calorific Value (MJ/kg)	Heat Energy Recovered (MJ/year)
Combustion	—	80% (Hebenstreit et al. 2014)	23.2	57,000
Pyrolysis	Bio-char	80% (Hebenstreit et al. 2014)	31.1	17,000
	Biogas[a]	85% (Evangelisti et al. 2015)	14.8	5,000
Anaerobic digestion	Biogas[b]	85% (Evangelisti et al. 2015)	55.5	35,000

[a] The composition of biogas produced through pyrolytic process, as well as its calorific value, was illustrated in Figure 15.10.
[b] The anaerobic digestion biogas consisted of CO_2 and CH_4 (56%–63%) (Lane 1983).

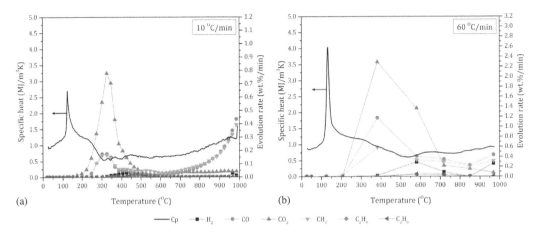

FIGURE 15.10 Specific heat and evolution rates of individual volatiles from coffee-ground pyrolysis with the temperature at heating rates of (a) 10°C/min. and (b) 60°C/min. (Reprinted from *J Anal. Appl. Pyrolysis.*, 110, Li, X. et al., Energy recovery potential analysis of spent coffee grounds pyrolysis products, 79–87, Copyright 2014, with permission from Elsevier.)

can be recovered from biomass will be 57,000 MJ by direct combustion, 22,000 MJ by combustion of the pyrolysis gas and solid products, or 35,000 MJ by AD biogas (Figure 15.10).

15.4.3 Energy Demand Offset and Greenhouse Gas Mitigation

With the renewable energy diversification approaches, the contribution of renewable energy sources on the building's energy demand offset is further estimated. For the aspect of electricity usage, more than 400 MWh/year electricity from the power grid of New South Wales in Australia can be replaced by the building's onsite electricity generator. The electricity produced by the BIPV approach and piezoelectric techniques have the potential to meet 34% of the total electricity needs of the building. Although the energy would be mainly generated by solar cells, the piezoelectric power-harvesting technology is newly emerging to the market with adequate space for further development and has been modelled to only cover 3.1% of the floor area, due to its current high cost. This technology is expected to make more contribution for future exploitation and with further reduction in its cost. On the other hand, the biomass-processing results indicated that by bioenergy conversion processes, the energy obtained from coffee grounds collected from the case study building alone would cover up to 10%–16% of building's annual heating energy consumption if the coffee waste is processed using an AD technique or subjected to biomass boilers for direct combustion, respectively; otherwise, it can cover 6% of the annual heating energy consumption, if the pyrolysis method is used.

The greenhouse gas mitigation potential of the renewable energy sources applied to the case building is evaluated based on the renewable energy production and emission factors obtained from the National Greenhouse Accounts (NGA) Factors Workbook (DCC 2013) and Green Star—Industrial v1 Greenhouse Gas Emissions Calculator Guide (GBCA 2013). The results indicate that a total amount of 434 tonnes of carbon dioxide equivalent can be offset by the application of renewable energy in the studied case building. The energy and greenhouse gas emission offset potential is summarized in Table 15.5.

TABLE 15.5

Building Energy and Greenhouse Gas Emissions Offset Potential by Integrating Renewable Energy Sources

Renewable Energy	PV and Piezoelectricity	Biomass
Annual Production	400,000 kWh/year[a]	57,000 MJ/year
Emission Factor[a]	0 kgCO$_2$-e/kWh	0.0018 kgCO$_2$-e/MJ[b]
GHG Mitigation (kgCO$_2$-e/year)	430,000	4,000
Energy Replacement Rate (%)	34.2	16.1

GHG: greenhouse gas; PV: photovoltaic.

[a] Photovoltaic and piezoelectric factors are presumed as no GHG emissions.

[b] Greenhouse Gas Emissions Factors for New South Wales in Australia from National Greenhouse Accounts (NGA) Factors Workbook (DCC, 2013).

15.5 CONCLUSIONS

This chapter bridges the knowledge gaps between the theoretical hypothesis and practical application of the renewable alternative energy sources. The empirical findings in this case study provide a solid understanding that the building's annual energy consumption profile can be improved by integrating renewable energy sources to the building. The building's energy security is improved by increasing the energy diversity and reducing peak energy demand. Considering the rapid increase in energy price, integration of green energy sources can be useful to avoid high operational costs.

In accordance with trends in building development, where modern buildings are becoming complex and multi-functional, the green technologies can either be integrated into pre-standing urban buildings or designed into new buildings as part of a shift toward a more-renewable, sustainable future. While the ideas and approaches are highly promising, they remain as components of a concept that is yet to be fully demonstrated in practice, due to investment costs and the very small energy generated. However, these technologies are expected to make more contributions for future exploitation and with further reduction in its cost, especially in the era of rapid breakthroughs on material developments, design innovations, and manufacturing revelations.

REFERENCES

Arjun, A. M., A. Sampath, S. Thiyagarajan, and V. Arvind. 2011. A novel approach to recycle energy using Piezoelectric Crystals, *International Journal of Environment and Sustainable Development*, 2: 488–492.

Authority, Olympic Delivery. 2012. Pedestrians to power walkway to London 2012 Olympic park, Accessed January 5, 2014. http://www.energyharvestingjournal.com/articles/pedestrians-to-power-walkway-to-london-2012-olympic-park-00004578.asp?sessionid=1.

Azevedo, J. A. R., and F. E. S. Santos. 2012. Energy harvesting from wind and water for autonomous wireless sensor nodes, *Circuits, Devices & Systems, IET*, 6: 413–420.

Battista, F., D. Fino, and G. Mancini. 2016. Optimization of biogas production from coffee production waste, *Bioresource Technology*, 200: 884–890.

Bean, B. 2013. 18,000 cups of coffee per week, Accessed March 21, 2014. http://www.staffnews.mq.edu.au/past_issues/past_stories/2010/18,000_cups_of_coffee_per_week.

Bedmutha, R., C. J. Booker, L. Ferrante, C. Briens, F. Berruti, K. K. C. Yeung, I. Scott, and K. Conn. 2011. Insecticidal and bactericidal characteristics of the bio-oil from the fast pyrolysis of coffee grounds, *Journal of Analytical and Applied Pyrolysis*, 90: 224–231.

BREEAM. 2013. BREEAM international new construction (NC), Accessed January 22, 2014. http://www.breeam.org/.

Cafiso, S., M. Cuomo, A. Di Graziano, and C. Vecchio. 2013. Experimental analysis for piezoelectric transducers applications into roads pavements, *Advances in Applied Materials and Electronics Engineering Ii*, 684: 253–257.

Chopra, K. L., P. D. Paulson, and V. Dutta. 2004. Thin-film solar cells: An overview, *Progress in Photovoltaics*, 12: 69–92.

Chow, T. T. 2010. A review on photovoltaic/thermal hybrid solar technology, *Applied Energy*, 87: 365–379.

DCC. 2013. National greenhouse accounts factors—July 2013, department of climate change and energy efficiency, Accessed April 7, 2014. http://www.climatechange.gov.au/sites/climatechange/files/documents/07_2013/national-greenhouse-accounts-factors-july-2013.pdf.

De Meester, S., J. Demeyer, F. Velghe, A. Peene, H. Van Langenhove, and J. Dewulf. 2012. The environmental sustainability of anaerobic digestion as a biomass valorization technology, *Bioresource Technology*, 121: 396–403.

de Wild-Scholten, M. J. 2013. Energy payback time and carbon footprint of commercial photovoltaic systems, *Solar Energy Materials and Solar Cells*, 119: 296–305.

Demirbas, A. 2011. Competitive liquid biofuels from biomass, *Applied Energy*, 88: 17–28.

Dincer, F. 2011. The analysis on photovoltaic electricity generation status, potential and policies of the leading countries in solar energy, *Renewable & Sustainable Energy Reviews*, 15: 713–720.

Dincer, I. 2000. Renewable energy and sustainable development: A crucial review, *Renewable & Sustainable Energy Reviews*, 4: 157–175.

Duarte, F., F. Casimiro, D. Correia, R. Mendes, and A. Ferreira. 2013. Waynergy people: A new pavement energy harvest system. *Proceedings of the Institution of Civil Engineers-Municipal Engineer*, 166(4), 250–256.

Dutton, A. G., J. A. Halliday, and M. J. Blanch. 2005. The feasibility of building-mounted/integrated wind turbines (BUWTs): Achieving their potential for carbon emission reductions, *Energy Research Unit, CCLRC*, 77–83.

Eriksson, S., H. Bernhoff, and M. Leijon. 2008. Evaluation of different turbine concepts for wind power, *Renewable and Sustainable Energy Reviews*, 12: 1419–1434.

Erturk, A., and D. J. Inman. 2011. Introduction to piezoelectric energy harvesting. In *Piezoelectric Energy Harvesting*, Hoboken, NJ: John Wiley & Sons.

Esen, M., and T. Yuksel. 2013. Experimental evaluation of using various renewable energy sources for heating a greenhouse, *Energy and Buildings*, 65: 340–351.

Evangelisti, S., P. Lettieri, R. Clift, and D. Borello. 2015. Distributed generation by energy from waste technology: A life cycle perspective, *Process Safety and Environmental Protection*, 93: 161–172.

GBCA. 2008. *The Dollars and Sense of Green Buildings: Building the Business Case for Green Commercial Buildings in Australia*. Sydney, Australia: Green Building Council of Australia (GBCA).

GBCA. 2013. Green Star–Ene-1 Greenhouse Gas Emissions, Green Building Council of Australia, Accessed April 7, 2014. http://www.gbca.org.au/green-star/queries-and-rulings/ene-1-greenhouse-gas-emissions/35198.htm.

Gilbert, J. M., and F. Balouchi. 2008. Comparison of energy harvesting systems for wireless sensor networks, *International Journal of Automation and Computing*, 5: 334–347.

Green, M. A., K. Emery, Y. Hishikawa, W. Warta, and E. D. Dunlop. 2012. Solar cell efficiency tables (version 39), *Progress in photovoltaics: Research and applications*, 20: 12–20.

Haapio, A., and P. Viitaniemi. 2008. A critical review of building environmental assessment tools, *Environmental Impact Assessment Review*, 28: 469–482.

Hebenstreit, B., R. Schnetzinger, R. Ohnmacht, E. Höftberger, J. Lundgren, W. Haslinger, and A. Toffolo. 2014. Techno-economic study of a heat pump enhanced flue gas heat recovery for biomass boilers, *Biomass and Bioenergy*, 71: 12–22.

Hoffmann, D., A. Willmann, R. Göpfert, P. Becker, B. Folkmer, and Y. Manoli. 2013. Energy harvesting from fluid flow in water pipelines for smart metering applications, *Journal of Physics: Conference Series*, 476(1): 012104.

Innowattech. 2010. Innowattech IPEG PAD harvests energy from passing trains, Accessed January 3, 2014. http://www.innowattech.co.il/technology.aspx.

Ishugah, T. F., Y. Li, R. Z. Wang, and J. K. Kiplagat. 2014. Advances in wind energy resource exploitation in urban environment: A review, *Renewable and Sustainable Energy Reviews*, 37: 613–626.

Kim, H. S., J. H. Kim, and J. Kim. 2011. A Review of piezoelectric energy harvesting based on vibration, *International Journal of Precision Engineering and Manufacturing*, 12: 1129–1141.

Komatsu, K., H. Yasui, R. Goel, Y. Y. Li, and T. Noike. 2011. Feasible power production from municipal sludge using an improved anaerobic digestion system, *Ozone-Science & Engineering*, 33: 164–170.

Krebs, F. C., J. Fyenbo, and M. Jorgensen. 2010. Product integration of compact roll-to-roll processed polymer solar cell modules: Methods and manufacture using flexographic printing, slot-die coating and rotary screen printing, *Journal of Materials Chemistry*, 20: 8994–9001.

Kwon, E. E., H. Yi, and Y. J. Jeon. 2013. Sequential co-production of biodiesel and bioethanol with spent coffee grounds, *Bioresource Technology*, 136: 475–480.

Lane, A. G. 1983. Anaerobic-Digestion of spent coffee grounds, *Biomass*, 3: 247–268.

Lee, W. L., and J. Burnett. 2008. Benchmarking energy use assessment of HK-BEAM, BREEAM and LEED, *Building and Environment*, 43: 1882–1891.

LEED. 2013. Projects earn points to satisfy green building requirements, Accessed January 22, 2014. http://www.usgbc.org/leed/rating-systems/credit-categories.

Li, D. H. W., T. N. T. Lam, W. W. H. Chan, and A. H. L. Mak. 2009. Energy and cost analysis of semi-transparent photovoltaic in office buildings, *Applied Energy*, 86: 722–729.

Li, X., and V. Strezov. 2014. Modelling piezoelectric energy harvesting potential in an educational building, *Energy Conversion and Management*, 85: 435–442.

Li, X., V. Strezov, and T. Kan. 2014. Energy recovery potential analysis of spent coffee grounds pyrolysis products, *Journal of Analytical and Applied Pyrolysis*, 110: 79–87.

Lin, Y. J., and H. T. Lin. 2011. Thermal performance of different planting substrates and irrigation frequencies in extensive tropical rooftop greeneries, *Building and Environment*, 46: 345–355.

Love, P. E. D., M. Niedzweicki, P. A. Bullen, and D. J. Edwards. 2012. Achieving the green building council of Australia's world leadership rating in an office building in Perth, *Journal of Construction Engineering and Management-Asce*, 138: 652–660.

Lunt, R. R., and V. Bulovic. 2011. Transparent, near-infrared organic photovoltaic solar cells for window and energy-scavenging applications, *Applied Physics Letters*, 98: 61.

Lunt, R. R., T. P. Osedach, P. R. Brown, J. A. Rowehl, and V. Bulovic. 2011. Practical roadmap and limits to nanostructured photovoltaics, *Advanced Materials*, 23: 5712–5727.

McKendry, P. 2002. Energy production from biomass (part 2): Conversion technologies, *Bioresource Technology*, 83: 47–54.

Peng, C. H., Y. Huang, and Z. S. Wu. 2011. Building-integrated photovoltaics (BIPV) in architectural design in China, *Energy and Buildings*, 43: 3592–3598.

Perez-Lombard, L., J. Ortiz, R. Gonzalez, and I. R. Maestre. 2009. A review of benchmarking, rating and labelling concepts within the framework of building energy certification schemes, *Energy and Buildings*, 41: 272–278.

Piezo-University. 2013. Basic designs of piezoelectric positioning drives/systems, Accessed February 24, 2014. http://www.physikinstrumente.com/en/products/primages.php?sortnr=400800.00&picview=2.

Prasad, D. K., and M. Snow. 2005. *Designing with Solar Power: A Source Book for Building Integrated Photovoltaics (BiPV)*. Mulgrave, Vic: Images Publication.

Pujol, D., C. Liu, J. Gominho, M. A. Olivella, N. Fiol, I. Villaescusa, and H. Pereira. 2013. The chemical composition of exhausted coffee waste, *Industrial Crops and Products*, 50: 423–429.

Ratanatamskul, C., G. Onnum, and K. Yamamoto. 2014. A prototype single-stage anaerobic digester for co-digestion of food waste and sewage sludge from high-rise building for on-site biogas production, *International Biodeterioration & Biodegradation*, 95, Part A: 176–180.

Raveendran, K., and A. Ganesh. 1996. Heating value of biomass and biomass pyrolysis products, *Fuel*, 75: 1715–1720.

Reed, R., A. Bilos, S. Wilkinson, and K. Schulte. 2009. International comparison of sustainable rating tools, *The Journal of Sustainable Real Estate*, 1: 1–22.

Roderick, Ya, D. McEwan, C. Wheatley, and C. Alonso. 2009. Comparison of energy performance assessment between LEED, BREEAM and green star. In *Eleventh International IBPSA Conference, Glasgow, Scotland, 27–30 July 2009*, pp. 1167–1176. Pune, India: Integrated Environmental Solutions Limited.

Saunders, T. 2008. A discussion document comparing international environmental assessment methods for buildings, Building Research Establishment (BRE), Watford. Available at: https://tools.breeam.com/filelibrary/International%20Comparison%20Document/Comparsion_of_International_Environmental_Assessment_Methods01.pdf (Accessed September 10, 2018)..

Sev, A. 2011. A comparative analysis of building environmental assessment tools and suggestions for regional adaptations, *Civil Engineering and Environmental Systems*, 28: 231–245.

Steven, R. A., and A. S. Henry. 2007. A review of power harvesting using piezoelectric materials (2003–2006), *Smart Materials and Structures*, 16(3): R1.

Strezov, V., T. J. Evans, and C. Hayman. 2008. Thermal conversion of elephant grass (Pennisetum Purpureum Schum) to bio-gas, bio-oil and charcoal, *Bioresource Technology*, 99: 8394–8399.

Tang, G., J. Q. Liu, H. S. Liu, Y. G. Li, C. S. Yang, D. N. He, V. Dzung Dao, K. Tanaka, and S. Sugiyama. 2011. Piezoelectric MEMS generator based on the bulk PZT/silicon wafer bonding technique, *Physica Status Solidi a-Applications and Materials Science*, 208: 2913–2919.

Tsang, S. W., and C. Y. Jim. 2011. Game-theory approach for resident coalitions to allocate green-roof benefits, *Environment and Planning A*, 43: 363–377.

Urbanetz, J., C. D. Zomer, and R. Ruther. 2011. Compromises between form and function in grid-connected, building-integrated photovoltaics (BIPV) at low-latitude sites, *Building and Environment*, 46: 2107–2113.

Vardon, D. R., B. R. Moser, W. Zheng, K. Witkin, R. L. Evangelista, T. J. Strathmann, K. Rajagopalan, and B. K. Sharma. 2013. Complete utilization of spent coffee grounds to produce biodiesel, bio-oil, and biochar, *ACS Sustainable Chemistry & Engineering*, 1: 1286–1294.

Yesil, A., and Y. Yilmaz. 2013. Review article: Coffee consumption, the metabolic syndrome and non-alcoholic fatty liver disease, *Alimentary Pharmacology & Therapeutics*, 38: 1038–1044.

Yilmaz, S., and H. Selim. 2013. A review on the methods for biomass to energy conversion systems design, *Renewable & Sustainable Energy Reviews*, 25: 420–430.

Index

Note: Page numbers in bold and italics refer to tables and figures, respectively.